# PARALLEL PROCESSING TECHNIQUES FOR SIMULATION

# APPLIED INFORMATION TECHNOLOGY

Series Editor:

M. G. SINGH
*UMIST, Manchester, England*

Editorial Board:

K. ASTROM
*Lund Institute of Technology, Lund, Sweden*

S. J. GOLDSACK
*Imperial College of Science and Technology, London, England*

M. MANSOUR
*ETH-Zentrum, Zurich, Switzerland*

G. SCHMIDT
*Technical University of Munich, Munich, Federal Republic of Germany*

S. SETHI
*University of Toronto, Toronto, Canada*

J. STREETER
*GEC Research Laboratories, Great Baddow, England*

A. TITLI
*LAAS, CNRS, Toulouse, France*

---

**PARALLEL PROCESSING TECHNIQUES FOR SIMULATION**
Edited by M. G. Singh, A. Y. Allidina, and B. K. Daniels

**INDUSTRIAL ARTIFICIAL INTELLIGENCE SYSTEMS**
Lucas Pun

# PARALLEL PROCESSING TECHNIQUES FOR SIMULATION

Edited by

## M. G. Singh
UMIST
Manchester, England

## A. Y. Allidina
Imperial Chemical Industries PLC
Northwich, Cheshire, England

and

## B. K. Daniels
National Computing Centre Ltd.
Manchester, England

PLENUM PRESS • NEW YORK AND LONDON

Library of Congress Cataloging in Publication Data

European Workshop on Parallel Processing Techniques for Simulation (1st: 1985:
University of Manchester Institute of Science and Technology)
   Parallel processing techniques for simulation.

   (Applied information technology)
   "Proceedings of the First European Workshop on Parallel Processing Techniques for Simulation, held October 28–29, 1985, at the University of Manchester Institute of Science
and Technology, Manchester, England"—T.p. verso.
   Includes bibliographies and index.
   1. Parallel processing (Electronic computers)—Congresses. 2. Digital computer simulation—Congresses. I. Singh, Madan G. II. Allidina, A. Y. III. Daniels, B. K. IV. Title. V.
Series.
QA76.5.E917   1985                         001.4′34                         86-22652

ISBN-13: 978-1-4684-5220-4          e-ISBN-13: 978-1-4684-5218-1
DOI:10.1007/ 978-1-4684-5218-1

Proceedings of the first European Workshop on Parallel Processing Techniques
for Simulation, held October 28–29, 1985, at the University of Manchester
Institute of Science and Technology, Manchester, England

© 1986 Plenum Press, New York

Softcover reprint of the hardcover 1st edition 1986

A Division of Plenum Publishing Corporation
233 Spring Street, New York, N.Y. 10013

PREFACE


    This volume provides the proceedings of the First European Workshop on
Parallel Processing Techniques for Simulation which was held at the end of
October 1985.  The Workshop was organized within the framework of a joint
project sponsored by the Commission of the European Communities under the
research part of the multiannual programme in the field of Data Processing
aming at promoting collaborative research work in the Community.  The
project involved collaborative work between the Complex Systems Group of
the Control Systems Centre at UMIST, the Systems Reliability Service of the
United Kingdom Atomic Energy Authority and the University of Bergamo, Italy.

    The aim of this project was to develop decomposition coordination
techniques which would be of help in the simulation of complex dynamical
systems on parallel processing facilities.  One of the major aims of the
Workshop was to report on the results produced within the project and to
try to relate these to the leading work going on in this field in other
centres of excellence.  With this in mind, the Proceedings Volume is split
up into a number of parts corresponding to the main sessions within the
Workshop programme.

    The first part comprises the report on the EEC project in terms of the
work done primarily in the U.K. between the CSG and the SRS and here the
4 papers provide on the one hand, an outline of the new techniques that
have been developed on the use of decomposition coordination techniques
for parallel simulation and on the other, we see the application of these
ideas to the nuclear safety code, RELAP, and to a reactor model.

    The next part deals with parallel system solvers where a number of
papers provide an outline of the research in this field.

    The next part deals with partitioning techniques and here four papers
highlight the aspects of partitioning.

    The next major area is concerned with distributed computing applica-
tions.

    The final part is concerned with parallel processing architectures and
here a number of new approaches are described.

    We believe that taken together, this Volume provides a coherent and up
to date account of the current state of the art of parallel processing
techniques for simulation.

We are most grateful to Mr. Desfosses and the EEC Commission for their financial support which made this project possible, and especially to Mr. Fangmeyer of the ISPRA Centre of the EEC for his continuous support and encouragement.

Manchester 1986

M.G. Singh
A.Y. Allidina
B.K. Daniels

CONTENTS

# DECOMPOSITION - COORDINATION TECHNIQUES FOR PARALLEL SIMULATION

K. Malinowski[*], A. Y. Allidina[+] and M.G. Singh[+]

[*]Dept. of Automatic Control, Technical University of Warsaw
Warsaw, Poland
[+]Control Systems Centre, UMIST, Manchester M60 1QD, U.K.

ABSTRACT

The paper investigates decomposition-coordination techniques which enable tasks to be performed in parallel using parallel-computing facilities when solving large sets of equations resulting from discretization of differential equations. Such an approach for system simulation can be useful in industries where it is vital to improve the speed of simulation.

## 1. INTRODUCTION

Simulation of the behaviour of multi-component dynamical systems described by large sets of ordinary differential equations (ODE) or by complicated partial differential equations (PDE) is amongst the most important and frequently required tools in modern decision making, training and control system design. Very fast simulation (solution) of dynamical systems is required in real-time simulators which are used as convenient, low-cost and effective tools to evaluate system design changes under different conditions, to evaluate operator workload and to train new operators. Even faster simulation (a few orders of magnitude faster than real-time) is required in, for example, on-line control mechanisms which involve solving mathematical models repetitively in order to investigate the impacts of different decisions. In particular, in the nuclear industry it is vital to have such a speed of simulation, so that on the onset of various accidents it would be possible to predict future reactions of the very fast and complex reactor system to different control actions. This would facilitate the choice of the most appropriate control action.

There are numerous ways of improving the simulation speed, such as:

(i)    The use of simplified mathematical models. This obvious possibility is sometimes forgotten, however in most cases a significant simplification of a set of equations describing a dynamical system is impossible due to accuracy requirements with respect to simulation results.

(ii)   Improvement of numerical techniques (e.g. integration algorithms).

1

(iii)  Optimisation of program codes and data bases.

(iv)  Design of dedicated devices such as analogue or hybrid simulators, special digital machines (e.g. dynamic differential analysers, special array processors, etc.).

(v)  The use of more powerful general purpose computers.

It should be noted that the last option (v), (together with (iii)), is perhaps the most appealing to an average user (the cost may be prohibitive). To date, the development of computer technology has resulted in more and more powerful central processing units which has justified this way of improving the simulation speed.

Another way of improving simulation speed consists of using suitable 'parallel-system-solvers' which can be applied on parallel computing facilities. This approach, however, has been less popular so far in practical applications. There are two main reasons for this. Firstly, the parallel machines were built of simple processing elements (e.g. 8-bit microprocessors) which could not compete in speed with the central processing units of available 'classical' computers, and, secondly, programming of parallel devices required knowledge of a given multiprocessor architecture, elaborate task allocation techniques, etc. This was not appealing to the users from different fields who preferred to use standard programming languages and did not want to be concerned with how the actual computing was organised and performed.

One can expect, however, a significant change in this situation over the next few years. It is believed that general purpose machines composed of clusters of very fast processing units with efficient communication systems and user-friendly programming languages will appear and be commercially available. In order to use the capabilities of these machines to the full extent with the aim of improving the simulation speed, it is necessary to develop numerical solution techniques involving parallel computing tasks of different levels of complexity.

The issue of parallelism can be addressed at many different levels (Burks, 1981; Schendel, 1981) with relation to different computer architectures and to concepts in parallel numerical methods. As far as architecture is concerned, one can think of parallelism at the instruction execution level (e.g. 'pipelining'), or of vector and array processors executing a stream of single instructions with multiple data (SIMD archi- tecture), or of special multiple instruction architectures (multi processing architectures) with several streams of instructions being executed simultaneously (MIMD) (Burks, 1981). One can have multi- processor or multi-computer networks having different configurations (e.g. common multi-bus systems (Arnold et al, 1983), or systems having a ring structure (Brasch and coworkers, 1981).

As far as concepts in parallel numerical mathematics are concerned, it is possible to consider parallel execution of elemental operations when evaluating mathematical expressions (e.g. when evaluating expressions like 'Horner scheme' with 'Log-sum-algorithm'). It is also possible to consider, at a higher level, inherent parallel operations in well estab- lished algorithms (Schendel, 1981). Finally, at a still higher level, one can specify larger tasks for separate computers to be executed in parallel while using decomposition or decomposition-coordination techniques. The execution of these tasks can be done in a synchronous or asynchronous mode. It is this last issue that we are concerned with in this paper , since we are interested in parallelism at a rather high level. This is related to the use of a multi-computer network rather than an array processor, for example.

2

In view of the above discussion this paper investigates decomposition-coordination techniques which enable tasks to be performed in parallel when solving large sets of nonlinear equations of a specific structure. This structure arises, for example, when solving a set of partial differential equations using a particular type of discretization scheme. In such an application it is necessary to solve the resulting equations at each time level, and therefore the developed method ('hierarchical system solver') needs to be used many times.

The paper is arranged as follows. In Section two, the problem is defined and the solution methodology of decomposition-coordination is discussed. In Sections three and four, two basic decomposition-coordinate techniques are adapted for solving the problem defined in Section two with attention being paid to effective coordination strategies. Some final remarks are given in Section five. The work reported here was carried out in the Control Systems Centre at UMIST under an EEC contract. The techniques are given in more detail in Allidina (ed.) (1984) and Malinowski et al (1985).

## 2. PROBLEM DEFINITION AND SOLUTION METHODOLOGY USING DECOMPOSITION-COORDINATION

A given simulation problem may consist largely of solving a set of ordinary differential equations or partial differential equations. In order to compute the solution by numerical techniques the equations need to be discretized (in time for ODEs, and in time and space for PDEs) and this leads to a system of difference equations. From the discretized equations the approximate values at the mesh points of the dependent variables can be computed. It may perhaps be possible to create such a system of difference equations (an integration scheme) in order to provide for parallel computing tasks of a large size. System partitioning and decomposition techniques can be useful at this stage (see e.g. Allidina et al, 1984). One of the possible applications of these techniques could be, for example, temporal decomposition. Let us assume that the problem to be solved consists of integrating the equations

$$\dot{x} = f(x) , \qquad x \in R^n$$

The integration is to be done over a given time horizon $[t_o, t_f]$ with the initial condition $x(t_o) = x_o$. This initial value problem is sequential in time by nature. An attempt to break this sequence could be, for example, as follows. The overall integration horizon can be split into, say, N parts $[t_{j-1}, t_j]$, $j = 1, \ldots, N$ where $t_N = t_f$, and N parallel problems can be defined:

$P_j$ : solve $\dot{x} = f(x)$ over $[t_{j-1}, t_j]$ with initial condition $x_{j-1}$ .

If we denote the solution of $P_j$ by $x^j(x_{j-1}, t)$, $t \in [t_{j-1}, t_j]$, then these solutions have to satisfy the conditions

$$x^j(x_{j-1}, t_j) = x_j \qquad j = 1, \ldots, N-1$$

The mode of coordination would consist then of updating the values of $x_j$, $j = 1, \ldots, N-1$ until the above conditions are fulfilled. Since for each collection of $x_j$, $j = 1, \ldots, N-1$, the problems $P_j$, $j = k, \ldots, N$ have to be solved (in parallel) at the k-th iteration of the coordinator, the above procedure can result in a speed-up of the simulation only if the number of iterations at the coordinating level is much less than N. If a sufficient number of parallel processors were available then one could solve simultaneously each of the problems $P_j$ for a number of different initial conditions, say $x_{j,k}$, $k = 1,\ldots,K$. Then the coordination problem is to satisfy the following conditions

$$x^1(x_o,t_1) = x_{1,k_1}, \quad x^2(x_{1,k_1},t_2) = x_{2,k_2},\ldots,$$

$$x^{N-1}(x_{N-2,k_{N-2}},t_{N-1}) = x_{N-1,k_{N-1}}$$

$x_{1,k_1},\ldots,x_{N-1,k_{N-1}}$      are a collection of intermediate conditions.

Similar ideas can be used to decompose the integration regions of PDEs. It should be observed, however, that for initial value problems consisting of large sets of equations, i.e. when n>>1, the above temperal decomposition cannot be expected to succeed. It should be noted that there are other parallel methods for the numerical solution of ordinary differential equations in which the discretization is done so as to provide for parallelism (e.g. Minanker and Liniger 1966 , Worland 1976 , Franklin 1978 , Katz et al 1977 ).

After these preliminary remarks concerning one possible use of decomposition at the discretization stage, let us consider the major case of interest in this paper which is the use of decomposition-coordination techniques after time and space discretization (where appropriate) of a set of differential equations. Whenever such discretization results in an explicit integration scheme then in order to compute the values of dependent variables at each subsequent time level one has to evaluate a set of explicit formulae. This can be done in parallel, for example by using equation segmentation (e.g. Franklin, 1978) and any special structure of the discretised equations can be useful in defining such parallel tasks so as to avoid excessive data transfer (communications) between the processors (e.g. Tao and Saeks, 1984). There is no scope, however, for the use of decomposition-coordination techniques. The situation changes when we deal with an implicit integration scheme (see, for example, Miranker, 1981) required, in particular for a stiff system of ODEs or when solving partial differential equations. Then, in order to advance the integration to the next time level it is necessary to solve a system of algebraic equations:

$$F(z) = 0 \tag{1}$$

where $Z \in R^{n_z}$ and $F: R^{n_z} \to R^{n_z}$:

Let us consider the case when equation (1) can be put into the following form:

$$F_i(x_i,y_i) = 0 \qquad i = 1,2,\ldots \nu \tag{2}$$

and

$$y_i = A_i x_{i-1} + B_i x_{i+1}, \quad i = 1,2,\ldots \nu \tag{3}$$

where $x_i \in R^{n_{x_i}}$, $y_i \in R^{n_{y_i}}$, $F_i: R^{n_{x_i}} \times R^{n_{y_i}} \to R^{n_{x_i}}$,

$A_1 = 0$ and $B_\nu = 0$.

Such equations can arise when solving for example, a set of partial-differential equations of the form:

$$\frac{\partial V}{\partial t} = f(\frac{\partial V}{\partial 1}, V, 1) \tag{4}$$

where $V(t,1) \in R^{n_V}$, $t \in R$ and $1 \in R$, with appropriate initial and boundary conditions. Assume that we use the discretization scheme depicted in Fig. 1, where $v_{ij}$ is an approximation to the variable $V$ at the ij-th grid point.

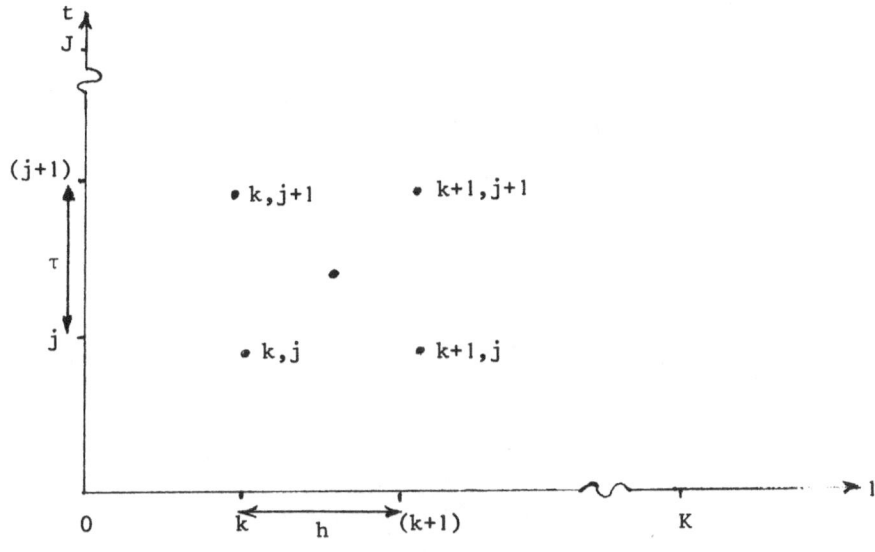

Fig. 1 Discretization scheme

The time and space derivatives are approximated by

$$\frac{\partial V}{\partial t} = \frac{1}{2\tau} (v_{k+1,j+1} - v_{k+1,j} + v_{k,j+1} - v_{k,j}) \tag{5}$$

$$\frac{\partial V}{\partial 1} = \frac{1}{2h} (v_{k+1,j+1} - v_{k,j+1} + v_{k+1,j} - v_{k,j}) \tag{6}$$

and $V$ itself is approximated by

$$V = \frac{1}{4} (v_{k+1,j+1} + v_{k+1,j} + v_{k,j+1} + v_{k,j}) \tag{7}$$

Suppose now that $v$ is known at all grid points at time level $j$, i.e. $v_{k,j}$ is known, for $0 \le k \le K$, that $v = [v^1, v^2]$ and that $v^1_{o,j+1}$ and $v^2_{K,j+1}$ represent boundary conditions which are known. (These are specified in such a way that the problem of solving equation (4) is well-posed.) The problem is then to find $v_{k,j+1}$ for $0 \le k \le K$. This amounts to solving a set of equations of the form (1), where

$$z = [v^2_{o,j+1}, v_{1,j+1}, \ldots, v_{K-1,j+1}, v^1_{K,j+1}] \tag{8}$$

The set of equations 11) in this case is partitioned in a natural way, as follows:

$$
\begin{aligned}
g_1(v^2_{o,j+1}, v_{1,j+1}) &= 0 \\
g_2(v_{1,j+1}, v_{2,j+1}) &= 0 \\
g_{k+1}(v_{k,j+1}, v_{k+1,j+1}) &= 0 \\
g_{K-1}(v_{K-2,j+1}, v_{K-1,j+1}) &= 0 \\
g_K(v_{K-1,j+1}, v^1_{K,j+1}) &= 0
\end{aligned}
\qquad\qquad (9)
$$

where each $g_i$ is of dimension $n_V$. By choosing particular space grid points $K_1, K_2, \ldots K_\nu$ it is possible to introduce the following variables:

$$
x_1 = (v^2_{o,j+1}, v_{1,j+1}, \ldots, v_{K_1-1,j+1}, v^1_{K_1,j+1})
$$

$$
y_1 = v^2_{K_1,j+1}
$$

$$
x_i = (v^2_{K_{i-1},j+1}, v_{K_{i-1}+1,j+1}, \ldots, v_{K_i-1,j+1}, v^1_{H_i,j+1}) \, ,
$$

$$
y_i = (v^1_{K_{i-1},j+1}, v^2_{K_i,j+1}) \, , \qquad i = 2,3,\ldots(\nu-1)
$$

$$
x_\nu = (v^2_{K_{\nu-1},j+1}, v_{K_{\nu-1}+1,j+1}, \ldots, v_{K_\nu-1,j+1}, v_{K_\nu,j+1})
$$

$$
y_\nu = v^1_{K_{\nu-1},j+1}, \qquad K_\nu = K
$$

By making use of the above variables, equation set (9) can be written in the form of equations (2) and 13), where $F_i(x_i, y_i) = 0$ represents the equations from $g_{K_{i-1}+1} = 0$ till $g_{K_i} = 0$. The variables $y_i$ have been chosen to be compatible with the original boundary conditions. In particular, this means that each group of equations (2) can be solved for $x_i$ given $y_i$ .

The example of partial-differential equations (equation 4), together with the particular discretization scheme was considered to show how the set of the equations (2 ) and (3) arise. Other implicit discretization schemes and two-point boundary value problem for ordinary-differential equations after discretization would also result with equations of the form (2) and (3) (with appropriate choice of $x_i$ and $y_i$).

Having defined our problem, we now consider ways of solving it with the use of a parallel computing facility (multi-computer network). One possible way of doing this would be to use successive linearisation. By this we mean that the equation set (1) or (2) can be linearised around a certain point and the resulting system of linear equations solved, after which the linearisation is repeated and so on. In such a case, in order to solve the set of linear equations, one can use suitable 'parallel-system-solvers'. There is a vast amount of literature on this subject, see for example Barlow and Evans (1982), Evans and Haghighi (1982) and Halada (1981). Another approach to solve the set of equations (2) and (3) is to use decomposition-coordination schemes. This involves decomposing the set of partitioned equations (2) and (3) into independent parallel problems by introducing suitable coordination variables. This approach goes along the lines of decomposition in large-scale optimization problems lHimmelblau (editor), 1973; Lasdon, 1970; Wismer (editor), 1971; Singh and Titli, 1978; Findeisen et al, 1980). It is possible to use well- established decomposing-coordination techniques after adapting them to solve the system of equations (2) and (3). It is necessary to use efficient coordination strategies in order that the solution of the sub- problems does not have to be repeated many times.

In this paper we are concerned with the latter approach of using decomposition-coordination. In the following sections two basic methods of this kind are adapted to solve the set of equations (2) and (3).

It should be noted at this point that the special structures of the set of algebraic equations have been exploited in the past in order to perform decompositions for serial processing (see, e.g., Himmelblau, 1973). The basic idea behind these decompositions is different from the approach used for decomposition of composite optimisation problems. In order to explain this let us consider the system of equations depicted in Fig. 2a as consisting of two subsets of equations coupled through interconnection variables $y_1$, $y_2$.

By 'tearing', say at the coupling associated with $y_1$ variable, (Fig. 2b), one obtains an acyclic partition of the system of equations. For given values of $y_1^o$, $F_1$ can be solved to produce $y_2$ and after that $F_2$ can be solved.

Then the value of $y_1^1$ would be known and the task of coordination would be to iterate on $y_1^o$ until $y_1^o = y_1^1$. Now, if $F_1$ and $F_2$ were not just coupled subsets of equations but the subproblems of an optimisation problem, the above approach would not lead in general to an overall optimal solution. In this case it would be necessary to 'tear' both couplings $y_1$ and $y_2$ in order to define two independent subproblems, (Fig. 2c). For solving the set of equations only it would be unreason- able to perform such a decomposition because it would result in a larger number of coordinating variables. However, if we want to produce parallel tasks for a general purpose multiprocessor of MIMD type, then we need to follow the decomposition idea as presented in Fig. 2c, hoping that the time saving obtained from solving the tasks in parallel will outweigh the extra computing effort associated with coordinating a larger number of variables.

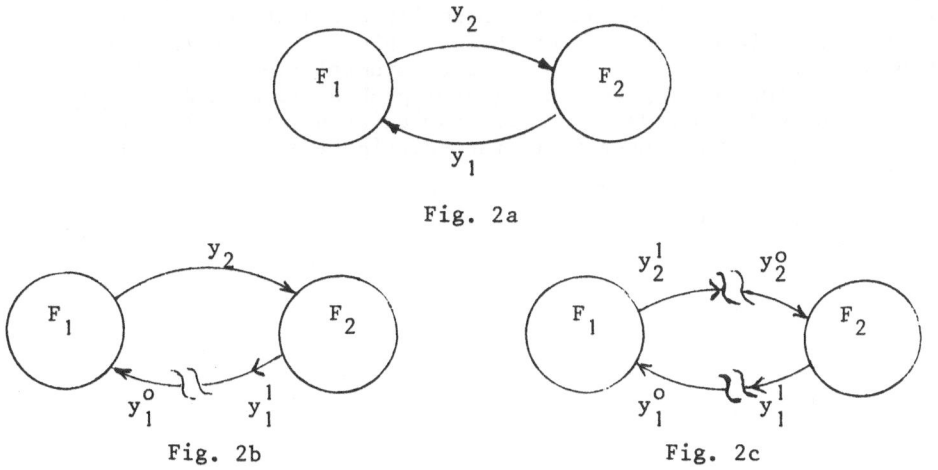

Fig. 2a

Fig. 2b                                   Fig. 2c

# 3. SOLUTION BY DECOMPOSITION-COORDINATION-DIRECT METHOD

One of the basic decomposition-coordination methods is the 'direct' method (e.g. Findeisen et al, 1980). In our case the procedure consists of treating the coupling variables $y_i$, $i = 1,2,\ldots\nu$, as the coordinating variables (hence the name 'direct'). For given $y_i$, $i = 1,2,\ldots\nu$, it is possible to define $\nu$ independent local problems to be solved in parallel. The overall procedure is as follows:

(i)     Local Problem i, $i = 1,2,\ldots\nu$.

For given $y_i$, solve $F_i(x_i,y_i) = 0$ for $\bar{x}_i\ (y_i)$.

(ii)    Coordinator Problem

Find $\hat{y} = (\hat{y}_1,\hat{y}_2,\ldots\hat{y}_\nu)$ such that

$$\hat{y}_i = A_i\bar{x}_{i-1}(\hat{y}_{i-1}) + B_i\bar{x}_{i+1}(\hat{y}_{i+1}), \quad i = 1,2,\ldots\nu \tag{10}$$

As far as the local problems are concerned, it is assumed that, given $y_i$, a solution $\bar{x}_i(y_i)$ exists for $F_i(x_i,y_i) = 0$. It is thus necessary, for example for the partial-differential-equations (4) considered, that the variables $y_i$ are chosen to be compatible with the original boundary conditions. $y_i$ should belong to such sets $Y_i$, that for $y_i \in Y_i$, these boundary conditions have physical meaning.

The local problems may be solved in different ways depending on the particular equation set $F_i(x_i,y_i) = 0$. For example, the Newton-Raphson procedure may be used:

$$x_i^{N+1} = x_i^N - \left( \left. \frac{\partial F_i}{\partial x_i} \right|_{x_i=x_i^N} \right)^{-1} F_i(x_i^N,y_i) \tag{11}$$

Starting from an initial guess $x_i^o$, equation (11) is used successively for $N = 1,2,\ldots.$ until convergence is achieved. It is of course assumed that $F_i$ is differentiable and that the Jacobian $[\partial F_i/\partial x_i]$ is non-singular. The convergence conditions for the above scheme have been extensively examined and are well known.

Since the solutions $\bar{x}_i(i)$ of the local problems are usually not available in explicit form, the coordinator problem must be solved iteratively. This implies that for each iteration at the coordination level resulting in new values for the variables $\hat{y}_i$, the local problems must be solved again. Therefore it is of crucial importance that the strategy used for solving the coordinator problem is as efficient as possible. We will consider two coordination strategies. In order to make the notation more convenient, the coordinating conditions (10) can be written as:

$$y = C\bar{x}(y) \tag{12}$$

where

$$y = (y_1\ y_2\ \ldots\ y\ ),\ \bar{x}(y) = (\bar{x}_1(y_1),\ \bar{x}_2(y_2)\ \ldots\ \bar{x}_\nu(y_\nu))$$

and

$$
C = \begin{pmatrix}
0 & B_1 & 0 & 0 & \cdots & 0 \\
A_2 & 0 & B_2 & 0 & \cdots & 0 \\
0 & A_3 & 0 & B_3 & \cdots & 0 \\
0 & & & A_{\nu-1} & 0 & B_{\nu-1} \\
0 & & & 0 & A_{\nu} & 0
\end{pmatrix}
$$

## Coordination strategy 1 - Relaxation

The coordination process has to be initiated with an initial guess $y^o$. In the partial-differential-equation example considered, the initial guess $y^o$ at the new time level could be taken as the value of $\hat{y}$ at the previous time level. Then at any iteration step M ($M \geq 0$), the local problems are solved for $\bar{x}(y^m)$. The coordinator then corrects the value of y according to:

$$
y^{M+1} = (1-\omega)y^M + \omega y^{-M+1} \tag{13}
$$

$$
y^{-M+1} = C\,\bar{x}(y^M) \tag{14}
$$

and $0 < \omega < 1$ is the relaxation parameter to be chosen so as to provide for the best convergence possible. In particular, for $\omega = 1$, we obtain a re-injection (or Jacobi type) iterative strategy. This coordination strategy is rather simple with the advantage that a small amount of information needs to be passed between the coordinator and the subsystems; however, the disadvantage is that many iterations may be required.

## Coordination strategy 2 - Newton-Raphson

Another strategy may be used for solving the coordinator problem of equation (12). Let us define:

$$
D(y) \triangleq y - C\,\bar{x}(y) \tag{15}
$$

Assuming that D is a differentiable function of y and $(\partial D(y)/\partial y)$ is non-singular, then at any iteration step the coordinator corrects the value of y according to:

$$
y^{M+1} = y^M - [\partial D(y)/\partial y\big|_{y=y^M}]^{-1} D(y^M) \tag{16}
$$

where

$$
\frac{\partial D(y)}{\partial y} = I - C\frac{\partial \bar{x}(y)}{\partial y} \tag{17}
$$

The differentiability of D is implied by that of $\bar{x}(\cdot)$, which can be attained under reasonable and well-known conditions (Implicit Function theorem). In order to compute the Jacobian $\partial D(y)/\partial y$ it is required to find

$$
(\partial \bar{x}(y)/\partial y) = \text{block-diag}\ (\partial \bar{x}_i(y_i)/\partial y_i) \tag{18}
$$

$\partial \bar{x}_i(y_i)/\partial y_i$ can be obtained in parallel at the subsystem level after the local problems have been solved. By differentiating equation (2) it can be seen that:

$$\frac{\partial \bar{x}_i(y_i)}{\partial y_i} = - \left( \frac{\partial F_i}{\partial x_i} \bigg|_{\bar{x}_i(y_i),y_i} \right) \frac{\partial F_i}{\partial y_i} \bigg|_{\bar{x}_i(y_i),y_i} \qquad (19)$$

The inverse of $(\partial F_i/\partial x_i)$ required in the above equation is in fact available from the last iteration at the subsystem level (see equation (11)).

It should be noted that it is not necessary to transfer the entire matrices $\partial \bar{x}_i(y_i)/\partial y_i$ from the subsystem to the coordinator, but only the appropriate part of the product $C \ \partial \bar{x}(y)/\partial y$ from each subsystem. Since the matrix $C$ is sparse, this would result in considerable saving in communication time between the subsystem and the coordinator.

A further point of interest is that if the functions $F_i$ are linear, then the above procedure requires only one iteration at the coordination level. The proposed scheme can then be considered as a particular parallel linear-system-solver.


4.  SOLUTION BY DECOMPOSITION-COORDINATION-DUAL METHOD

Another basic decomposition-coordination technique is the 'dual' method (Lasdon, 1970; Singh and Titli, 1978; Findeisen et al, 1980). The method requires introducing extra variables (duel variables) which are used as the coordinating instruments. The dual method cannot, however, be applied directly to solve the equations (2) and (3), even if these are linear. This is unlike the case of solving a large set of inequalities where the dual approach can be directly and successfully applied (Suri, 1981). The reason that, for a system of *equations*, the dual method is not directly applicable lies in the inherent lack of strict convexity properties (Allidina et al, 1982).

In order to apply the idea of decomposition by dual variables, the problem as given by equations (2) and (3) is transformed into an artificial optimization problem with a convenient performance index. This performance index can be chosen to satisfy any desired requirements (strong convexity, differentiability, etc.), since we are really inter-ested only in finding a point satisfying equations (2) and (3). The sufficient conditions under which the problem of solving equations (2) and (3) can be transformed into an optimization problem with adequate convexity properties for the dual method to be applicable, have been given in Singh et al (1983).

The performance index $J(.)$ must have the form:

$$J(x,y) = \sum_{i=1}^{\nu} J_i(x_i,y_i) \qquad (20)$$

In order to be able to work out efficient algorithms for solving the local and coordinator problems, and to have strong convexity, the functions $J_i(.,.)$ are chosen as follows:

$$J_i(x_i,y_i) = \frac{1}{2}||x_i||^2 + \frac{1}{2}||y_i||^2 \qquad (21)$$

Then the performance $J(.,.)$ (see equations (20) and (21)) is to be minimised subject to equations (2) and (3).

Suppose now that we will relax the coupling constraints (3) by forming the Lagrangian:

10

$$L(x,y,\lambda) \;=\; \sum_{i=1}^{\nu} \left| \frac{1}{2}||x_i||^2 + \frac{1}{2}||y_i||^2 + \lambda_i^T (y_i - A_i x_{i-1} + B_i x_{i+1}) \right| \tag{22}$$

where the $\lambda_i$'s are vectors of Lagrangian multipliers, $\lambda_i \in R^{n_{y_i}}$. The above Lagrangian can be separated as follows:

$$L(x,y,\lambda) \;=\; \sum_{i=1}^{\nu} L_i(x_i,y_i,\lambda) \tag{23}$$

where

$$L_i(x_i,y_i,\lambda) \;=\; \frac{1}{2}||x_1||^2 + \frac{1}{2}||y_i||^2 + \lambda_i^T y_i - \lambda_{i+1}^T A_{i+1} x_i + \lambda_{i-1}^T B_{i-1} x_i \tag{24}$$

and where $\quad \lambda^T \;=\; [\lambda_1^T, \lambda_2^T, \ldots, \lambda_\nu^T] \tag{25}$

Now for a given , we can define  local problems which can be solved in parallel.

Local Problem i,  i = 1,2,...ν.

$$\left. \begin{array}{l} \underset{x_i, y_i}{\text{Min}} \;\; L_i(x_i,y_i,\lambda) \\[20pt] \text{subject to } F_i(x_i,y_i) = 0 \end{array} \right\} \tag{26}$$

The solution $(\bar{x}_i(\lambda), \bar{y}_i(\lambda))$ of the i-th local problem is assumed to be unique, which is so if  proper conditions are satisfied (Findeisen et al, 1980). It should be noted that it is possible to fulfill these conditions for a wide class of nonlinear constraints $F_i(.,.) = 0$.

The solution of the local problems will be approached through the necessary conditions. For every local problem i, these necessary conditions are:

$$\frac{\partial L_i}{\partial x_i} + \mu_i^T \frac{\partial F_i}{\partial x_i} \;=\; 0 \tag{27}$$

$$\frac{\partial L_i}{\partial y_i} + \mu_i^T \frac{\partial F_i}{\partial y_i} \;=\; 0 \tag{28}$$

$$F_i(x_i,y_i) \;=\; 0 \tag{29}$$

In view of equation (24), equations (27)-(29) become:

$$x_i^T - \lambda_{i+1}^T A_{i+1} - \lambda_{i-1}^T B_{i-1} + \mu_i^T \frac{\partial F_i}{\partial x_i} \;=\; 0 \tag{30}$$

$$y_i^T + \lambda_i^T + \mu_i^T \frac{\partial F_i}{\partial y_i} \;=\; 0 \tag{31}$$

$$F_i(x_i,y_i) \;=\; 0 \tag{32}$$

The above equations can be solved for $x_i, y_i$ and $\mu_i$ in different ways. Details are given in Malinowski et al, (1985). It should be noted that for the linear case, the solution techniques discussed in the reference cited would give the solution in one step.

11

## Coordinator Problem

Find $\hat{\lambda} = (\hat{\lambda}_1, \hat{\lambda}_2, \ldots \hat{\lambda}_\nu)$ such that

$$\bar{y}_i(\hat{\lambda}) = A_i \bar{x}_{i-1}(\hat{\lambda}) + B_i \bar{x}_{i+1}(\hat{\lambda}) \tag{33}$$

where $(\bar{x}_i(\lambda), \bar{y}_i(\lambda))$ is the solution of the i-th local problem.

Thus, the task of the coordinator consists of finding $\lambda$ so as to satisfy equation (33) for $i = 1, 2, \ldots \nu$, which may be written as

$$y(\lambda) - C\, x(\lambda) = 0 \tag{34}$$

where $C$ is as defined before (see equation (12)), and $x(\lambda)$, $y(\lambda)$ represent the collection of local problem solutions $x_i(\lambda), y_i(\lambda)$. If we define a function $\phi(\lambda)$ ('dual function') as

$$\phi(\lambda) = \sum_{i=1}^{\nu} L_i(x_i(\lambda), y_i(\lambda), \lambda) \tag{35}$$

then satisfying equation (34) may be viewed as maximising $\phi$ with respect to $\lambda$ since

$$\left(\frac{\partial \phi(\lambda)}{\partial \lambda}\right)^T = y(\lambda) - Cx(\lambda) \tag{36}$$

The updates of $\lambda$ for maximising $\phi$ can be obtained by some hill-climbing procedure. Since the gradient of $\phi$ is readily available after solving the local problems for given $\lambda$, one may use gradient hill-climbing procedures in order to maximise $\phi$ and thus to solve equation (34).

However, since each evaluation of $(\partial \phi(\lambda)/\partial \lambda)$ requires the set of local problems to be solved, then it may be better to update $\lambda$ by a more efficient (Newton-Raphson type) procedure even at the expense of some extra computation. A Newton-Raphson type of procedure requires the Hessian of $\phi$ or at least some reasonable approximation of the Hessian to be computed. If the Hessian $H(\lambda)$ is available, then the coordinator can correct the value of $\lambda$ according to:

$$\lambda^{M+1} = \lambda^M - H(\lambda^M)^{-1} G(\lambda^M) \tag{37}$$

where

$$G(\lambda) = \left(\frac{\partial \phi}{\partial \lambda}\right)^T \tag{38}$$

and

$$H(\lambda) = \left(\frac{\partial G(\lambda)}{\partial \lambda}\right) = \frac{\partial((\partial \phi / \partial \lambda)^T)}{\partial \lambda} \tag{39}$$

The matrix $H(\lambda)$ consists of the blocks $(\partial^2 \phi / \partial \phi_i \partial \lambda_j)$. Now, from equation (36), since $(\partial \phi / \partial \lambda_i) = (y_1(\lambda) - A_i x_{i-1}(\lambda) - B_i x_{i+1}(\lambda))$, it follows that:

$$\frac{\partial^2 \phi}{\partial \lambda_i \partial \lambda_j} = 0 \quad \text{for } j \neq i-2, i-1, i, i+1, i+2 \tag{40}$$

In view of equation (40), $H$ would be of the form:

$$H = \begin{pmatrix} H_{11} & H_{12} & H_{13} & 0 & \cdots & 0 \\ H_{21} & H_{22} & H_{23} & H_{24} & \cdots & 0 \\ H_{31} & H_{32} & H_{33} & H_{34} & & 0 \\ 0 & H_{42} & H_{43} & H_{44} & & 0 \\ \vdots & & & & & \vdots \\ 0 & 0 & \cdots\cdots\cdots\cdots & & & H_{\nu\nu} \end{pmatrix} \tag{41}$$

where $\quad H_{ij} = \dfrac{\partial^2 \phi}{\partial\lambda_i \partial\lambda_j}$ , $\quad i,j = 1,2,\ldots\nu$.

It should be noted that $H_{ij}(\lambda) = H^T_{ji}(\lambda)$ for $j = i+1, i+2$. The development of the blocks $(\partial^2\lambda/\partial\lambda_i\partial\lambda_j)$ is given in Malinowski et al, (1985). It is apparent from the equations given there that a considerable effort would be required to set-up the exact Hessian, and this would not help our objective of solving the original system of equations as fast as possible. A suitable approximation is considered which decreases the required computation in setting up the Hessian. It is emphasised that the approximation is such that it provides an exact expression of the Hessian when the equations $F_i(x_i, y_i) = 0$ are linear.

It should be noted that some of the matrices involved in the setting up of the (approximate) Hessian are extremely sparse. Further, some other matrices that are also required for the Hessian would be available at the Local Problems. These features should obviously be taken into account during implementation.

Now, since an approximation is made in setting up the matrix $H(\lambda)$ ($= \partial(\partial\phi/\partial\lambda)^T$) according to the equations given in Malinowski et al (1985), it may perhaps be better to relax the updating formula for $\lambda$ given by equation (37) to the following:

$$\lambda^{M+1} = \lambda^M - \varepsilon[H(\lambda^M)]^{-1} G(\lambda^M) \tag{42}$$

where $o < \varepsilon \le 1$.

As mentioned earlier, the approximate procedure does provide the exact Hessian of $\phi$ in the linear case, in which case $H(\lambda) = H = $ constant. In this case, by taking $\varepsilon = 1$, the overall set of equations could be solved in one iteration.

Finally, it may be worthwhile to compare the results obtained when equation (42) is used for updating $\lambda$ with those obtained when a simple gradient update as given below is used,

$$\lambda^{M+1} = \lambda^M + G(\lambda^M) \tag{43}$$

It is of course also possible to use a conjugate gradient algorithm with a suitable effective linear search.

## 5. GENERAL COMMENTS AND CONCLUSIONS

In this paper, we have considered the problem of solving large sets of equations as given in Section 2. It has been shown that such equations may arise, for example, when simulating a system described by partial-differential equations. If it is necessary to simulate such a system

13

rapidly, then clearly it is desirable to split the overall problem into sub-problems each of which may be solved in parallel using a multi-component network. For this purpose, two decomposition-coordination techniques (the Direct method and the Dual method) have been adapted to solve the problem defined in Section 2.

The Direct method appears to be less complicated than the Dual method in relation to our problem. In particular, the local problems for the Direct method are simpler, and it has been easier to develop a reasonable coordination strategy. A general disadvantage of the Direct method is the difficulty in determining a feasible set of coordinating inputs $y$. However, if in any particular case, the variables $y_i$ are chosen as some physically meaningful quantities, then it should be possible to find a feasible set of the $y_i$'s. For example, when solving a set of partial-differential-equations describing gas flow in a pipeline, the variables $y_i$ are chosen as the boundary conditions for the segments of the pipeline, and the problem cited above does not appear.

A general comment as far as the Dual method in particular is concerned is that the matrix inverses implied in the development of the Hessian are not strictly necessary (Malinowski et al, 1985). This is to say, that when for example solving a set of equations $Ax = b$, it is not necessary to invert the matrix $A$. Any fast algorithm (making use of the sparsity of A) for finding x may be used. If numerical robustness is an issue, and the relevant matrix $A$ is symmetric and positive definite, then an appropriate procedure (e.g. Cholesky factorization) may be used for solving the equation $Ax = b$. Note that the matrices A are symmetric and positive definite (or at least non-negative definite), while $H(\lambda)$ is symmetric and should be negative definite. If $H(\lambda)$ is not negative definite, then the artificial optimisation problem created does not exhibit the required convexity properties. The Dual method is then not appropriate.

Finally, if the equations $F_i(x_i,y_i) = 0$ are linear and of the form $K_i x_i + L_i y_i - b_i = 0$, $i = 1,2,\ldots\nu$, then together with the coupling equations $y_i = A_i x_{i-1} + B_i x_{i+1}$, $i = 1,2,\ldots\nu$, the entire set of equations may be written as $AX = b$, where $X = (x,y)$ and A has a special structure. In such a case, the Direct method and the Dual method with the Newton-Raphson Algorithm used as the coordination strategy may be interpreted as special sparse matrix techniques for inverting A. The Newton-Raphson algorithm provides convergence in one step in the linear case.

The methods have been applied in the simulation of gas flow through a pipeline (Crorkin et al, 1985) and on reactor models (Allidina et al, 1985; Buro et al, 1985; Wang et al, 1985).

REFERENCES

Allidina, A.Y., Malinowski, K. and Singh, M.G., 1982, A note on parallel processing techniques for algebraic equations, ordinary differential equations and partial differential equations, Control Systems Centre Report No. 568, UMIST.

Allidina, A.Y, ed., 1984, Development of hierarchical techniques for the simulation of large scale systems with particular application to the nuclear industry, EEC Project Phase 1 Report, May, 1984.

Allidina, A.Y., Cook, R., Malinowski, K., Plowman, S., Singh, M.G., 1984, Real time simulation study for fast breeder dynamics and control, Interim Report No. 2, December 1984.

Arnold, C.P., Michael, I.P. and Michael, B.D., 1983, An efficient parallel algorithm for the solution of large sparse linear matrix equations, IEEE Transactions on Computers, vol. C-32, No. 3.

Barlow, R.H. and Evans, D.J., 1982, Parallel algorithms for the iterative solution to linear systems, The Computer Journal, vol. 25, No. 1, 1982.

Blech, R.A. and Arpasi, D.J., 1985, Hardware for a real-time multi-processor simulator, Distributed Simulation 85, San Diego, California, January 1985.

Brash, F.M. Jr., Van Ness, J.E. and Kang, S.C., Design of Multi-processor Structures for Simulation of Power-System Dynamics, Report, Electric Power Research Institute, Palo Alto, California 94304.

Burks, A.W., 1981, Programming and structure changes in parallel computers, Proceedings CONPAR 81, Springer-Verlag, Lecture Notes in Computer Science, vol. III, edited by W. Handler.

Crorkin, W., Allidina, A.Y., Malinowski, K. and Singh, M.G., 1985, "Decomposition-coordination Techniques for Parallel Simulation, Part 2", Large Scale Systems, North Holland.

Evans, D.J. and Haghighi, R.S., 1982, Parallel iterative methods for solving linear equations, Intern. J. Computer Math., vol. II, pp. 247-284.

Findeisen, m., Bailey, F.N., Brdys, M., Malinowski, K., Tatjewski, P. and Wozniak, A., 1980, "Control and Coordination in Hierarchical Systems", International Series on Applied Systems Analysis vol. 9, John Wiley and Sons.

Franklin, M.A., 1978, Parallel solution of ordinary differential equations, IEE Trans. on Computers, vol. C-27, No. 5.

Halada, L., 1981, A parallel algorithm for solving band systems and matrix inversion, Proceedings CONPAR 81, Springer-Verlag, Lecture Notes in Computer Science, vol. III, edited by W. Handler

Himmelblau, D.M., ed., 1973, "Decomposition of Large Scale Systems (collection of articles on decomposition and coordination techniques", American Elsevier, New York.

Katz, I.N., Franklin, M.A. and Sen, A., 1977, Optimally stable parallel predictors for Adams-Moulton correctors, Comp. & Maths. with Appls., vol. 3, pp. 217-233.

Lasdon, L.S., 1970, "Optimisation Theory for Large Systems", MacMillan, London.

Malinowski, K., Allidina, A.Y., Singh, M.G. and Crorkin, W., 1985, "Decomposition-coordination Techniques for Parallel Simulation - Part 1", Large Scale Systems, North Holland.

Miranker, W.L. and Liniger, W., 1967, Parallel methods for the numerical integration of ordinary differential equations, Math. Comput., vol. 21, pp. 303-320.

Miranker, W.L., 1981, Numerical methods for stiff equations, Mathematics and Its Applications, 15, D. Reidel Publishing Co.

Schendel, U., 1981, On basic concepts in parallel numerical mathematics, Proc. CONPAR 81, Springer-Verlag, Lecture Notes in Computer Science, vol. III, edited by W. Handles.

Singh, M.G., Allidina, A.Y. and Malinowski, K., 1983, Hierarchical simulation techniques, MECO 83, Athens, July.

Singh, M.G. and Titli, A., 1978, "Systems: Decomposition, Optimisation and Control", Pergamon Press, Oxford.

Suri, R., 1981, "Resource Management Concepts for Large Systems", Pergamon Press, Oxford.

Tao, H.M. and Saeks, R., 1984, Parallel System Simulation, IEEE Trans. on SMC, vol. SMC-14, No. 2.

Travassos, R. and Kaufman, H., 1980, Parallel algorithms for solving
    nonlinear two-point boundary-value problems which arise in
    optimal control, J. of Optimisation Theory & Applications,
    vol. 30, No. 1, January.
Wismer, D.A., 1971, Distributed multilevel systems, in "Optimisation
    Methods for Large Scale Systems", edited by D.A. Wismer,
    McGraw-Hill, New York.
Worland, P.B., 1976, Parallel methods for the numerical solution of
    ordinary differential equations, IEEE Trans. on Computers, Oct.

APPLICATION OF PARALLEL PROCESSING TO RELAP

R. Buro

Control Systems Centre, University of Manchester Institute

of Science and Technology, Manchester M60 1QD

ABSTRACT

Given a simulation problem, decomposition-coordination techniques can be used to enhance parallelism in order to speed-up the simulation. This paper describes the use of such techniques on a simplified reactor model. In particular, two decomposition-coordination techniques are applied to the fluid system of the model. The decomposition of the fluid system from the heat system, and further decomposition of the latter is also described.

With the eventual aim of investigating the possibility of speeding up the Nuclear code RELAP V, a discussion of the possible decomposition in RELAP V is also given in this paper.

1. INTRODUCTION

This paper is concerned with the possible application of two decomposition-coordination techniques to the RELAP 5 nuclear safety code. Section 2 describes the two techniques in detail. Section 3 demonstrates the results obtained when applying the above methods to a benchmark model which is described by a set of Partial Differential Equations. Two forms of discretization are applied, one leading to a set of nonlinear algebraic equations, the other to a set of linear algebraic equations. Section 4 discusses the application of the two decomposition-coordination techniques to the RELAP 5/Mod 1 nuclear safety code. This section is also concerned with describing the natural parallel components and areas of explicit coupling are pointed out.

2. DECOMPOSITION-COORDINATION TECHNIQUES

The Dual and Direct Decomposition-Coordination Techniques

A simulation problem can consist of solving a set of Partial-Differential Equations, Ordinary Differential Equations, Algebraic Equations or a mixture of the above, together with appropriate initial and boundary conditions. The objective is to find a solution over a certain time horizon $[t_o, t_f]$.

In this paper the decomposition-coordination techniques are applied to systems described by algebraic equations which arise from discretizing the original set of equations, i.e.

SYSTEM OF
ODE's AND PDE's $\longrightarrow$ SYSTEM OF LINEAR OR
NONLINEAR ALGEBRAIC EQN's

It should be observed that one can attempt to enhance parallelism either after a particular discretization scheme has been used or before such a discretization is done. It will be seen that if the Dual and Direct methods are applied after a particular discretization scheme is implemented, the accuracy of the final solution will not be altered as a result of using parallel computing.

Let us assume that, after discretization and given initial conditions and appropriate boundary conditions, the problem consists of having to solve a set of equations of the form:

$$F(z) = 0 \quad \text{where} \quad z \in R^M \quad \text{and} \quad F : R^M \to R^M \tag{1}$$

To decompose the above set of algebraic equations into subsets (subsystems) we will consider the following partitioned form (Allidina et al, 1984) of equation (1):

$$F_i(x_i, y_i) = 0 , \quad i = 1, 2, \ldots, \nu \tag{2}$$

and

$$y_i = A_i x_{i-1} + B_i x_{i+1} , \quad i = 1, 2, \ldots, \nu \tag{3}$$

where $x_i \in R^{n_{x_i}}$, $y_i \in R^{n_{y_i}}$, $F_i : R^{n_{x_i}} \ R^{n_{y_i}} \to R^{n_{x_i}}$

and $A_i = 0$, $B_\nu = 0$.

This paper is not concerned with the partitioning of equation (1) as such (Lei, 1985). It suffices to say that the partitioning should be effected in such a way so as either to minimize the number of inter-connecting variables

$$\sum_{k=1}^{\nu} y_k$$

or to reduce the 'degree of coupling' between the connected subsystems.

Having redefined the problem as in equations (2) and (3), we now consider ways of solving it with the use of a parallel computing facility (MIMO machine). One possible way of doing this would be to use successive linearization. One could then use a suitable 'parallel-system solver' (e.g. Evans and Haghighi, 1982; Halada, 1981). Another approach is to use decomposition-coordination schemes. This involves having independent parallel problems by introducing suitable coordination variables, based on the set of partitioned equations (2) and (3). Two methods will be considered.

2.1 The Direct Method

Here we treat the variables $y_i$, $i = 1, 2, \ldots \nu$ as the coordinating variables. Thus, for given $y_i$, $i = 1, 2, \ldots \nu$ it is possible to define $\nu$ independent local problems to be solved in parallel. The overall procedure is as follows.

(i)    Local Problem i,   $i = 1, 2, \ldots \nu$

For a given $y_i$, solve $F_i(x_i, y_i) = 0$ for $\bar{x}_i(y_i)$. If the local algebraic equations are nonlinear, an iterative scheme will need to be used to solve for $\bar{x}_i(y_i)$. For example, Newton-Raphson can be used to solve the local level as follows (provided $F_i$ is differentiable and that the Jacobian is non-singular):

$$x_i^{J+1}(y_i) = x_i^J(y_i) - \left[\frac{\partial F_i}{\partial x_i}\bigg|_{x_i = x_i^J}\right]^{-1} \cdot F_i(x_i^J, y_i)$$

with the error criterion being given by

$$||x_i^{J+1} - x_i^J|| = \sum_{k=1}^{n_{x_i}} \frac{|x_{i,k}^{J+1} - x_{i,k}^J|}{n_{x_i}} \leq \epsilon$$

If the local problem is linear, then it can be solved in one step. Sparse matrix techniques should be used wherever possible.

(ii) Coordinator Problem

Find $\hat{y} = (\hat{y}_1, \hat{y}_2, \ldots, \hat{y}_\nu)$ such that $\hat{y}_i = A_i \bar{x}_{i-1}(\hat{y}_{i-1}) + B_i \bar{x}_{i+1}(\hat{y}_{i+1})$.
The coordination process has to be initiated with an initial guess $y^o$ which can be taken as the value of $\hat{y}$ at the previous time level. The coordinator can correct for the value of $\hat{y}$ in two ways:

(A)   Relaxation

$$y^{M+1} = (1-w)y^M + w\,\bar{y}^{M+1} \tag{4}$$

where     $\bar{y}^{M+1} = [y_1^{M+1}, y_2^{M+1}, \ldots, y^{M+1}]^T$

and     $y_i^{M+1} = A_i \bar{x}_{i-1}(y_{i-1}^M) + B_i \bar{x}_{i+1}(y_{i+1}^M)$

which can be written as

$$\bar{y}^{M+1} = C\,\bar{x}(y^M)$$

where     $\bar{x}(y^M) = [x_1(y^M), x_2(y^M), \ldots, x_\nu(y^M)]^T$

and C has the following form:

$$C = \begin{bmatrix} 0 & B_1 & 0 & 0 & \ldots & 0 \\ A_2 & 0 & B_2 & 0 & \ldots & 0 \\ 0 & A_3 & 0 & B_3 & \ldots & 0 \\ 0 & & & & & \\ 0 & \ldots & & A_{\nu-1} & 0 & B_{\nu-1} \\ 0 & \ldots & & 0 & A_\nu & 0 \end{bmatrix}$$

and  w  is the relaxation parameter.

This coordination strategy is rather simple but has the advantage that a small amount of information needs to be passed between the coordinator and the subsystems, and the disadvantage that many iterations are required.

(B) Newton-Raphson

Define $D(y) = y - C\bar{x}(y)$ . Then the coordinator can update $y$ according to

$$y^{M+1} = y^M - \left[\frac{\partial D(y)}{\partial y}\bigg|_{y=y^M}\right]^{-1} . D(y) \qquad (5)$$

where

$$\frac{\partial D(y)}{\partial y} = I - C \frac{\partial \bar{x}(y)}{\partial y} \qquad (6)$$

To compute the Jacobian in (5) we need to find

$$\frac{\partial \bar{x}(y)}{\partial y} = \text{block diag}\left[\frac{\partial \bar{x}_i(y_i)}{\partial y_i}\right] \qquad (7)$$

where

$$\frac{\partial \bar{x}_i(y_i)}{\partial y_i} = -\left[\frac{\partial F_i}{\partial x_i}\bigg|_{\bar{x}_i(y_i).y_i}\right]^{-1} \frac{\partial F_i}{\partial y_i}\bigg|_{\bar{x}_i(y_i).y_i} \qquad (8)$$

Equation (8) can be solved for $\partial \bar{x}_i(y_i)/\partial y_i$ at the local levels using any sparsity in $[\partial F_i/\partial x_i]$ and only the elements which contribute to $\partial D(y)/\partial y$ in (6) need be sent to the coordinator. In the coordinator, equation (5) can also be solved using any sparsity present in the Jacobian.

## 2.2 The Dual Method

This method requires introduction of additional variables which are used as the coordinating variables. The method is another basic decomposition-coordination technique e.g. Singh and Titli, 1978; Findeisen et al, 1980).

To apply decomposition by dual variables we transform the problem given by equations (2) and (3) into an artificial optimization problem with a convenient performance index. The performance index chosen is of the form

$$J(x,y) = \sum_{i=1}^{\nu} J_i(x_i,y_i)$$

To satisfy strong convexity and differentiability, let

$$J_i(x_i,y_i) = \frac{1}{2}||x_i||^2 + \frac{1}{2}||y_i||^2 \qquad (9)$$

The optimization problem becomes that of minimizing (9) subject to (2) and (3). The solution is approached by forming the Lagrangian by relaxing the coupling constraints (3) and rearranging so that:

$$L(x,y, ) = \sum_{i=1}^{\nu} L_i(x_i,y_i,\lambda)$$

where $\lambda$ is a vector Lagrange multipliers, and

$$L_i(x_i, y_i, \lambda) = \frac{1}{2}||x_i||^2 + \frac{1}{2}||y_i||^2 + \lambda_i^T y_i - \lambda_{i+1}^T A_{i+1} x_i - \lambda_{i-1}^T B_{i-1} x_i$$

The procedure for solution is as follows:

(i)  Local problem i,   i = 1,2,...$\nu$

$$\begin{array}{c} \text{Min} \\ x_i, y_i \end{array} L_i(x_i, y_i, )$$

subject to  $F_i(x_i, y_i) = 0$

The solutions of the local problems are obtained from the necessary conditions for stationarity:

$$x_i^T - \lambda_{i+1}^T A_{i+1} - \lambda_{i-1}^T B_{i-1} + \mu_i^T \frac{\partial F_i}{\partial x_i} = 0 \tag{10}$$

$$y_i^T + \lambda_i^T + \mu_i^T \frac{\partial F_i}{\partial y_i} = 0 \tag{11}$$

$$F_i(x_i, y_i) = 0 \tag{12}$$

If (12) is linear, the above equations can be solved directly for $\mu_i$ and back substitution would yield $x_i$ and $y_i$. However, if (12) is nonlinear an iterative scheme will be required to solve equations (10)-(12).  One such method is as follows.  $x_i$ and $y_i$ are first expressed as functions of $\mu$ , then substituted into (12).  The resulting function is then linearized about $\mu_i^N$ .  Solution of the linearized equation gives $\mu_i^{N+1}$ and back substition yields $x_i^{N+1}$ and $y_i^{N+1}$ .

(ii)  Coordinator problem

Find  $\hat{\lambda} = (\hat{\lambda}_1, \hat{\lambda}_2, ... \hat{\lambda}_\nu)$  such that

$$\bar{y}_i(\hat{\lambda}) = A_i \bar{x}_{i-1}(\hat{\lambda}) + B_i \bar{x}_{i+1}(\hat{\lambda}) \tag{13}$$

Now let

$$\phi(\lambda) = \sum_{i=1}^{\nu} L_i(x_i(\lambda), y_i(\lambda), \lambda)$$

Maximizing the above with respect to $\lambda$ satisfies (13), since

$$\frac{\partial \phi(\lambda)}{\partial \lambda} = y(\lambda) - Cx(\lambda)$$

To find the maximum, a gradient hill-climbing technique can be used:

$$\lambda^{M+1} = \lambda^M + \alpha G(\lambda)$$

where  $G(\lambda) = \partial\phi/\partial\lambda$ where  $\alpha$  is a parameter to be chosen.

Apart from the simple gradient update, we can also use a more efficient Newton-Raphson type procedure which requires the elements $\partial^2\phi/\partial\lambda_i \partial\lambda_j$  of the Hessian.  In order to compute the Hessian it is necessary to have the following derivatives:

$$\frac{\partial y_i}{\partial \lambda_j} \quad \text{and} \quad \frac{\partial x_i(\lambda_i)}{\partial \lambda_j} \quad\quad \begin{array}{l} \text{for } j = i-1, i, i+1 \\ \text{and } i = 1,2,...\nu \end{array}$$

The above are obtained by differentiating the necessary conditions for the local problems (10),(11) and (12). For example, differentiating with respect to $\lambda_{i-1}$ gives:

$$\frac{\partial x_i(\lambda)}{\partial \lambda_{i-1}} + \left[\frac{\partial}{\partial x_i}\left[\frac{\partial F_i^T}{\partial x_i}\mu_i\right]\right]\frac{\partial x_i(\lambda)}{\partial \lambda_{i-1}} + \left[\frac{\partial}{\partial y_i}\left[\frac{\partial F_i^T}{\partial x_i}\mu_i\right]\right]\frac{\partial y_i(\lambda)}{\partial \lambda_{i-1}}$$
$$+ \frac{\partial F_i^T}{\partial x_i}\cdot\frac{\partial \mu_i(\lambda)}{\partial \lambda_{i-1}} = B_{i-1}^T$$

$$\frac{\partial y_i(\lambda)}{\partial \lambda_{i-1}} + \left[\frac{\partial}{\partial x_i}\left[\frac{\partial F_i^T}{\partial y_i}\mu_i\right]\right]\frac{\partial x_i(\ )}{\partial \lambda_{i-1}} + \left[\frac{\partial}{\partial y_i}\left[\frac{\partial F_i^T}{\partial y_i}\mu_i\right]\right]\frac{\partial y_i(\lambda)}{\partial \lambda_{i-1}}$$
$$+ \frac{\partial F_i^T}{\partial y_i}\cdot\frac{\partial \mu_i(\lambda)}{\partial \lambda_{i-1}} = 0$$

To simplify the above equations we approximate the Hessian in such a way that if $F_i(x_i,y_i) = 0$ were linear, the Hessian would be exact. Thus, we drop the terms involving the matrices

$$\frac{\partial}{\partial x_i}\left[\frac{\partial F_i^T}{\partial x_i}\mu_i\right] \quad \text{and} \quad \frac{\partial}{\partial y_i}\left[\frac{\partial F_i^T}{\partial x_i}\mu_i\right]$$

and the diagonal terms will become:

$$\frac{\partial^2 \phi}{\partial \lambda_i \partial \lambda_i} = -A_i A_i^T + A_i\left(\frac{\partial F_{i-1}}{\partial x_{i-1}}\right)^T J_{i-1}(\lambda)^{-1}\left(\frac{\partial F_{i-1}}{\partial x_{i-1}}\right)A_i^T$$
$$- B_i B_i^T + B_i\left(\frac{\partial F_{i+1}}{\partial x_{i+1}}\right)^T J_{i+1}(\lambda)^{-1}\left(\frac{\partial F_{i+1}}{\partial x_{i+1}}\right)B_i^T$$
$$- I + \left(\frac{\partial F_i}{\partial y_i}\right)^T J_i(\lambda)^{-1}\left(\frac{\partial F_i}{\partial \lambda_i}\right)$$

The off-diagonal terms can also be simplified in a similar way.

The elements of the Hessian are evaluated at the local levels and no inversion of matrices is necessary. The coordinator updates $\lambda$ according to:

$$\lambda^{M+1} = \lambda^M - \epsilon H(\lambda^M)^{-1} G(\lambda^M)$$

where $G(\lambda^M) = (\frac{\partial \phi}{\partial \lambda})^T$.

The Hessian is block five-diagonal and if convergence at the coordinator is to be achieved, the Hessian must be negative-definite.

Summarizing both the Direct and Dual methods, it can be seen that if the local $F_i(x_i,y_i)$ are linear, convergence would be obtained in one iteration for a Newton-Raphson type coordinator update.

22

## 3. APPLICATION TO A BENCHMARK MODEL

The system simulated consists of fluid flowing along an open ended pipe, as shown in Fig. 1.

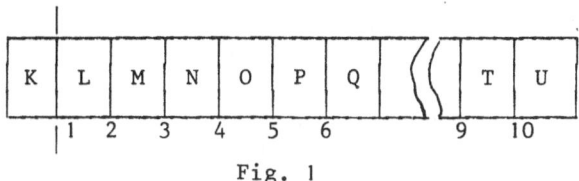

Fig. 1

The system is described by a set of nonlinear hyperbolic partial differential equations. These equations are:

Continuity 
$$\frac{\partial \rho}{\partial t} + \frac{\partial (\rho v)}{\partial x} = 0$$

Thermal Energy Eqn. 
$$\frac{\partial (\rho u)}{\partial t} + \frac{\partial (\rho v u)}{\partial x} = -P \frac{\partial v}{\partial x} + Q + \rho v^2 FW$$

Momentum 
$$\rho \frac{\partial v}{\partial t} + \frac{\rho}{2} \frac{\partial v^2}{\partial x} = -\frac{\partial P}{\partial x} + \rho Bx - \rho v FW$$

Two forms of discretization were tried on the above set of PDE's, resulting in:

(a) A set of nonlinear algebraic equations. Fluid was taken as methane gas and the state equation was assumed to be ideal. All properties were defined at cell centres.

(b) A set of linear algebraic equations. Flow was taken as water and the state equation was obtained by linearizing the dependent properties around the previous time level.

Vector properties such as velocity are defined at the junctions of the meshes in Fig. 1, i.e. at points 1,2,3, etc., while scalar properties such as temperature, density and pressure are defined at the cell centres, i.e. points K,L,M,N, etc.

Thus, for case (A):

$$Y_i = \begin{bmatrix} \rho_{i-1,k}^{j+1} \\ T_{i-1,k}^{j+1} \\ \overline{\phantom{v}} \\ v_{i+1,1}^{j+1} \end{bmatrix}$$

Elements from subsystem i-1

Element from subsystem i+1

where k is the number of cells in subsystem i and

$$X_i = \begin{bmatrix} \rho_{i,1}^{j+1}, \rho_{i,2}^{j+1}, \ldots, \rho_{i,k}^{j+1}, & T_{i,1}^{j+1}, T_{i,2}^{j+1}, \ldots, T_{i,k}^{j+1}, & v_{i,1}^{j+1}, v_{i,2}^{j+1}, \ldots, v_{i,k}^{j+1} \end{bmatrix}$$

and for case (B)

$$Y_i = \begin{bmatrix} p^{j+1}_{i-1,k} \\ - \overline{\phantom{p}} \overline{\phantom{j+1}} \overline{\phantom{p}} - \\ p^{j+1}_{i+1,k} \end{bmatrix} \qquad \begin{array}{l} \text{Elements of subsystem i-1} \\[1.5em] \text{Elements of subsystem i+1} \end{array}$$

and

$$X_i = \begin{bmatrix} p^{j+1}_{i,1}, p^{j+1}_{i,2}, \ldots, p^{j+1}_{i,k-1}, p^{j+1}_{i,k} \end{bmatrix}$$

The results of cases (A) and (B) are not compared with each other. All results were achieved using a serial computer.

Figure 2 shows the results for the nonlinear set of algebraic equations Case (A). It must be pointed out that no use is made of sparse matrix routines. Figure 3 shows the results for Case (B). CPU time is plotted versus the number of subsystems for the Direct Method. It can be seen that when the number of subsystems is greater than three no further speed up would be obtained if a parallel processing facility were used. This is due to the increased computational effort required at the coordinator. Multi-level hierarchies may aleviate this.

## 4. POSSIBLE APPLICATION OF DECOMPOSITION-COORDINATION TECHNIQUES TO RELAP 5/MOD 1

Figure 4 shows a schematic representation of a typical RELAP 5/Mod 1 nodalization. The diagram shows that primary and secondary cooling systems are simulated as is the core with heat structures. The pressurizer, pump and valves are also simulated (RELAP 5/Mod 1 Manual, 1982).

It can immediately be seen that the secondary cooling system (nodal volumes 500-555) can be simulated in parallel to the primary cooling system. The pressurizer (volumes 400-420) can also be treated in parallel (volumes 225-240). In addition, the emergency core cooling systems (volumes 300-330 and volumes 335-365) could also be simulated in parallel to the primary cooling system.

At this stage it should be noted that to obtain maximum efficiency by utilizing the parallelism in the system, the tasks for each processor must be approximately equal. Thus, it is necessary to distribute the computational tasks evenly amongst the processors.

To show the parallelism in a nuclear reactor system in a more appropriate form, a simplified model of the reactor was developed. This is shown in Fig. 5.

The model considered consists of four main features: a Fluid Flow model, a Heat Structures model, a Reactor Kinetics model and a Pump model. In the RELAP 5/Mod 1 code, the coupling between fluid and heat structures is explicit, as is the coupling between the Heat Structures and the Reactor Kinetics. This implies that each of the three components can be advanced in parallel with an exchange of information being made after every time step. If, in addition, the coupling between the components is weak it may become possible to interchange information only after several time step advancements (Birta, 1980; Franklin, 1978). In RELAP the heat structure sections can themselves be advanced independently, as axial heat flow in the core is assumed negligible.

Apart from the natural parallelism in the system, decomposition-

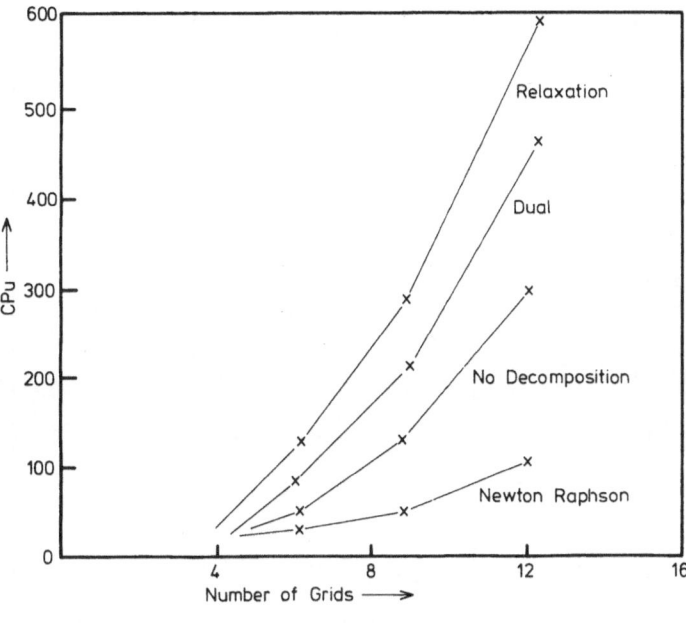

Fig. 2

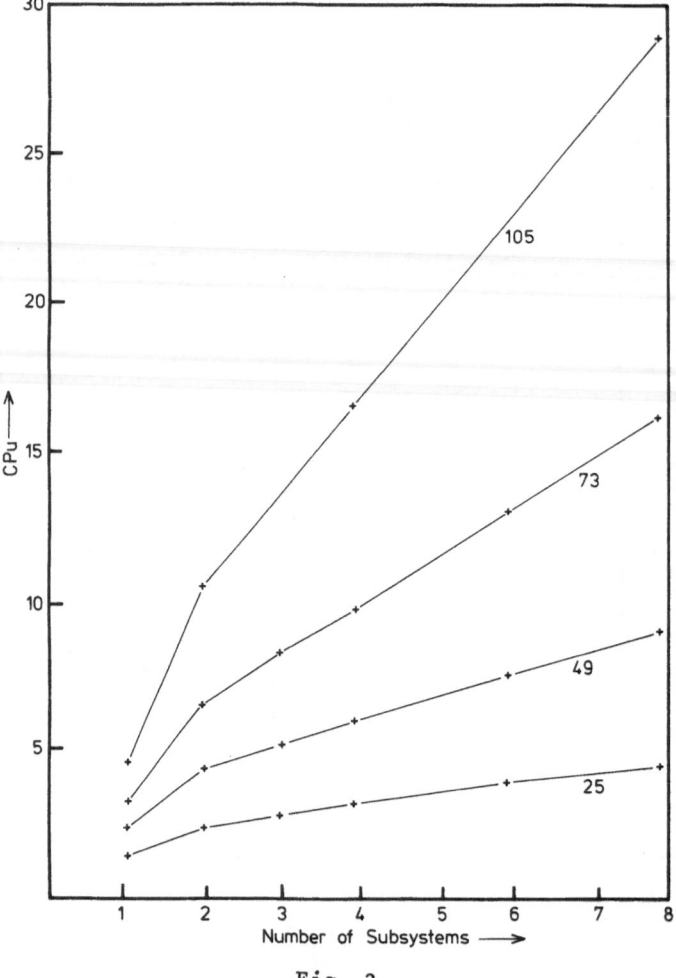

Fig. 3

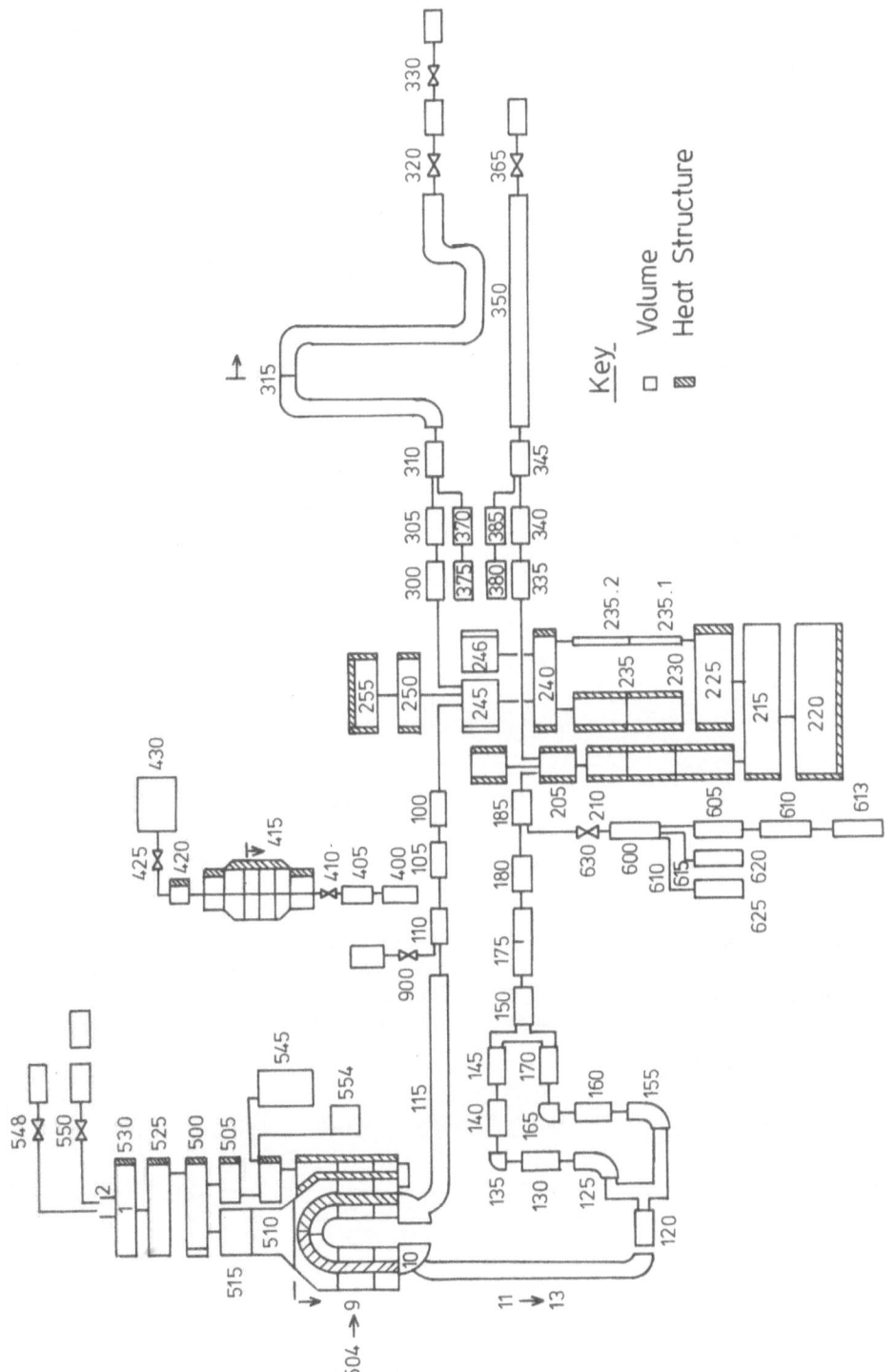

Fig. 4  RELAP V/MOD 1 NODALIZATION FOR LOFT TEST

Key

□  Volume

▨  Heat Structure

27

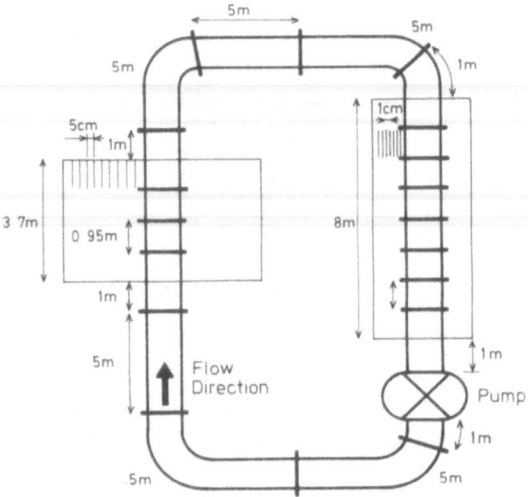

Fig. 5 Simplified Reactor Model

Closed circuit pipe is assumed constant area. Diameter of pipe = 10cm.
Velocity of flow at steady state = 4.8ms$^{-1}$.
Core contains 4 volumes each of 10 nodes.
Heat Exchanger contains 8 volumes each of 10 nodes.

coordination techniques can be applied to the fluid system. In RELAP the partial Differential Equations describing the fluid flow are discretized so that a linear system of equations results effectively making the Direct Method act as a particular parallel system solver. In the simplified model of Fig. 5, the system of equations is in fact a tri-diagonal system of equations and the use of the direct method would not produce great speed-up. However, the more generalized system of Fig. 4 would lead to a fluid system equation which was not diagonally bounded, and the Direct and Dual methods could produce considerable speed-up if implemented on a parallel processing facility.

## 5. ACKNOWLEDGMENT

The author would like to thank Mr. W. Crorkin for his contribution to this paper.

## REFERENCES

Allidina, A.Y., ed., 1984, Development of hierarchical techniques for the simulation of large scale systems with particular application to the nuclear industry, EEC Project, Phase I Report, May 1984.

Lei, S., Allidina, A.Y. and Malinowski, K., 1985, Clustering technique for rearranging ODE systems, Control Systems Centre Report 635, UMIST.

Evans, D.J. and Haghighi, R.S., 1982, Parallel iterative methods for solving linear equations, Intern. J. Computer Math., vol. II, pp. 247-284.

Halada, L., 1981, A parallel algorithm for solving band systems and matrix inversion, Proceedings CONPAR 81, Springer-Verlag, Lecture notes in Computer Science, vol. III, edited by Handler.

Singh, M.G. and Titli, A., "Systems, Decomposition, Optimization and Control", Pergamon Press, Oxford.

RELAP 5/Mod 1 Code Manuals, vol. 1, 2 and 3, U.S. Dept. of Energy, 1982.

Birta, L.D., 1980, A quasi-parallel method for the simulation of loosely coupled continuous subsystems, Mathematics and Computers in Simulation XXII, pp. 189-199.

# CLUSTERING TECHNIQUE FOR REARRANGING ODE SYSTEMS

S. Lei,  A.Y. Allidina and  K. Malinowski

Control Systems Centre, University of Manchester Institute

of Science and Technology, P.O. Box 88, Manchester, U.K.

## ABSTRACT

This report presents a technique which makes use of the concept of single and double connected clusters to rearrange a large system of ODEs into a 'nearly' block-diagonal form. The eventual aim is to partition the large system of ODEs into subsystems with few interactions between them. This is useful, for example, when employing parallel processing together with decomposition techniques for simulating large dynamic systems. Based on the analysis presented, an algorithm has been implemented which provides an automatic procedure for clustering system variables in a desired form.

## 1. INTRODUCTION

When dealing with large scale systems it is often necessary to partition the system into subsystems in order to cope with, for example, the problem of dimensionality and/or to use parallel processing.  The performance of the solution technique depends upon, amongst other things, the manner in which the partitioning is done.

System partitioning is required in many areas, such as in the study of electrical power networks, ordered Gaussian elimination on sparse systems, circuit design, application of hierarchical optimisation techniques (see, e.g., Lasdon, 1970), etc. The issues related to problem partitioning have been considered by many authors. For example, Carre (1968) presents a method for calculating load-flows by partitioning systems into trees in networks. Ogbuobiri et al (1970) give a survey of sparsity-directed decomposition for Gaussian elimination. The survey reviews some early work concerned with ordering schemes but they require a special structure of the system to be partitioned. Several other schemes for problem partitioning are given in Himmelblau (ed.) (1972, but these are mostly concerned with acyclic partitions and tearing methods for solving sets of equations using sequential processing. In the recent work by Siljak et al (1982), Sezer and Siljak (1984), and Vannelli and Vidyasagar (1984), several concepts and methods for system partitioning were presented. In particular, Sezer and Siljak (1984) consider the so-called 'ε-decompositions' in which weights are assigned to interactions and the system partitions are carried out so that the interactions between them are below a specified threshold.  The above approaches are oriented towards different applications, for example decentralised control design.

A particular problem of interest is the simulation of a large scale dynamic system (Malinowski et al, 1984) by solving large sets of Ordinary Differential Equations (Allidina and coworkers 1985; Singh and others 1983, 1985).  In order to speed up the simulation of such a system it is desirable to use parallel processing which requires the system to be partitioned into subsystems.  Let us assume that the set of ordinary differential equations is:

$$\dot{x} = f(x) , \qquad x = [x_1, \ldots, x_N]^T \qquad\qquad (1)$$

and we would like to partition it into the following subsets:

$$\dot{x}^i = f_i(x^i, \bar{x}^i) , \qquad i = 1, 2, \ldots, \nu \qquad\qquad (2)$$

where $\bar{x}^i$ represents those variables that are contained in the i-th block of equations and are not included in the subvector $x^i$.  The variables $\bar{x}^i$ therefore represent the interactions between the i-th subsystem and the rest of the system.  There are different criteria that can be used in attempting to carry the partition discussed above (see, e.g. Singh and coworkers 1985).  One particular criterion is to have the number of interactions as small as possible.  Now, in the simulation problem mentioned earlier, the subsystems can, for example, be solved in parallel on processors.  The advantage of having a small number of interaction variables is that the size of the coordination problem becomes smaller making the overall solution more efficient.

In many practical cases an efficient partitioning can be done "by hand", taking into account the physical features of the system.  It is desirable, however, to have an automatic procedure that can carry out the partitioning with a small number of interactions.  Such a procedure could consist of the following two steps:  (1)  rearrangement of the components of the vector  x ;  (2)  actual partitioning, i.e. splitting the rearranged vector  x  into subvectors  $x^i$  (where  $x = [x^{1T}, \ldots, x^{\nu T}]^T$) corresponding to a desired number of subsystems of desired sizes.  The second step is, of course, more straightforward, since the splitting is to be done without any further reordering of variables.  It is the first step which should provide a proper sequence of variables and equations.

The basic idea behind the method presented in this report is as follows.  We do not want, since it may not be clear at this preliminary stage, to specify *a priori* either the number of subsystems or their sizes. Having in mind the potential objective of minimising the number of inter- action inputs to the subsystems, we want to develop an automatic procedure which would be capable of rearranging (renumbering) the variables  $x_i$, together with associated equations, so as to expose the system structure and make the partitioning decisions (regarding number and sizes of sub- systems) easier.

In order to explain this idea we first associate with a given large system of ODEs (eqn. (1)) a square binary occurrence matrix $P = [p_{ij}]$, such that

$$p_{ii} = 1 \quad \text{by definition}$$

$$p_{ij} = 1 \quad \text{if variable } x_j \text{ appears in the i-th equation}$$

$$p_{ij} = 0 \quad \text{otherwise.}$$

Now we would like to rearrange the variables $x_i$ within the vextor $x$, together with the respective equations, so that the nonzero entries in the permuted matrix $\tilde{P} = [\tilde{p}_{ij}]$ are grouped in clusters along the main diagonal.

$$
P = \begin{array}{c} \\ \text{eq.1} \\ \text{eq.2} \\ \text{eq.3} \\ \text{eq.4} \\ \text{eq.5} \\ \text{eq.6} \end{array}
\begin{array}{cccccc} x_1 & x_2 & x_3 & x_4 & x_5 & x_6 \\ 1 & 0 & 1 & 0 & 0 & 0 \\ 0 & 1 & 0 & 1 & 1 & 0 \\ 0 & 0 & 1 & 0 & 0 & 1 \\ 0 & 1 & 0 & 1 & 1 & 0 \\ 0 & 1 & 0 & 0 & 1 & 1 \\ 1 & 0 & 1 & 0 & 0 & 1 \end{array}
\quad , \quad
\tilde{P} = \begin{array}{c} \\ \text{eq.1} \\ \text{eq.3} \\ \text{eq.6} \\ \text{eq.5} \\ \text{eq.2} \\ \text{eq.4} \end{array}
\begin{array}{ccc|ccc} x_1 & x_3 & x_6 & x_5 & x_2 & x_4 \\ 1 & 1 & 0 & 0 & 0 & 0 \\ 0 & 1 & 1 & 0 & 0 & 0 \\ 1 & 1 & 1 & 0 & 0 & 0 \\ \hline 0 & 0 & 1 & 1 & 1 & 0 \\ 0 & 0 & 0 & 1 & 1 & 1 \\ 0 & 0 & 0 & 1 & 1 & 1 \end{array}
$$

Fig. 1

For example, in Fig. 1 an initial occurrence matrix P is shown, together with matrix P which represents the structure of the system after suitable rearrangement of variables. It is easy to observe from P that the associated system of equations can be split into two third-order subsystems with only one interaction from subsystem 1 to subsystem 2 ($\tilde{p}_{43} = 1$). For a large system of equations the task of grouping the nonzero elements of P into clusters along the main diagonal of P is by no means a trivial one. Also, it cannot be uniquely quantitatively defined (e.g. in terms of minimising the number of interactions), since we would not like to specify *a priori* the number and the sizes of subsystems.

The permutations of the occurrence matrices in order to obtain a desired rearrangement of equations and variables or to extract from those matrices valuable information regarding system structural properties, are commonly used in graph-theoretical analysis of dynamic systems, sets of linear or nonlinear equations, etc. One could consider here the method mentioned before of Vannelli and Vidyasagar (1984), or the approaches presented in Siljak and coworkers (1982), aimed at decomposing the system into hierarchically ordered (acyclic) input reachable (or input-output reachable) partitions. In particular, Vannelli and Vidyasagar (1984) consider partitioning of a weighted graph into k subgraphs and they maximise the total weight in each subgraph and an algorithm is given for the case of equipartitioning.

At this point it is worthwhile mentioning that the acyclic partitions as considered by Siljak and coworkers (1982) can be vital for stability stability analysis, decentralised controller design, or for solving large sets of equations on a single processor computer (by sequentially solving subsequent complexes of equations (Himmelblau (ed.), 1972). Yet in many cases there may not exist nontrivial acyclic partitions of a given set of equations, and also this kind of partitioning may not necessarily lead to a small number of interactions between the subsystems.

The method presented in this report can be regarded as a general purpose algorithm for obtaining 'nearly' block-diagonal structure of a given system.

The report is organised as follows. In Section 2 some preliminary notions are introduced. Then in Sections 3 and 4 the method is presented. In particular, Section 3 deals with identifying completely independent subsystems if these exist. In Section 4 the method for determining groups of clusters of variables ('connected' variables) is presented, together with the general rules for 'chaining' these groups (providing a required permuted form $\tilde{P}$ of the occurrence matrix) is described. Finally, in Section 5 the examples are given. They demonstrate that the developed computer program can handle quite easily the rearrangement of variables for systems of large dimensions.


## 2. PRELIMINARY NOTIONS AND BASIC IDEAS

Given a set of ordinary differential equations (eqn. (1)), and the binary occurrence matrix $P = [p_{ij}]$ related to this set of equations, (let us remember that $p_{ij} = 1$ if variable $x_j$ enters the i-th equation) one may introduce a corresponding digraph (ordered graph) $G = (V, E_G)$, where $V$ is a set of vertices $v_i$ representing the variables $x_i$, $i = 1,\ldots,N$, (and the associated equations), and $E_G$ is a set of edges $(v_j, v_i)$ directed from vertex $v_j$ to vertex $v_i$. An edge $(v_j, v_i)$, $i \neq j$, exists if $p_{ij} = 1$. With a given occurrence matrix $P$ and hence with a given digraph $G$, we can also associate the single connectivity (s-connectivity) N by N matrix $S = [s_{ij}]$, where $s_{ij} = 1$ if $p_{ij} = 1$ or $p_{ji} = 1$ and $s_{ij} = 0$ otherwise. It is clear that $S = S^T$. The s-connectivity matrix $S$ corresponds to a graph $G_S = (V_1, E_{GS})$, where an (nondirected) edge $e_{ij} \in E_{GS}$ between the vertices $v_i$ and $v_j$ exists whenever $s_{ij} = 1$. Similarly, we can define a double-connectivity (d-connectivity) matrix $D = [d_{ij}]$, such that

$$d_{ij} = 1 \quad \text{if} \quad p_{ij} = 1 \quad \text{and} \quad p_{ji} = 1$$

$$d_{ij} = 1 \quad \text{otherwise,}$$

d-connectivity matrix $D$ $(D = D^T)$ corresponds then to a graph $G_D = (V, E_{GD})$, where a non-directed edge $e_{ij} \in E_{GD}$ between the vertices $v_i$ and $v_j$ exists whenever $d_{ij} = 1$.

The subset $V^k$ of the set of vertices $V$ is called a 'single-connected cluster' (sc-cluster) if for every $v_i, v_j \in V^k$ we have $s_{ij} = 1$. The sc-cluster $V^k \subset V$ is said to be a 'maximal sc-cluster' if it is not a proper subset of any other sc-cluster. Similarly, we define a 'double-connected cluster' (dc-cluster) as the subset $V^\ell \subset V$ in which for every $v_i, v_j \in V^\ell$ $d_{ij} = 1$. The dc-cluster $V^\ell$ is a maximal dc-cluster if it is not a proper subset of any other dc-cluster.

In order to illustrate the above notions let us consider the following system of ODEs:

$$
\begin{aligned}
\dot{x}_1 &= f_1(x_1,x_2,x_4,x_5) \\
\dot{x}_2 &= f_2(x_2,x_3,x_4) \\
\dot{x}_3 &= f_3(x_2,x_3) \\
\dot{x}_4 &= f_4(x_1,x_4,x_5) \\
\dot{x}_5 &= f_5(x_1,x_2,x_4,x_5)
\end{aligned}
\qquad
P =
\begin{bmatrix}
1 & 1 & 0 & 1 & 1 \\
0 & 1 & 1 & 1 & 0 \\
0 & 1 & 1 & 0 & 0 \\
1 & 0 & 0 & 1 & 1 \\
1 & 1 & 0 & 1 & 1
\end{bmatrix}
$$

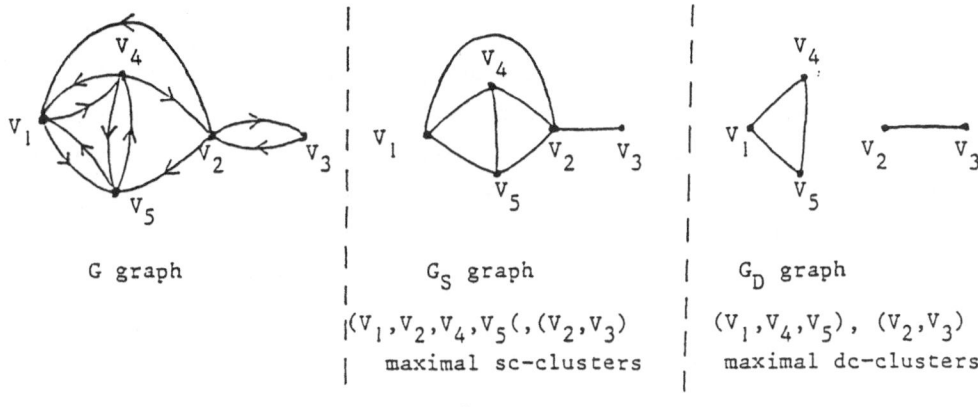

G graph

$G_S$ graph

$(v_1, v_2, v_4, v_5(, (v_2, v_3)$
maximal sc-clusters

$G_D$ graph

$(v_1, v_4, v_5), (v_2, v_3)$
maximal dc-clusters

Fig. 2

In what follows we will also need the notions of a complete non-directed graph $G_C = (V, E)$ with an edge between every pair of vertices $v_i$ and $v_j$ and of complementary graphs $G_S'$ and $G_D'$, where $G_S' = (V, E\backslash E_{GS})$, $G_D' = (V, E\backslash E_{GD})$. For example, in graph $G_S'$ an edge between the vertices $v_i$ and $v_j$ exists when $s_{ij} = 0$. Complementary graphs $G_S'$ and $G_D'$ correspond to a complementary s-connectivity matrix $S' = [s_{ij}']$, where $s_{ij}' = 1$ when $s_{ij} = 0$ ($s_{ij}' = 0$ when $s_{ij} = 1$), and, respectively, to a complementary d-connectivity matrix $D' = [d_{ij}']$, where $d_{ij}' = 1$ when $d_{ij} = 0$ and $d_{ij}' = 0$ when $d_{ij} = 1$.

Finally, we define by $A_i$ the set of all edges in a complete graph $G_C$ which are connected to a vertex $v_i$ and by $A_i'$ the complementary set of edges, i.e. $A_i' = E\backslash A_i$.

In Sections 4 and 5 the above graphs and sets will be used to develop the clustering technique leading to a proper rearrangement of system variables. The basic idea of the method is the following. The algorithm will first of all find all maximal sc-clusters and then, within them, all maximal dc-clusters. The maximal dc-clusters will be used to rearrange the variables within the sc-clusters and then the sc-clusters will be 'chained' properly - resulting finally in a desired permutation of the occurrence matrix. Before the above procedure is applied it is first necessary to examine whether the system of equations can be split into two or more independent subsystems.


3.  IDENTIFICATION OF INDEPENDENT SUBSYSTEMS

The logical first step when given a large set of equations (1) is to check whether this system contains any disconnected subsystems - the components in terms of graph theory. There are several algorithms available for finding the components in a digraph. For example, the 'fussion' method which replaces with one (fuses) any two connected vertices. Here we present a simple algorithm which 'fuses' the rows of an occurrence matrix P. This algorithm, coded in FORTRAN, is a part of the software described in this report. The algorithm is as follows:

(i)   Set NxN matrix $B = b_{ij}$ such that $b_{ij} = j \cdot P_{ij}$.
      Assign label (-) to each row of B.   Set $\ell = 0$.

(ii)  Change $\ell$ to $\ell+1$. Put the first row of B with label (-) in the $\ell$-th row of matrix C and change the label of this row of B to (+).

(iii)   Compare the $\ell$-th row of C with every row of B with label (−).  If
        for, say, the k-th row of B, $c_{\ell j} = b_{kj} > 0$  for at least one
        $j \in \{1,\dots,N\}$ then replace each element of the $\ell$-th row of C with
        $\bar{c}_{\ell i} = \max(c_{\ell i}, b_{ki})$, $i = 1,\dots,N$, and change the label of the k-th
        row of B to (+).  After all rows of B with labels (−) have been
        compared with the $\ell$-th row of C, proceed to the next step.

(iv)    Go to step (iii) unless all rows of B have labels (+), in which
        case there are $\ell$ disconnected subsystems and each row of C con-
        tains the numbers of equations (and variables) forming an
        independent subsystem.


It is clear that further on we can consider each disconnected subsystem separately and thus in the next sections we assume that the
system of equations (1) does not contain independent subsystems.


## 4.   METHOD FOR FINDING sc-CLUSTERS AND dc-CLUSTERS

Consider first the graph $G_S$,  together with the corresponding
s-connectivity matrix  S  and the complementary s-connectivity matrix S'
in order to establish an algorithm for finding the maximal sc-clusters.
The formula on which the algorithm is based can be obtained by representing the edge set $E_{GS}$ of $G_S$ as follows:

$$E_{GS} = E \setminus \bigcup_{(i,j)\in W} (A_i \cap A_j) = \left( \bigcup_{(i,j)\in W} (A_i \cap A_j) \right)' \qquad (3)$$

where $(i,j) \in W$ if $s'_{ij} = 1$ $(s_{ij} = 0)$.  The above formula can be
easily explained if we observe that the intersection $A_i \cap A_j$ represents
an edge in a complete graph $G_C$ (see Section 2) between the vertices $v_i$
and $v_j$ .  Thus, equation (3) means that $E_{GS}$ is obtained by removing
those edges from a complete set  E  which do not exist in $E_{GS}$ .  Before
we make further transformations of equation (3) let us observe that the
following set

$$A'_{k_1} \cap \dots \cap A'_{k_n} \qquad (4)$$

represents all those edges in a complete graph  $G_C$,  which *are not*
*connected* to any of the vertices  $v_{k_1},\dots,v_{k_n}$ .  Thus, if $A'_{k_1} \cap \dots \cap A'_{k_n} \subset$
$E_{GS}$  then the set  $V \setminus \{v_{k_1},\dots,v_{k_n}\}$  is the sc-cluster.  Also, to each
sc-cluster  $V^*$  there corresponds a set of edges  $A'_{k_1} \cap \dots \cap A'_{k_n}$  such
that  $v_{k_1} \notin V^*$ .

Equation (3) can now be transformed into the following form (using
De Morgan's laws),

$$E_{GS} = \bigcap_{(i,j)\in W} (A'_i \cup A'_j) \qquad (5)$$

The order of the union and intersection operations can then be changed by the following procedure. First we transform equation (5) to

$$E_{GS} = \bigcap_{i \in \{1,\ldots,N\}} [A_i' \cup ( \bigcap_{j \in J_i} A_j' ] \tag{6}$$

where $j \in J_i$ when $s_{ij}' = 1$ $(s_{ij} = 0)$ .

Let us examine now the intersections of subsequent pairs of the terms in square brackets in equation (6). For example, consider

$$[A_1' \cup ( \bigcap_{j \in J_1} A_j')] \cap [A_2' \cup ( \bigcap_{j \in J_2} A_j')]$$

$$= A_1' \cap A_2' \cup ( \bigcap_{j \in J_2 \cup \{1\}} A_j') \cup ( \bigcap_{j \in J_1 \cup \{2\}} A_j') \cup ( \bigcap_{j \in J_1 J_2} A_j') \tag{7}$$

From the components of the union on the right-hand side of equation (7) we drop out those sets which are contained in or are the same as one of the remaining terms (since they are absorbed by these terms). For example, if the second term is $A_1' \cap A_2' \cap A_p' \cap \ldots \cap A_r'$ , then this term is absorbed by $A_1' \cap A_2'$, etc. After this elimination procedure is performed for all pairs of the terms in square brackets in equation (6), the intersections between the reduced terms (7) are performed in the same way and so on until we obtain $E_{GS}$ represented in the form:

$$E_{GS} = (A_{k_1}' \cap \ldots \cap A_{k_n}') \cup (A_{r_1}' \cap \ldots \cap A_{r_m}') \cup (A_{p_1}' \cap \ldots \cap A_{p_\ell}') \tag{8}$$

$$k_i, r_i, p_i \in \{1,\ldots,N\}$$

As already observed, to each of the above terms $(A_{k_1}' \cap \ldots \cap A_{k_n}'$ ,etc.) corresponds an sc-cluster of vertices. Because of the absorbtion procedure used, the sets $V \{v_{k_1},\ldots,v_{k_n}\}$, $V \{v_{r_1},\ldots,v_{r_m}\}$, etc., are also the *maximal sc-clusters*.

The above procedure can be efficiently put in an algorithmic form and implemented on a computer, since all we need to do in fact is to remember which indices are related to the sets $A_i'$ in subsequent terms representing the intersections of $A_i's$ . In particular, the sets $J_i$ in equation (6) correspond to the rows of the complementary s-connectivity matrix $S'$ and, for example, intersection $\bigcap_{j \in J_i} A_j'$ can be represented by the i-th row of $S'$.

In general, whenever we consider the intersection of two terms, $(A_{q_1}' \cap \ldots \cap A_{q_n}')$ and $(A_{z_1}' \cap \ldots \cap A_{z_\ell}')$, then this intersection is represented by a Boolean sum of two binary N-vectors with unit entries at the positions $q_1,\ldots,q_n$ and $z_1,\ldots,z_\ell$ respectively. Thus, for example, the terms in brackets on the right-hand side of equation (2) are represented by the four binary N-vectors with unit elements in the positions indicated by the indices of sets $A_j'$ appearing in these intersections. These vectors are compared as follows: the binary N-vector $\underline{a}$ is absorbed by the binary N-vector $\underline{b}$ and is dropped out if $a_i \geq b_i$, $i = 1,\ldots,N$ . If $a_i \leq b_i$ for $i = 1,\ldots,E$, then $\underline{b}$ is absorbed by $\underline{a}$ and $\underline{b}$ is dropped out. Otherwise, both vectors are retained.

When performing the intersections of two unions, say $(B_1 \cup B_2 \cup B_3) \cap (B_4 \cup B_5)$, where each $B_i$ is represented by a binary N-vector, we create in the above way six binary N-vectors corresponding to $B_1 \cap B_4$, $B_1 \cap B_5$, etc.

The details of the above procedure, which has been implemented in FORTRAN, are explained in terms of an example in Section 5.

Once the maximal sc-clusters are determined they are put in a descending order according to the number of vertices contained in them. Only as many of the maximal sc-cluster in this list are retained as is necessary to include all the vertices in V. Then each of the retained maximal sc-clusters can be examined as a separate 'system' to determine the maximal dc-clusters. The procedure is identical to that described above for finding sc-clusters, except that the d-connectivity matrix of this new 'system' is used (in place of the s-connectivity matrix).

The procedure given in this section enables us to obtain sc-clusters and dc-clusters within each sc-cluster. What is required now is to 'chain' these clusters. There are many ways in which this may be done, based on the following general rules:

(1)  Within each sc-cluster the vertices corresponding to dc-clusters are grouped together. If there are common vertices between any two dc-clusters then the vertices in these dc-clusters are arranged so that the common vertices are grouped together.

(2)  Each sc-cluster is placed so as to have the largest number of common elements with its neighbours. The vertices within the sc-clusters are then arranged so that the common elements are grouped together without destroying the dc-clusters as far as possible.

The algorithm for 'chaining' has been implemented based on the above two rules. Precise details of the algorithm are not given here as it represents only one possibility of doing this.

5.  EXAMPLES

Example 1

Let us consider first, for the purpose of illustrating the described method, a single example of a system of ODEs with the occurrence matrix P given in Fig. 1, repeated here for convenience.

$$P = \begin{pmatrix} 1 & 0 & 1 & 0 & 0 & 0 \\ 0 & 1 & 0 & 1 & 1 & 0 \\ 0 & 0 & 1 & 0 & 0 & 1 \\ 0 & 1 & 0 & 1 & 1 & 0 \\ 0 & 1 & 0 & 0 & 1 & 1 \\ 1 & 0 & 1 & 0 & 0 & 1 \end{pmatrix}$$

First of all, it is easy to check (e.g. by using the procedure described in Section 3), that there are no disconnected subsystems. The next step is to look for the maximal sc-clusters. To begin with we form the s-connectivity matrix S and its complementary matrix S',

$$S = \begin{pmatrix} 1 & 0 & 1 & 0 & 0 & 1 \\ 0 & 1 & 0 & 1 & 1 & 0 \\ 1 & 0 & 1 & 0 & 0 & 1 \\ 0 & 1 & 0 & 1 & 1 & 0 \\ 0 & 1 & 0 & 1 & 1 & 1 \\ 1 & 0 & 1 & 0 & 1 & 1 \end{pmatrix} ; \quad S' = \begin{pmatrix} 0 & 1 & 0 & 1 & 1 & 0 \\ 1 & 0 & 1 & 0 & 0 & 1 \\ 0 & 1 & 0 & 1 & 1 & 0 \\ 1 & 0 & 1 & 0 & 0 & 1 \\ 1 & 0 & 1 & 0 & 0 & 0 \\ 0 & 1 & 0 & 1 & 0 & 0 \end{pmatrix}$$

Using the matrix $S$ we obtain, according to equation (6),

$$
\begin{aligned}
EG_S = \ & [A_1' \cup (A_2' \cap A_4' \cap A_5')] \cap [A_2' \cup (A_1' \cap A_3' \cap A_6')] \cap \\
& [A_3' \cup (A_2' \cap A_4' \cap A_5')] \cap [A_4' \cup (A_1' \cap A_3' \cap A_6')] \cap \\
& [A_5' \cup (A_1' \cap A_3')] \cap [A_6' \cup (A_2' \cap A_4')]
\end{aligned}
\tag{9a}
$$

Then we perform the intersections of the pairs of unions on the right-hand side of equation (9a):

$$
\begin{aligned}
EG_S = \ & [(A_1' \cap A_2') \cup (A_1' \cap A_3' \cap A_6') \cup (A_2' \cap A_4' \cap A_5') \cup \\
& (A_1' \cap A_2' \cap A_3' \cap A_4' \cap A_5' \cap A_6')] \cap \\
& [(A_3' \cap A_4') \cup (A_1' \cap A_3' \cap A_6') \cup (A_2' \cap A_4' \cap A_5') \cup \\
& (A_1' \cap A_2' \cap A_3' \cap A_4' \cap A_5' \cap A_6')] \cap \\
& [(A_6' \cap A_6') \cup (A_2' \cap A_4' \cap A_5') \cup (A_1' \cap A_3' \cap A_6') \cup \\
& (A_1' \cap A_2' \cap A_3' \cap A_4')]
\end{aligned}
\tag{9b}
$$

Within each term in the square brackets we drop out the sets which can be absorbed by the other elements in the union in this bracket. Then, since there is an odd number of terms in square brackets in equation (9b), we work out the intersection of the first two terms and leave the third term unchanged. Thus,

$$
\begin{aligned}
EG_S = \ & [(A_1' \cap A_2' \cap A_3' \cap A_4') \cup (A_1' \cap A_2' \cap A_3' \cap A_6') \cup \\
& (A_1' \cap A_2' \cap A_4' \cap A_5') \cup (A_1' \cap A_3' \cap A_4' \cap A_6') \cup \\
& (A_1' \cap A_3' \cap A_6') \cup (A_1' \cap A_2' \cap A_3' \cap A_4' \cap A_5' \cap A_6') \cup \\
& (A_2' \cap A_3' \cap A_4' \cap A_5') \cup (A_1' \cap A_2' \cap A_3' \cap A_4' \cap A_5' \cap A_6') \cup \\
& (A_2' \cap A_4' \cap A_5')] \cap [(A_5' \cap A_6') \cup \\
& (A_2' \cap A_4' \cap A_5') \cup (A_1' \cap A_3' \cap A_6') \cup (A_1' \cap A_2' \cap A_3' \cap A_4')]
\end{aligned}
\tag{9c}
$$

After deleting the sets in the first square brackets which are absorbed by the other sets in the brackets, we perform finally the final intersection of the sets in square brackets,

$$EG_S = (A_1' \cap A_2' \cap A_3' \cap A_4' \cap \overline{A_5'} \cap A_6') \cup (A_1' \cap A_2' \cap A_3' \cap A_4' \cap \overline{A_5'}) \cup$$
$$(A_1' \cap A_2' \cap A_3' \cap A_4' \cap \overline{A_6'}) \cup (A_1' \cap A_2' \cap A_3' \cap A_4') \cup$$
$$(A_1' \cap A_3' \cap A_5' \cap \overline{A_6'}) \cup (A_1' \cap A_2' \cap A_3' \cap A_4' \cap \overline{A_5'} \cap \overline{A_6'}) \cup$$
$$(A_1' \cap A_3' \cap A_6') \cup (A_1' \cap A_2' \cap A_3' \cap A_4' \cap \overline{A_6'}) \cup$$
$$(A_2' \cap A_4' \cap A_5' \cap \overline{A_6'}) \cup (A_2' \quad A_4' \quad A_6') \cup$$
$$(A_1' \cap A_2' \cap A_3' \cap A_4' \cap \overline{A_5'} \cap \overline{A_6'}) \cup (A_1' \cap A_2' \cap A_3' \cap A_4' \cap \overline{A_5'})$$

$$= (A_1' \cap A_2' \cap A_3' \cap A_4') \cup (A_1' \cap A_3' \cap A_6') \cup (A_2' \cap A_4' \cap A_5')$$

Hence the maximal sc-clusters are

$$\{v_5, v_6\} \quad \{v_2, v_4, v_5\}, \quad \{v_1, v_3, v_6\}$$

Now we see that the two larger clusters contain all the vertices of the system and hence we can drop the sc-cluster $\{v_5, v_6\}$. Then, in each of the two remaining sc-clusters, we find by the same procedure (using d-connectivity matrices) the dc-clusters which are:

$$\{v_3, v_6\}, \{v_1\} \qquad \text{in the sc-cluster} \quad \{v_1, v_3, v_6\}$$
$$\{v_2, v_4\}, \{v_2, v_5\} \qquad \text{in the sc-cluster} \quad \{v_2, v_4, v_5\}$$

As far as chaining is concerned, we arrange the vertices within each sc-cluster so as to group together the vertices forming dc-clusters. We obtain thus,

$$\{v_1, v_3, v_6\} \quad \text{and} \quad \{v_4, v_2, v_5\}$$

We see now that there are no common elements between the sc-clusters. In order to obtain a reasonable chain, we locate some vertex $v_i$ from the set $\{v_1, v_3, v_6\}$ that is connected to some $v_j$ from the set $\{v_1, v_2, v_5\}$. There must, of course, be at least one such pair of $v_i$ and $v_j$ since there are no disconnected subsets. In our case $v_5$ and $v_6$ are connected and so the final arrangement of the vertices is as follows:

$$v_1, \ v_3, \ v_6, \ v_5, \ v_2, \ v_4$$

which corresponds to the permuted occurrence matrix $\tilde{P}$ given below:

$$\tilde{P} = \begin{pmatrix} 1 & 1 & 0 & 0 & 0 & 0 \\ 0 & 1 & 1 & 0 & 0 & 0 \\ 1 & 1 & 1 & 0 & 0 & 0 \\ 0 & 0 & 1 & 1 & 1 & 0 \\ 0 & 0 & 0 & 1 & 1 & 1 \\ 0 & 0 & 0 & 1 & 1 & 1 \end{pmatrix}$$

Example 2

We now consider a larger example of dimension 40x40 and give the

results as obtained from the implemented FORTRAN program. The occurrence
matrix P corresponding to the initial ordering of the vertices $v_1, v_2, \ldots$
$\ldots, v_{40}$ is as follows:

The final arrangement of the vertices is:

$$v_9, v_1, v_{12}, v_{31}, v_{23}, v_{14}, v_{24}, v_5, v_{25}, v_{13}, v_6, v_{15},$$

$$v_{35}, v_{39}, v_{26}, v_{16}, v_{27}, v_{29}, v_7, v_{32}, v_{40}, v_{38}, v_{19},$$

$$v_4, v_{20}, v_3, v_{36}, v_{10}, v_{21}, v_{34}, v_{30}, v_{11}, v_{11}, v_2, v_{33},$$

$$v_{22}, v_8, v_{17}, v_{18}, v_{28}, v_{37} \ .$$

The corresponding occurrence matrix is:

$$\tilde{P} =$$

```
1 1 0 1 0 0 0 0 1 0 0 0 0 0 0 0 0 0 0 0 0 0 0 0 0 0 0 0 0 0 0 0 0 0 0 0 0 0 0 0
1 1 1 0 1 0 0 0 0 0 0 0 0 0 0 0 0 0 0 0 0 0 0 0 0 0 0 0 0 0 0 0 0 0 0 0 0 0 0 0
1 0 1 1 0 0 0 0 0 0 0 0 0 0 0 0 0 0 0 0 0 0 0 0 0 0 0 0 0 0 0 0 0 0 0 0 0 0 0 0
0 1 0 1 1 0 0 1 0 0 0 0 0 0 0 0 0 0 0 0 0 0 0 0 0 0 0 0 0 0 0 0 0 0 0 0 0 0 0 0
0 1 1 1 1 0 0 0 0 0 0 0 0 0 0 0 0 0 0 0 0 0 0 0 0 0 0 0 0 0 0 0 0 0 0 0 0 0 0 0
0 0 0 0 0 1 0 0 1 0 0 1 0 0 0 0 0 0 0 0 0 0 0 0 0 0 0 0 0 0 0 0 0 0 0 0 0 0 0 0
0 0 0 0 0 1 1 0 1 0 0 0 0 0 0 0 0 0 0 0 0 0 0 0 0 0 0 0 0 0 0 0 0 0 0 0 0 0 0 0
0 0 0 0 0 1 1 1 1 1 0 0 0 0 0 0 0 0 0 0 0 0 0 0 0 0 0 0 0 0 0 0 0 0 0 0 0 0 0 0
0 0 0 0 0 1 1 1 1 0 0 0 0 0 0 0 0 0 0 0 0 0 0 0 0 0 0 0 0 0 0 0 0 0 0 0 0 0 0 0
0 0 0 0 0 0 1 1 1 1 0 0 0 0 0 0 0 0 0 0 0 0 0 0 0 0 0 0 0 0 0 0 0 0 0 0 0 0 0 0
0 0 0 0 0 0 0 1 0 1 1 1 0 0 0 0 0 0 0 0 0 0 0 0 0 0 0 0 0 0 0 0 0 0 0 0 0 0 0 0
0 0 0 0 0 0 0 0 1 1 1 1 0 0 0 0 0 0 0 0 0 0 0 0 0 0 0 0 0 0 0 0 0 0 0 0 0 0 0 0
0 0 0 0 0 0 0 0 0 1 1 1 0 0 0 0 0 0 0 0 0 0 0 0 0 0 0 0 0 0 0 0 0 0 0 0 0 0 0 0
0 0 0 0 0 0 0 0 1 0 0 1 1 1 0 0 0 0 0 0 0 0 0 0 0 0 0 0 0 0 0 0 0 0 0 0 0 0 0 0
0 0 0 0 0 0 0 0 0 0 0 1 1 1 1 0 0 0 0 0 0 0 0 0 0 0 0 0 0 0 0 0 0 0 0 0 0 0 0 0
0 0 0 0 0 0 0 0 0 0 0 0 1 1 1 0 0 0 0 0 0 0 0 0 0 0 0 0 0 0 0 0 0 0 0 0 0 0 0 0
0 0 0 0 0 0 0 0 0 0 1 0 0 1 1 1 0 0 0 0 0 0 0 0 0 0 0 0 0 0 0 0 0 0 0 0 0 0 0 0
0 0 0 0 0 0 0 0 0 0 0 0 0 0 1 0 1 0 0 0 0 0 0 0 0 0 0 0 0 0 0 0 0 0 0 0 0 0 0 0
0 0 0 0 0 0 0 0 0 0 0 0 0 0 0 1 1 0 1 0 0 0 0 0 0 0 0 0 0 0 0 0 0 1 0 0 0 0 0 0
0 0 0 0 0 0 0 0 0 0 0 0 0 0 0 1 1 1 0 1 0 0 0 0 0 0 0 0 0 0 0 0 0 0 0 0 0 0 0 0
0 0 0 0 0 0 0 0 0 0 0 0 0 1 0 0 1 1 1 1 1 0 0 0 0 0 0 0 0 0 0 0 0 0 0 0 0 0 0 0
0 0 0 0 0 0 0 0 0 0 0 0 0 0 0 0 1 0 1 0 0 0 0 0 0 0 0 0 0 0 0 0 0 0 0 0 0 0 0 0
0 0 0 0 0 0 0 0 0 0 0 0 0 0 0 0 0 0 0 1 0 1 1 0 0 0 0 0 0 0 0 0 0 0 0 0 0 0 1 0
0 0 0 0 0 0 0 0 0 0 0 0 0 0 0 0 0 0 1 0 0 0 1 1 1 1 1 0 0 0 0 0 0 0 0 0 0 0 0 0
0 0 0 0 0 0 0 0 0 0 0 0 0 0 0 0 0 0 0 0 1 1 1 1 1 0 0 0 0 0 0 0 0 0 0 0 0 0 0 0
0 0 0 0 0 0 0 0 0 0 0 0 0 0 0 0 0 0 0 0 0 1 0 0 0 0 0 0 0 0 0 0 0 0 0 0 0 0 0 0
0 0 0 0 0 0 0 0 0 0 0 0 0 0 0 0 0 0 0 0 0 0 0 1 1 0 0 0 0 0 0 0 0 0 0 0 0 0 0 0
0 0 0 0 0 0 0 0 0 0 0 0 0 0 0 0 0 0 0 0 0 0 0 1 1 1 0 1 0 0 0 0 0 0 0 0 0 0 0 0
0 0 0 0 0 0 0 0 0 0 0 0 0 0 0 0 0 0 0 0 0 0 1 0 1 1 1 1 0 0 0 0 0 0 0 0 0 0 0 0
0 0 0 0 0 0 0 0 0 0 0 0 0 0 0 0 0 0 0 1 0 0 0 1 1 1 1 1 0 0 0 0 0 0 0 0 0 0 0 0
0 0 0 0 0 0 0 0 0 0 0 0 0 0 0 0 0 0 0 1 0 0 0 1 0 1 1 0 0 0 0 0 0 0 0 0 0 0 0 0
0 0 0 0 0 0 0 0 0 0 0 0 0 0 0 0 0 0 0 0 0 0 0 0 0 1 1 1 0 0 0 0 0 0 0 0 0
1 0 0 0 1 0 0 0 0 0 0 0 0 0 0 0 0 0 0 0 0 0 0 0 0 0 1 1 0 1 0 0 0 0 0 0 0
0 0 0 0 0 0 0 0 0 0 0 0 0 0 0 0 0 0 0 0 0 0 0 0 0 1 1 1 1 0 0 0 0 0
0 0 0 0 0 0 0 0 0 0 0 0 0 0 0 0 0 0 0 0 0 0 0 0 0 1 0 1 1 1 0 0 0 0 0
0 0 0 0 0 0 0 1 0 0 0 0 0 0 0 0 0 0 0 0 0 1 0 0 0 1 0 1 1 1 1 0 0 0 0 0
0 0 0 0 0 0 0 0 0 0 0 0 0 0 0 0 0 0 0 0 0 0 0 0 0 0 0 0 1 1 1 1 0
0 0 0 0 0 0 0 0 0 0 0 0 0 0 0 0 0 0 0 0 0 0 0 0 0 0 0 0 1 1 0 1 1
0 0 0 0 0 0 0 0 0 0 0 0 0 0 0 0 0 0 0 0 0 0 0 0 0 0 0 0 0 1 1 0 0
0 0 0 0 0 0 0 0 0 0 0 0 0 0 0 0 0 0 0 0 0 0 0 0 0 0 0 0 0 1 1 1
0 0 0 0 0 0 0 0 0 0 0 0 0 0 0 0 0 0 0 0 0 0 0 0 0 0 0 0 1 1 0 1
```

It can be seen from the above matrix that the clusters of unity elements are grouped along the diagonal. It is easy now to split the system into subsystems which would have a small number of interactions. For example, if we split the system into two equal subsystems, we then have 5 interactions as opposed to over 60 in the initial arrangement of the vertices with the subsystems $(v_1, v_2, \ldots, v_{20})$ and $(v_{21}, v_{22}, \ldots, v_{40})$.

## 6. CONCLUSIONS

In this report, a technique has been presented for clustering the system variables in such a way that the splitting of the system into subsystems with a small number of interactions is made easier. This was demonstrated in the examples.

It is worth noting that the technique presented in this paper can also be used (directly or after some modifications) to rearrange the system variables taking into account the strength of the interactions, represented for example by the elements of the Jacobian matrix. In particular, the technique presented can be directly used to obtain $\varepsilon$-decompositions where we want to identify subsystems lif any) amongst which the interactions are below a certain threshold ($\varepsilon$). Similarly, using the clustering technique, it is possible to obtain clusters of variables depending upon the strength of their mutual interactions.

REFERENCES

1.  A.Y. Allidina, S. Lei and L. Wang, Hierarchical simulation techniques for ODE systems, Distributed Simulation 85, San Diego, USA. Simulation Series, vol. 15, No. 2 (A Society for Computer Simulation publication), (1985).
2.  B.A. Care, Solutions of load-flow problems by partitioning systems into trees, IEEE Trans. on Power Apparatus and Systems, (1968).
3.  L. Himmelblau (ed.), "Decomposition of Large Scale Problems", North Holland, Amsterdam, (1972).
4.  L.S. Lasdon, "Optimization Theory for Large Systems", MacMillan, London, (1970).
5.  Malinowski, K., A.Y. Allidina, M.G. Singh and W. Crorlein, Decomposition-coordination techniques for parallel simulation - Part I, Control Systems Centre Report No. 599, UMIST. To appear in Journal of Large Scale Systems.
6.  E.C. Ogbuobiri, m.F. Tinney, and J.W. Walker, Sparsity-directed decomposition for Gaussian elimination on matrices, IEEE Trans. on Power Apparatus and Systems, 141-150, vol. PAS-89, No. 1, January 1970.
7.  D.F. Robinson and L.R. Foulds, "Digraphs, Theory and Techniques", Gordon and Breach, New York.
8.  M.E. Seser and D.D. Siljak, Nested $\varepsilon$-decompositions of complex systems, Proceedings of the 9th IFAC Congress, Budapest, Hungary, (1984).
9.  Siljak, D.D., V. Pichai and M.E. Seser, Graph-theoretic analysis of dynamic systems, Report DE-AC037ET29138-35, University of Santa Clara, Santa Clara, USA.
10. M.G. Singh, A.Y. Allidina and K. Malinowski, Hierarchical simulation techniques, Proc. of MECO 83, Athens, Greece, (1985).
11. M.G. Singh, A.Y. Allidina and K. Malinowski, Parallel simulation methods for dynamical systems: issues and challenges, Proc. of 2nd International Symposium on Systems Analysis and Simulation, Berlin, GDR, August 1985.
12. A. Vanelli and M. Vidyasagar, Three approximate solution techniques for the optimal K-decomposition problem, Proc. of the International Conference for Systems, Man and Cybernetics, Delhi, India, (1984).

MULTI-LEVEL HIERARCHICAL STRUCTURES FOR THE SOLUTION OF

LARGE SETS OF ORDINARY DIFFERENTIAL EQUATIONS

L. Wang[*], A.Y. Allidina[*], K. Malinowski[+] and M.G. Singh[*]

[*]Control Systems Centre, UMIST, Manchester M60 1QD

[+]Dept. of Automatic Control, Technical University of
Warsaw, Warsaw, Poland

ABSTRACT

   This paper is concerned with the rapid solution of large sets of
ordinary differential equations using parallel computing.  Parallel
computing facilities can be used for this purpose if, for example, due to
structural properties it is possible to use problem partitioning and
related decomposition methods.  It is shown in this paper how such
methods can be used to organise the overall computation task into multi-
level hierarchies.  A particular example is considered and results
regarding the evaluation of the possible speed-up for different structures
are given.

1.  SNTRODUCTION

   The problem of rapid simulation of large scale plants is amongst the
most important problems in modern engineering.  For example, real-time
simulators are required for a convenient, low-cost and effective
evaluation of system design changes as well as for operator training.  In
many situations it is crucial to simulate the system behaviour as fast as
possible (perhaps many times faster than real-time) in order to be able
to predict future reactions of the system to different possible control
actions.

   One obvious way of improving the simulation speed is to use more
powerful computers and/or to introduce more efficient simulation
algorithms for serial processing.  Another possibility of making
simulation faster and cost effective is emerging with the rapid development of microcomputer technology and consists of using parallel
processing facilities which would require special software for simulation.
What is also required is algorithms that enable parallel tasks to be
specified for the parallel processing facility.

   As far as simulation is concerned, a large class of problems involves
the solution of sets of differential equations describing the behaviour
of many Engineering Systems.  In this paper we are concerned with solving
large sets of Ordinary Differential Equations (ODEs).  As far as the
numerical solution of a set of ODE's is concerned, it is possible to
develop many types of parallel techniques.  First of all it should be

45

observed that one can attempt to enhance parallelism either after a particular discretisation scheme has been used or before such a discretisation is done. To illustrate the first approach, let us suppose that some standard discretisation has been used resulting, for example, in a sequence of algebraic expressions to be evaluated at each consecutive time level (as in explicit Runge-Kutta routines for instance). Then it is possible to evaluate these expressions by using the equation segmentation method (Franklin, 1978) (where the vector of formulas for computing the right-hand side of the ODE's is split into segments being evaluated in parallel).

On the other hand, there are some parallel methods for the numerical solution of ODE's in which the discretisation is done so as to develop a 'parallel front' of computation (e.g. Miranker and Liniger, 1966; Worland 1976; Franklin, 1978). One can mention, for example, some modified versions of block methods for parallel applications (Worland, 1976). Most of the existing approaches do not take into account the structure and the size of the set (system) of ODE's. In this paper, we consider decomposition-coordination for enhancing parallelism after discretisation while taking into account the system structure prevalent in large scale systems.

Systems of ODE's which result from the modelling of complex processes usually have a specific structure and numerical properties. In particular, a large set of ODE's can often be partitioned into interconnected subsets where the number of interconnecting variables is significantly less than the number of system variables. The number of coupling variables depends on the manner in which the partitioning is done as well as on the number of subsystems. The objective of the partitioning is to specify independent computing tasks with the use of decomposition and coordination techniques.

Another feature of large sets of ODE's resulting from the modelling of complex processes is 'stiffness'. That is, different parts of the process may have greatly different dynamics. This difficulty is usually handled by special implicit integration schemes (e.g. Gear, 1971; Miranker 1981) which may also offer some potential for the application of parallel computing facilities for the case in which the large set of ODE's has a special structure. For example, the inherent matrix inversions that are required can be carried out in parallel. In general, these implicit techniques result in systems of linear or nonlinear algebraic equations to be solved at each consecutive time level. Parallel computing facilities can be used for this purpose if, for example, due to structural properties it is possible to use problem partitioning and related decomposition methods, as discussed in this paper. Such methods can be used to organise the overall computation task into multi-level hierarchies.

The paper is organised as follows. In Section 2, system partitioning together with a basic decomposition-coordination technique is discussed. Special system structures and multi-level hierarchies are discussed in Section 3. In Section 4 some practical issues related to the use of decomposition techniques are discussed. A realistic model from the nuclear industry is considered in Section 5, and it is shown how multi-level hierarchies may be applied. While multi-level hierarchies are discussed in general in Section 3, a specific structure including decision on number of subsystems, etc., can only be discussed in terms of a particular application. This is shown in Section 5. Some concluding remarks are also given in this section.

## 2. SYSTEM PARTITIONING AND DECOMPOSITION-COORDINATION

Let us consider the case when the simulation problem consists of solving a large set of ODE's, say

$$\dot{x} = \bar{f}(x,t), \quad x \in R^m \tag{1}$$

The objective is to find a numerical solution of the above set of equations over a certain time horizon $[t_0, t_f]$. Since we are considering a simulation problem, initial conditions $(x(t_0) = x_0)$ are assumed to be given. The desire is to obtain the solution as fast as possible given a multi-processor system (e.g. a partitioned ring structure) or a multi-computer network. If the set of equations (1) describes a large scale system, then, as stated before, it may be worthwhile to exploit the structural properties of this set of equations. To present the Decomposition-Coordination technique in this context let us assume that the above set can be partitioned into, for example, the following $\nu$ subsets:

$$\dot{x}_i = f_i(x_i, u_i, t), \quad i = 1, 2, \ldots \nu \tag{2}$$

where $x_i \in R^{m_i}$, and $u_i \in R^{r_i}$ is an interaction input to the i-th subsystem, and is given by

$$u_i = C_i y = \sum_{j=1}^{\nu} C_{ij} y_j \tag{3}$$

The vector $y \in R^l$ is composed of subsystem interaction outputs $y_1, \ldots y_\nu$ (i.e. $y^T = [y_1^T, \ldots, y_\nu^T]$, $y_i \in R^{l_i}$), where

$$y_i = g_i(x_i) \tag{4}$$

Other special structures will be discussed in Section 3.

The partitioning as given by equations (2)-(4) has to be consistent in the sense that solving these equations would be equivalent to solving equation (1). While performing the partitioning as described above, several objectives should be met. In particular, it is worth mentioning the following ones:

(i) the coupling variables should be rather few relative to the overall system dimension, i.e. we would like to have dim u << dim x, where

$$u^T = [u_1^T, \ldots, u_\nu^T]$$

(ii) the subsystems should, if possible, be of an equal and moderate computational complexity in order to be able to allocate the work load evenly between the processors and to solve each of the sub-systems efficiently.

It is very difficult, if not impossible, to develop a general purpose automatic procedure for system partitioning which would be capable of choosing the number of subsystems, their structure, interaction couplings, etc. Since the first objective can be defined quantitatively in terms of minimising the number of interactions, then a logical first step would be to develop an algorithm oriented towards minimising the number of coupling variables. Such an automatic, general purpose technique is presented in Lei et al,(1985).

In general, the manner in which a numerical solution may be found for equations (2)-(4) depends on the integration scheme (discretisation) employed. The use of implicit schemes results in a set of algebraic equations to be solved at each time level and does not give rise to naturally occurring independent subsystems. However, decomposition-coordination can be used to specify independent tasks.

In order to demonstrate the application of such techniques, let us consider, for example, the use of an implicit Euler-scheme on the set of equations (2)-(4). This results in having to solve a set of algebraic equations of the form:

$$x_{i,n+1} = x_{i,n} + f_i(x_{i,n+1}, u_{i,n+1}, t_{n+1})h \qquad (5)$$

$$u_{i,n+1} = C_i y_{n+1} \qquad (6)$$

and

$$y_{i,n+1} = g_i(x_{i,n+1}) \qquad (7)$$

$$i = 1,2,\ldots\nu$$

where $(.)_n$ represents the value of the variable $(.)$ at the time-level $t_n$, and $h = (t_{n+1} - t_n)$. The aim is to solve the above set of algebraic equations for $x_{i,n+1}$, $i = 1,2,\ldots\nu$ in parallel, using decomposition-coordination techniques (Malinowski et al, 1984; Allidina et al, 1985). One basic method that can be considered consists of treating the inter-action inputs $u_{i,n+1}$, $i = 1,2,\ldots\nu$ as 'coordinating' variables (Direct Method). For such coordinating variables it is possible to define $\nu$ independent 'local' problems and a coordinator problem:

Local Problem i, $i = 1,2,\ldots\nu$

For given $u_{i,n+1}$, we need to solve the following equation for $x_{i,n+1}$:

$$x_{i,n+1} = x_{i,n} + f_i(x_{i,n+1}, u_{i,n+1}, t_{n+1})h \qquad (8)$$

Equation (8) may be solved, for example, by Relaxation or Newton-Raphson.

Relaxation

$$x_{i,n+1}^{k+1} = w_1 x_{i,n+1}^k + (1-w_1)(x_{i,n} + f_i(x_{i,n+1}^k, u_{i,n+1}, t_{n+1})h) \qquad (9)$$

where $w_1$ is the Relaxation parameter.

Newton-Raphson

Let

$$J_{1i} = x_{i,n+1} - x_{i,n} - f_i(x_{i,n+1}, u_{i,n+1}, t_{n+1})h \qquad (10)$$

Then a solution for $x_{i,n+1}$ may be obtained from:

$$x_{i,n+1}^{k+1} = x_{i,n+1}^k - \left(\frac{\partial J_{1i}}{\partial x_i}\right)_{x_{i,n+1}^k} . J_{1i}(x_{i,n+1}^k) \qquad (11)$$

where

$$\frac{\partial J_{1i}}{\partial x_i} = (I - \frac{\partial f_i}{\partial x_i} . h) \qquad (12)$$

48

Equations (9) or (11) can be used repeatedly for $k = 0,1,2,\ldots$ until, for example,

$$||J_{1i}|| < \epsilon_1 \quad \text{or} \quad ||x_{i,n+1}^{k+1} - x_{i,n+1}^{k}|| < \epsilon_1 \tag{13}$$

where $\epsilon_1$ is a specified error tolerance. Let the solution of the i-th local problem be $\bar{x}_{i,n+1}$, which is a function of $u_{i,n+1}$. This can then be used to obtain the interaction output variable $y_i$ according to

$$\bar{y}_{i,n+1}(u_{i,n+1}) = g_i(\bar{x}_{i,n+1}(u_{i,n+1})) \tag{14}$$

It is clear that the solution of the $\nu$ local problems can be obtained in parallel on a suitable number of parallel processors.

## Coordinator Problem

The coordination task is to find the variables $u_i$ such that equation (6) is eventually satisfied. It is necessary to solve the coordinator problem iteratively. This means that for each iteration at the coordination level resulting in new values for the variables $u_i$, the local problems must be solved again. It is therefore important that the strategy used for the coordinator problem is as efficient as possible. From equation (6), $i = 1,2,\ldots\nu$, we have the coordination condition:

$$u_{n+1} = C\,\bar{y}_{n+1}(u_{n+1})$$

where

$$u_{n+1} = [u_{1,n+1}^T, u_{2,n+1}^T, \ldots, u_{\nu,n+1}^T]^T$$

$$\bar{y}_{n+1}(u_{n+1}) = [\bar{y}_{1,n+1}^T(u_{1,n+1}), \ldots, \bar{y}_{\nu,n+1}^T(u_{\nu,n+1})]$$

and

$$C = [C_1^T, C_2^T, \ldots, C_\nu^T]^T$$

Now, two iterative coordination strategies may be considered:

## Relaxation

$$u_{n+1}^{k'+1} = w_c u_{n+1}^{k'} + (1-w_c)C\bar{y}_{n+1}^{-k'} \tag{15}$$

where $w_c$ is the Relaxation parameter and $k'$ is the iteration index.

The local solutions $\bar{y}_{i,n+1}^{-k'} = \bar{y}_{i,n+1}(u_{i,n+1}^{k'})$ have to be obtained for $u_{i,n+1} = u_{i,n+1}^{k'}$ and then they must be collected in the coordinator.

## Newton-Raphson

Let

$$J_c = u_{n+1} - C\,\bar{y}_{n+1}(u_{n+1}) \tag{16}$$

This gives the following strategy:

$$u_{n+1}^{k'+1} = u_{n+1}^{k'} - \left(\frac{\partial J}{\partial u}\right)_{u_{n+1}^{k'}}^{-1} \cdot J_c(u_{n+1}^{k'}) \tag{17}$$

where $\quad \dfrac{\partial J}{\partial u} = I - c\dfrac{\partial \bar{y}}{\partial u}$ $\hspace{4cm}$ (18)

The index n+1 is dropped for convenience. Also,

$$\frac{\partial \bar{y}}{\partial u} = \text{block diag}\left(\frac{\partial \bar{y}_i}{\partial u_i}\right)$$

where

$$\frac{\partial \bar{y}_i}{\partial u_i} = \frac{\partial g_i}{\partial x_i}\frac{\partial \bar{x}_i}{\partial u_i}$$

and since

$$\frac{\partial \bar{x}_i}{\partial u_i} = \frac{\partial f_i}{\partial x_i} \cdot \frac{\partial \bar{x}_i}{\partial u_i} \cdot h + \frac{\partial f_i}{\partial u_i} h$$

Then

$$\frac{\partial \bar{x}_i}{\partial u_i} = \left[I - \frac{\partial f_i}{\partial x_i} \cdot h\right]^{-1} \frac{\partial f_i}{\partial u_i} \cdot h$$

Thus

$$\frac{\partial \bar{y}_i}{\partial u_i} = \frac{\partial g_i}{\partial x_i} \cdot \left[I - \frac{\partial f_i}{\partial x_i} \cdot h\right]^{-1} \frac{\partial f_i}{\partial u_i} \cdot h \tag{19}$$

The above derivative $(\partial \bar{y}_i/\partial u_i)$ can be evaluated at the 'local level' at $\bar{x}_{i,n+1}^{k'}(u_{i,n+1}^{k'})$, $\bar{u}_{i,n+1}^{k'}$, and then communicated to the coordinator. The coordinator iterations (equation (15) or (17)) are carried on until

$$||u_{n+1}^{k'+1} - u_{n+1}^{k'}|| < \varepsilon_o \tag{20}$$

where $\varepsilon_o$ is a specified error tolerance at the coordinator level.

The Relaxation strategy (equation (15)) is rather simple and thus requires less computation per iteration compared to the Newton-Raphson strategy (equation (17)); however, the disadvantage is that many iterations may be required.

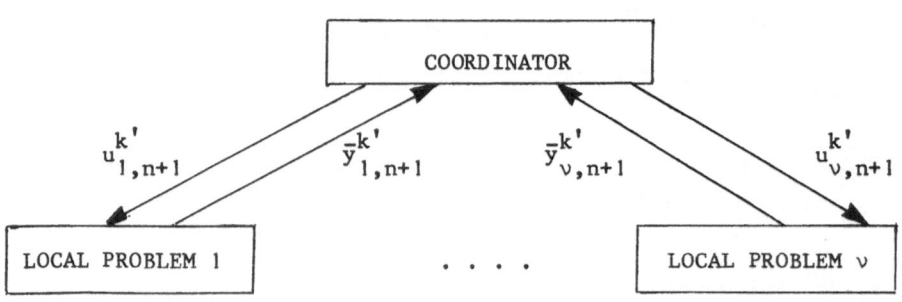

Fig. 1

## 3.  SPECIAL SYSTEM STRUCTURES AND MULTILEVEL HIERARCHIES

In the previous section a basic general form of system partition was presented, and a basic two-level hierarchical arrangement of computing tasks ($\nu$ parallel tasks of solving Local Problems (eqn. (8)) and the coordination task) was described. The structural properties of a system to be simulated, as well as the availability of a particular parallel computing facility, can make it worthwhile to exploit the system structure in more detail (using the general partitioning ideas presented in the previous section) and/or to consider the use of multilevel hierarchies.

As far as special system structures are concerned, let us consider, for example, the following system configuration, depicted in Fig. 2, where the interaction inputs to the subsystems are indicated by arrows. It can be seen that the subsystems 2,...,5 are coupled through the subsystem 1. Now, suppose that this subsystem is rather small in size (dimension) when compared to the remaining subsystems. Then the following question can be asked: is it better to consider all interaction inputs depicted in Fig. 2 as the coordinating variables (i.e. according to the basic approach presented in Section 2) or is it preferable to treat as the coordinating variables only the interaction inputs to subsystem 1?

To consider such a situation in the general case, let us assume that the large set of ODE's can be partitioned as follows:

$$\dot{x}_i = f_i(x_i, u_i, t) , \qquad i = 1,\ldots,\nu \tag{21}$$

$$\dot{y} = g(y,z,t) \tag{22}$$

where
$$u_i = d_i(y) \tag{23a}$$

$$z = h(x) , \qquad x = [x_1^T,\ldots,x_\nu^T]^T \tag{23b}$$

Thus z is the interaction input to the 'coupling subsystem' described by equation (22). Now, after a given implicit discretisation scheme is used on equations (21)-(23), then at time level n+1 we can either use $u_{i,n+1}$, $i = 1,2,\ldots\nu$ __and__ $z_{n+1}$ as the coordinating variables, or use only $z_{n+1}$ (or only $u_{i,n+1}$, $i = 1,2,\ldots\nu$) as the coordinating variables. In the former case the coupling subsystem (22) is treated simply as one of the parallel subsystems considered in Section 2. In the latter case the subsystem (22) is advanced in time (e.g. by solving a set of nonlinear equations) either before (or after the remaining independent subsystems (21) are advanced (on $\nu$ parallel processors). In this case the number of interactions to be handled by the coordinator is decreased, but the solution of subsystem (22) cannot be done in parallel with the other subsystems.

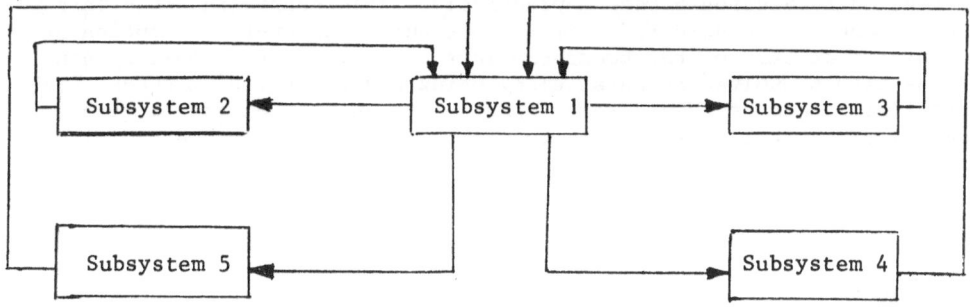

Fig. 2

The task of solving this subsystem may be regarded as a part of the coordinator task*. It is impossible in general to determine which of the above approaches is preferable, as it may depend upon numerous issues, for example, the size and numerical properties of the coupling subsystem with respect to subsystems (21), the number of parallel processors available, the rate at which data is transfered between the processors, etc. In Section 5 the details regarding the two possible approaches discussed above are highlighted by an example system.

The two-level hierarchical method described in the previous section leads usually to specifving $\nu$ parallel computing tasks of large size (singularity (Singh et al 1985)). The most straightforward scheduling (allocation) of these tasks is to allocate them to $\nu$ processors.

There are, however, two important issues to be observed here. First of all, if there are more processors available than the number of subsystems then the remaining processors will be kept idle while the local problems are solved. Secondly, it may easily happen that the solution of some of the sub-problems will take more time than others, which will also result in some processors being partially idle. One obvious possibility is to split the system into more subsystems. This would, however, increase the number of coordinating variables, the workload on the coordination and the amount of data to be transferred to and from the coordinator. The other possibility is to introduce multilevel hierarchies. That is, to treat each subsystem as a set of the form of equation (1) and to use the partitioning and coordination technique to advance this subsystem in time on several parallel processors (for example, by splitting Local Problems (8)). The details of such an approach are the same as those described in Section 2. Thus, the use of the two-level technique at the system-level and at the subsystem level (for all or for some of the subsystems) leads to a multilevel hierarchy of tasks, as depicted in Fig. 3.

It is clear that each sub-problem at, say, level 2 can be split further if desired, into a two level hierarchy, and so on. It should be observed that the use of multilevel hierarchies may be enhanced by a hierarchically structured parallel computing system. For example, a system with a partitioned ring structure (Brasch et al, 1981) consists of, say, n branches with several processors in each branch communicating through a multibus. These branches are connected to a communication ring on which several messages can be put and circulated around. Then, for example, it might be useful to associate each two-level numerical scheme (levels 1 and 2 in Fig. 3) with one branch of processors and to use the communications over the ring in order to organise the coordination at level 0.

The possible advantages of using multilevel hierarchies are investigated in Section 5 when a practical example consisting of a large system of ODE's is considered. It is worthwhile, however, to mention at this point that each of the Local Problems, as given for example by eqn. (8), may also be solved in parallel by using other parallel methods, such as parallel linear-system-solvers, parallel techniques on the level of elemental arithmetic operations, etc. This paper is not concerned with these possibilities.

---

*It should be observed that such an approach is not proper in the case of multilevel optimisation, when the coupling subsystem has its own performance function.

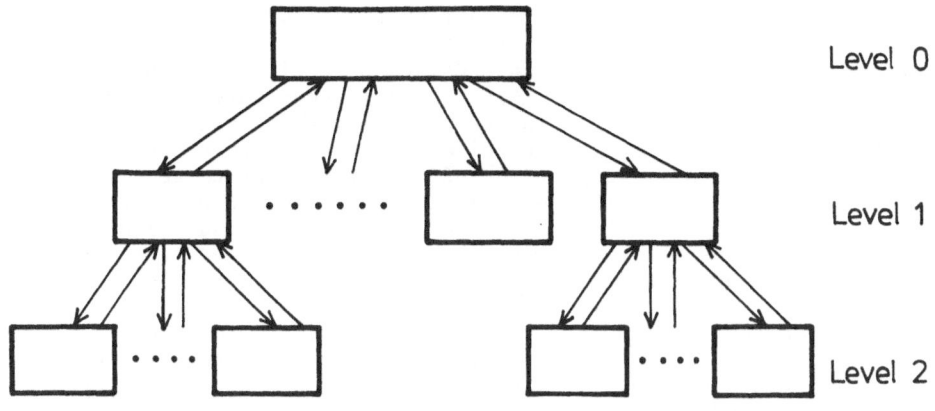

Level 0

Level 1

Level 2

Fig. 3

## 4. PRACTICAL ISSUES RELATED TO THE USE OF DECOMPOSITION TECHNIQUES

An efficient implementation of any parallel technique requires numerous important aspects to be taken into account. In particular, for the decomposition-coordination method considered in this paper, the following issues are essential:

(i)   numerical procedures for solving local problems and the coordination strategies,

(ii)  accuracy requirements with respect to solutions computed at different levels of a hierarchical scheme (see Fig. 3),

(iii) evaluation of the 'speed-up' that can be achieved by using the decomposition-coordination methods on a parallel facility with respect to 'classical' solution techniques.

As far as the first issue (i) is concerned, it is observed that the use of hierarchical techniques offers the possibility of using different solution techniques for different subproblems. Also, since the subproblems will be of small dimension, there is a possibility of obtaining analytical solutions of these subproblems or at least to derive analytical expressions which make the numerical solutions of subproblems more efficient. For example, it may be possible to derive analytical expressions for the inverses of the local Jacobians in equation (11).

The coordination strategy represents a key factor in any hierarchical technique, since at each iteration step of this strategy the local problem must be solved again. It may seem that the strategy with a larger rate of convergence (thus requiring less iterations) would be preferable to other strategies. However, one should also be aware of the fact that the computing at the coordination level is performed between lower level solutions and therefore an elaborate (complicated) coordination strategy should be avoided. The compromise can again only be established for a particular application.

The second major issue (ii) stated above is concerned with the specification of solution accuracies required at different levels (e.g.

the specification of $\varepsilon_1$ and $\varepsilon_o$ in equations (13),(20)). It is apparent and can be theoretically explained (Malinowski, 1983) that in order to obtain a solution with a prescribed accuracy of a given problem (for example, as defined by equations (5)-(7)) it is necessary to solve both the local problems and the coordinator problem with certain accuracies. Moreover, in order to solve the coordinator problem with a desired accuracy, it is necessary to provide sufficiently accurate solutions of the local problems. It is also evident that more accurate solutions require more iterations to be done at both levels, and thus require longer computing time. Thus, before a simulation routine is ready for an efficient on-line application, it is important to determine by off-line experiments and with appropriate techniques the accuracies required at all levels of the decomposition-coordination scheme. This is in order to provide for the fastest possible simulation and at the same time produce sufficiently accurate results. The method for selecting optimal accuracy parameters at different levels of the hierarchical technique considered in this paper is currently being investigated.

Before a given parallel simulation method is implemented on a parallel computing facility, it is necessary to evaluate the possible speed-up which can be achieved by the use of this method. The speed-up factor S is usually defined as the ratio of the sequential processing time to the processing time required when the method is applied on a parallel machine. It is of course desirable to evaluate the speed-up which could be achieved by using a given parallel simulation method on a parallel machine with respect to the most efficient sequential numerical routine implemented on a 'classical' computer, costing the same as the considered parallel system. Such an evaluation is, however, rather difficult. Thus, in this paper when investigating the speed-up the former more restrictive definition of the speed-up factor is used. It should be mentioned, however, that the numerical experience available so far shows (see, for example, Crorkin et al, 1985; Allidina et al, 1984) that the decomposition-coordination techniques, when used to solve systems of nonlinear or linear equations and implemented on a sequential computer, require about the same, or sometimes even shorter, computing time than the typical 'classical' methods. Therefore, if a significant reduction of computing time can be achieved by implementing a decomposition method on a parallel machine, then this technique may well be useful in practical applications. It would be unreasonable to expect the speed-up factor to be equal to the number of subsystems (local problems) and processors used. The uneven distribution of computing tasks between parallel processors, the time required to perform the coordination and to transfer data between the processors, will have an effect of reducing the speed-up factor. In modern multiprocessor systems the interprocessor communications are usually rather fast and so it may be reasonable that an initial evaluation of speed-up is neglecting the overhead due to data transfers. In order to account for an uneven task distribution and the coordinating actions a simple 'book-keeping' procedure can be used. This basically involves summing up the time units required for elemental operations performed during the computing. In the example investigated in the next section the time-unit corresponds to one addition of two real numbers in floating-point arithmetic. The multiplication and division of such numbers can be assumed to consume on an average minicomputer (e.g. PERKIN-ELMER) about 8 and 20 time units, respectively.

## 5. SIMULATION EXAMPLE AND CONCLUDING REMARKS

In order to illustrate the use of decomposition-coordination and multilevel hierarchies, a realistic model from the nuclear industry is considered. The model is of order 151 and is of the form:

$$\dot{x} = f(x,u) = C_1(x).x + C_2(x).u \tag{24}$$

$$\dot{y} = g(y,z,v_m) = C_3(z).y + C_4.z \tag{25}$$

$$z = h(x) \tag{26}$$

where $x = [x^1, x^2, \ldots x^{144}]^T$, $y = [y^1, y^2, \ldots, y^7]^T$ and $u = y^1$.

The variable $v_m$ represents an external input. It can be seen now that if the matrices $C_1, C_2$ and $C_3$ are held constant over each time-step (with their values being calculated using x and z at the beginning of the time-step), then the set of equations (24) and (25) can be regarded as linear over that time-step. (Note that $C_4$ and $C_5$ are constant anyway.) This is as done in practice when solving the above model. Simulation results are given here for this case and for the case when these system equations are considered to be nonlinear, as given in (24) and (25). For both cases, different multilevel hierarchies are considered. For each hierarchy considered, an evaluation of the speed-up that may be obtained when using an appropriate parallel-processing facility is given. It is emphasized that communication times (resulting from data transfer between processors) are not taken into account. The speed-up evaluation is done using the 'book-keeping' procedure described in Section 4.

Now, in order to find a solution of the set of equations (24-26) it is necessary to use some integration (discretisation) scheme. In this paper we do not consider the merits of different integration schemes as our main aim is to show, given a particular integration scheme, how two-level or multi-level structures can be useful in speeding up simulation, when used in conjunction with a parallel processing facility. Thus, for the problem being considered (equations (24)-(26)), an implicit two-point discretisation scheme is used. (Such a discretisation scheme, when used on, for example, the equation $\dot{w} = a(w)$ results in

$$w_{n+1} = w_n + a.(w_{n+1} + w_n)/2.H$$

where $w_n$ is the value at the n-th time-step and H is the step length.)

As mentioned earlier, the system equations are first solved with the matrices $C_1, C_2$ and $C_3$ held constant over each time step. In this case, therefore, the two-point discretisation is applied on the remaining terms in equations (24) and (25), giving:

$$x = x_n + [C_1(x_n).(\frac{x+x_n}{2}) + C_2(x_n).(\frac{u+u_n}{2})].H \tag{27}$$

$$y = y_n + [C_3(z_n).(\frac{y+y_n}{2}) + C_4(\frac{z+z_n}{2})].H \tag{28}$$

$$z = h(x) \tag{29}$$

These equations have to be solved for x,y,z and u in order to obtain the values $x_{n+1}$, $y_{n+1}$, $z_{n+1}$ and $u_{n+1}$ ($u_{n+1} = y^1_{n+1}$) at the time level (n+1).

As far as the solution is concerned, the following situations are considered. The cases are presented below, but the actual results are presented later (in Table 1).

## Case L1 (basic)

Here, the partitioning implied by equations (27)-(29) is made use of directly. The following strategy is adopted. For some given value of the variable u, say $u^k$, equation (27) is solved for the variable x. This solution for x is then used in (29) to obtain z, which in turn is used in equation (28) to obtain a value of y. Using this value of y, the variable $u = y^1$ is compared with its previous value $u^k$. If these are within a specified tolerance, then a solution for x,y has been found, otherwise the value for u needs to be updated and the procedure repeated.

The updating of u is carried out using Relaxation, as discussed in Section 2. Relaxation is also used for the solution of equations (27) and (28). The overall procedure is shown in Fig. 4.

The tasks discussed above and shown in Fig. 4 are carried out sequentially. The time required according to the 'book-keeping' procedure for a simulation horizon of a given length of 100 steps is noted. The corresponding times for the other cases to be discussed below will be compared with this time, as the structure is the most reasonable for a sequential implementation. (Note that the variable u is the coordination variable in terms of the discussion in Section 4).

## Case L2

Considering the system equations, in particular equation (24), it turns out that this equation can be partitioned into three subsystems without any interactions (see equation (21)). This may be detected by making use of the knowledge of the system, or by using a suitable procedure (Lei et al, 1985). Thus, the vector x can be split into three parts, $x = \begin{bmatrix} x_1^T, x_2^T, x_3^T \end{bmatrix}^T$. The dimensions of $x_1, x_2$ and $x_3$ are 69, 41 and 34 respectively. It is clear that the subsets of equation (27) corresponding to $x_1, x_2$ and $x_3$ can be solved in parallel. The implementation is depicted in Fig. 5.

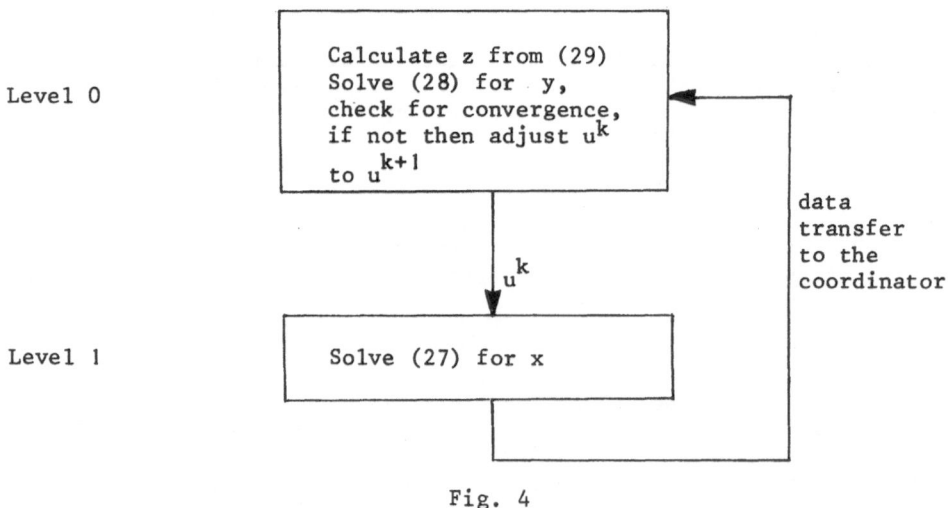

Level 0
```
Calculate z from (29)
Solve (28) for y,
check for convergence,
if not then adjust u^k
to u^{k+1}
```
$u^k$

Level 1
```
Solve (27) for x
```

data transfer to the coordinator

Fig. 4

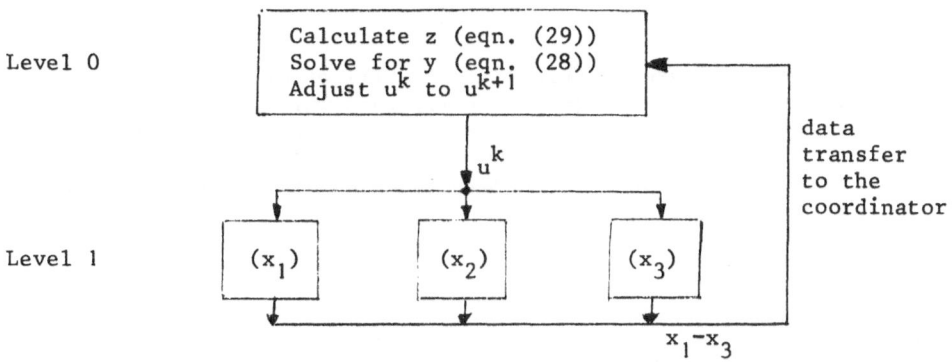

Level 0

Calculate z (eqn. (29))
Solve for y (eqn. (28))
Adjust $u^k$ to $u^{k+1}$

data
transfer
to the
coordinator

$u^k$

Level 1

$(x_1)$    $(x_2)$    $(x_3)$

$x_1 - x_3$

Fig. 5

If an appropriate parallel facility is used, then the total time
required for each iteration, neglecting the communication time, would be:

$$\left( \begin{array}{l} \text{time required for} \\ \text{carrying out task} \\ \text{at level 0} \end{array} \right) + \left( \begin{array}{l} \text{time required for the} \\ \text{longest task out of the} \\ \text{parallel tasks at level 1} \end{array} \right)$$

The total time using the 'book-keeping' procedure is recorded, with the
simulation horizon the same as in Case L1.

Case L3

It can be observed that out of the three parallel tasks at level 1
in Fig. 5, the subsystem corresponding to $x_1$ requires the most time. This
is because it is of dimension 69 as compared to 41 and 34 for the other
two subsystems respectively. This is also indicated by the 'book-keeping'
procedure. It is thus desirable to split the first subsystem into two by
introducing an additional coordinating variable w (see Fig. 6).

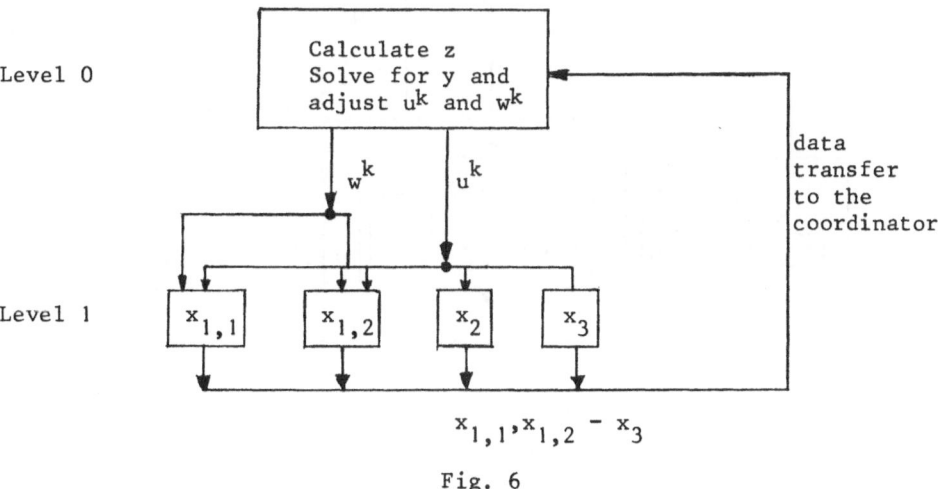

Level 0

Calculate z
Solve for y and
adjust $u^k$ and $w^k$

data
transfer
to the
coordinator

$w^k$    $u^k$

Level 1

$x_{1,1}$    $x_{1,2}$    $x_2$    $x_3$

$x_{1,1}, x_{1,2} - x_3$

Fig. 6

## Case L4

   The subsystems in case L3 (Fig. 6) are still of high dimension.  In case L4, we split the system (equation (27)) into more subsystems (21 subsystems) introducing the required number of coordinating variables ($w_1$,... ...,$w_{18}$,u).  The system is such that the 21 subsystems have at most one interaction variable between each other.  These subsystems (see Fig. 7) are of dimension 6 or 8.

## Case L5

   A further split into a larger number of subsystems (up to a maximum of 144) was investigated but no significant improvement in speed-up was obtained.  This is because the coordination task contributes significantly to the overall time required, since it includes the solution of equation (28).  Thus, to obtain any significant improvements in the speed-up factor, it is necessary to move the solution of y (equation (28)) out of the coordinator and place it as a subsystem task that may be solved in parallel with the other subsystems.  This, of course, requires z to be an additional coordination variable as depicted in Fig. 8.

   As can be seen in Table 1, this results in a considerable improvement in the speed-up factor.

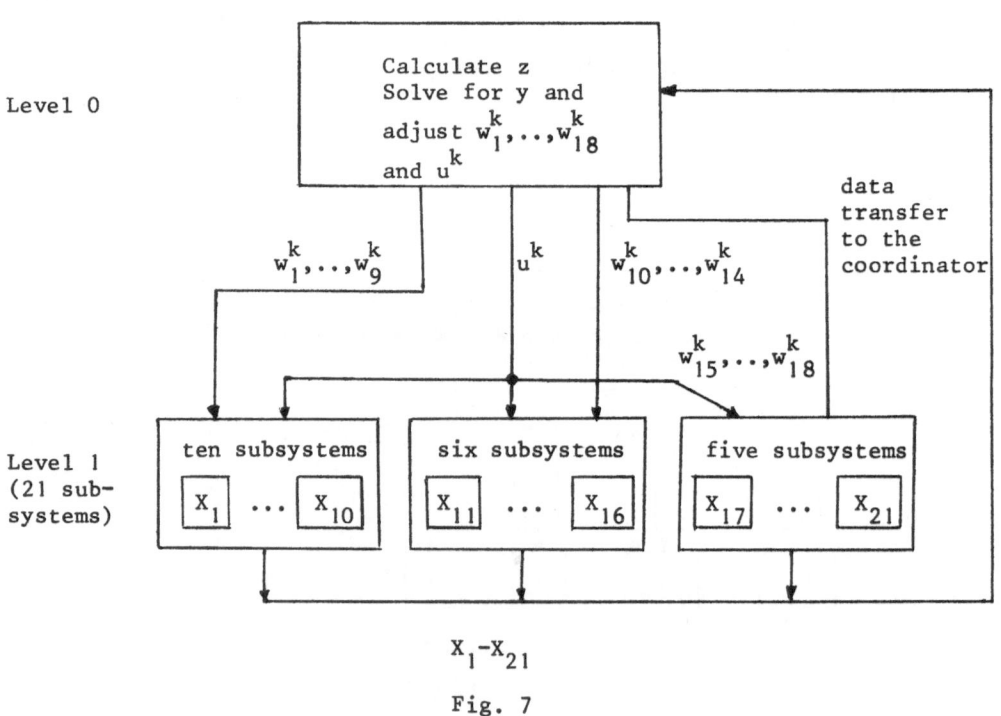

Fig. 7

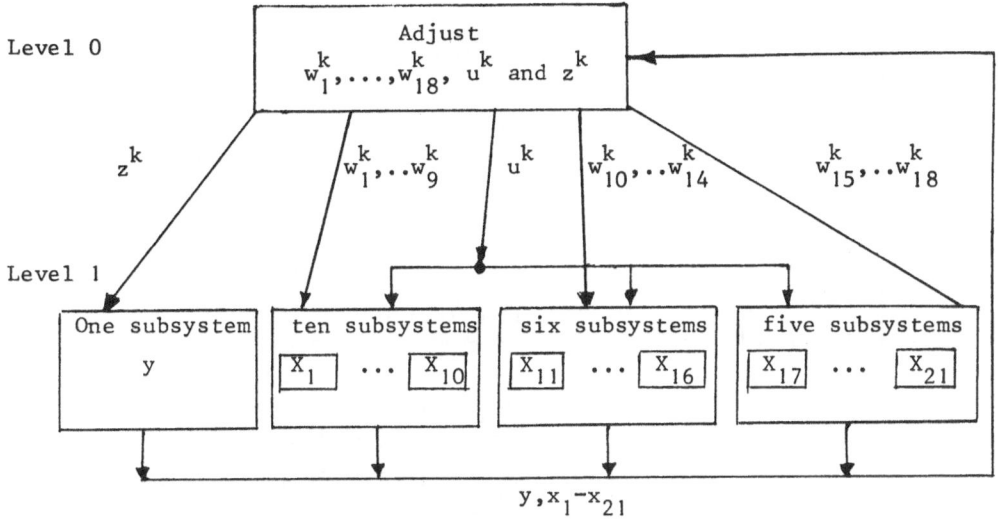

Level 0

Adjust
$w_1^k, \ldots, w_{18}^k$, $u^k$ and $z^k$

$z^k$

$w_1^k, \ldots w_9^k$    $u^k$    $w_{10}^k, \ldots w_{14}^k$    $w_{15}^k, \ldots w_{18}^k$

Level 1

| One subsystem | ten subsystems | six subsystems | five subsystems |
| y | $\boxed{X_1}$ ... $\boxed{X_{10}}$ | $\boxed{X_{11}}$ ... $\boxed{X_{16}}$ | $\boxed{X_{17}}$ ... $\boxed{X_{21}}$ |

$y, x_1 - x_{21}$

Fig. 8

## Case L6

It is interesting now to consider multilevel structures. For example, from case L2 (Fig. 5) the three subsystem tasks can themselves be split into two levels. With reference to Fig. 8, the variables $w_1, \ldots w_{18}$ are removed from the main coordinator at Level 0, and for a fixed value of $u$ the three groups of variables $(w_1, \ldots, w_9)$, $(w_{10}, \ldots, w_{14})$ and $(w_{15}, \ldots, w_{18})$ are iterated in a 'lower level coordinator' lsee Fig. 9). After convergence at this lower level, the values of $x$ are used to calculate $z$ (from equation (29)) which is compared with its previous value. As can be seen in Table 1, no significant improvement in the speed-up factor is obtained. However, when the sub-problems within this structure can be split further

## Case L7

In this case, the single sub-problem in case L6 (Fig. 9) at Level 1 pertaining to $y$ is split into two levels by introducing additional coordinating variables $v_1, \ldots, v_7$. The equations concerned with the variable $x$ (i.e. equation (27)) are split into a three-level structure by introducing additional coordinating variables $t_1, \ldots, t_{126}$. In all, 131 subsystems are considered. This number of subsystems was considered taking into account the structure of the system. From the 131 subsystems there are 126 subsystems at Level 3 and 5 subsystems (together with 18 coordinators) at Level 2. This structure is depicted in Fig. 10.

With this structure a significant improvement in the speed-up factor was obtained assuming, of course, that there are enough parallel processors available. It should be remembered that the data-transfer (communication) times were not taken into account. On the other hand, it can be observed from Fig. 10 that these communications could be conveniently arranged within a hierarchically structured multiprocessor facility.

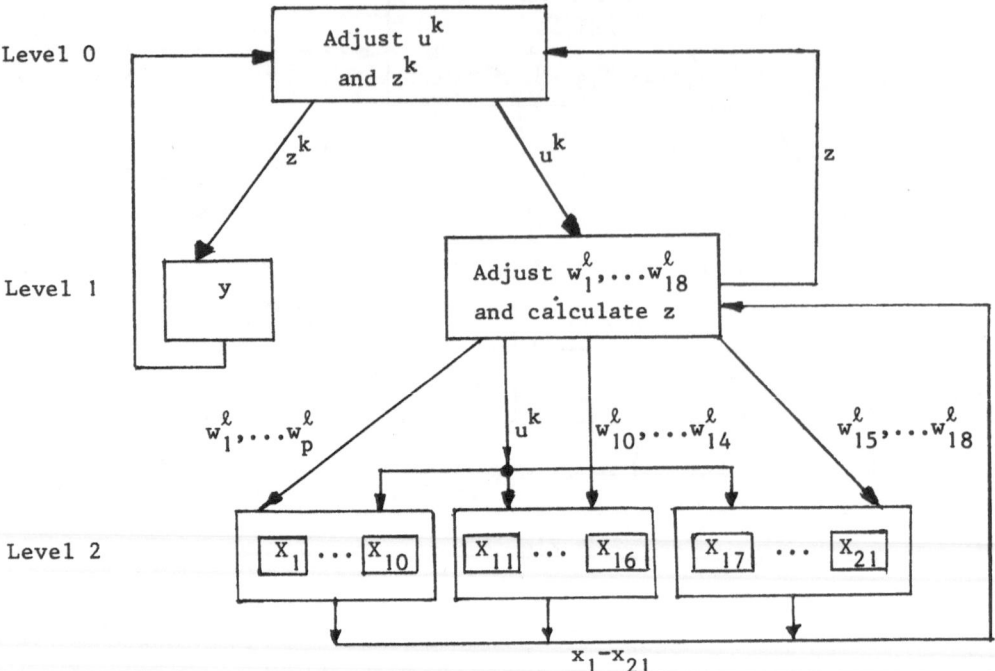

Fig. 9

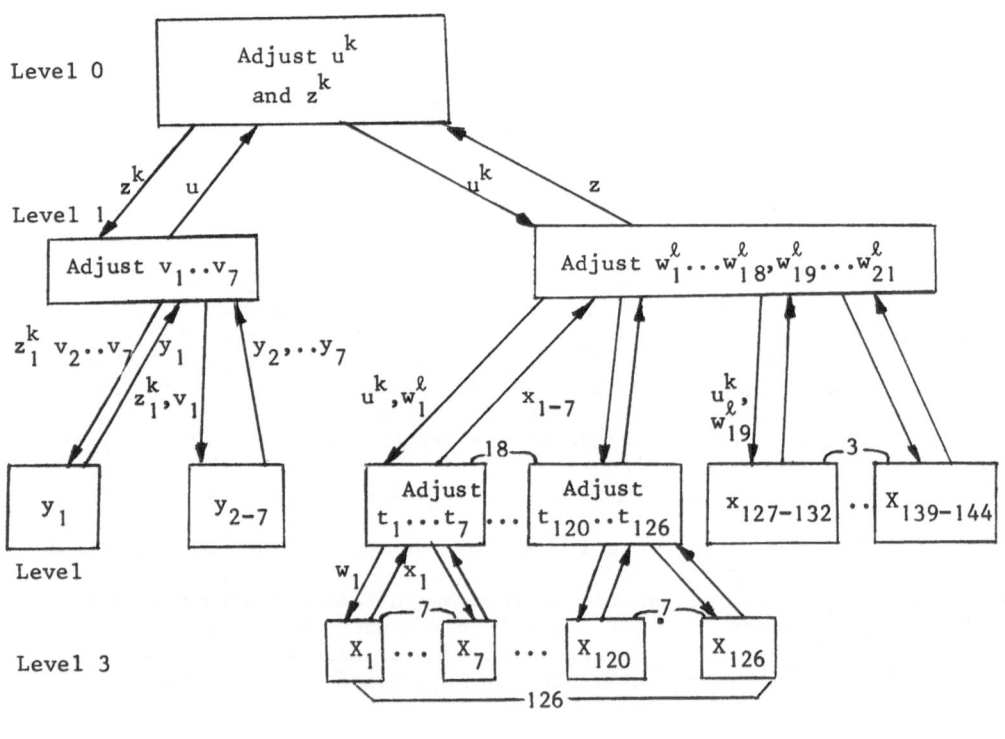

Fig. 10

60

TABLE 1

| Case | Structure | Speed-up Factor (over 100 integration steps) |
|------|-----------|----------------------------------------------|
| L1 | basic, sequential - Fig. 4 | 1 |
| L2 | two-level, 3 subsystems solution of eqn. (28) at Level 0 - Fig. 5 | 2.0 |
| L3 | two-level, 4 subsystems solution of eqn. (28) at Level 0 - Fig. 6 | 3.6 |
| L4 | two-level, 21 subsystems solution of eqn. (28) at Level 0 - Fig. 7 | 9.5 |
| L5 | two-level, 22 subsystems solution of eqn. (28) at Level 1 - Fig. 8 | 16 |
| L6 | three-level, 22 subsystems - Fig. 9 | 17 |
| L7 | four-level, 131 subsystems - Fig. 10 | 60 |

The simulation for all the cases was performed over 100 integration steps with transient conditions in order to have a more meaningful evaluation of the speed-up factors.

It should be noted that in cases L1 to L7, as mentioned earlier, linear equations (27) and (28) were being solved at each time step. In order to examine the nonlinear situation, the structures L1 to L7 were repeated as N1 to N7 with the nonlinear equations, which are:

$$x = x_n + [C_1(\frac{(x+x_n)}{2}).(\frac{x+x_n}{2}) + C_2(\frac{(x+x_n)}{2}).(\frac{u+u_n}{2})].H \qquad (30)$$

$$y = y_n + [C_3(\frac{(z+z_n)}{2}) (\frac{y+y_n}{2}) + C_4(\frac{z+z_n}{2})].H \qquad (31)$$

For such equations the application of the parallel-system-solvers available in the literature requires successive linearisations, while the decomposition-coordination discussed in this paper can be used directly in order to define parallel tasks. In fact, these tasks are as in cases L1 to L7.

The purpose of this exercise was to examine whether the results regarding speed-up are sensitive to the linearity of discretized equations. The results for cases N1 to N7, summarised in Table 2, are similar to the results obtained for L1 to L7.

TABLE 2

| Case | Structure | Speed-up Factor (over 100 integration steps) |
|------|-----------|----------------------------------------------|
| N1 | basic, sequential | 1 |
| N2 | two-level, 3 subsystems, solution of eqn. (31) at Level 0 | 2.1 |
| N3 | two-level, 4 subsystems, solution of eqn. (31) at Level 0 | 3.5 |
| N4 | two-level, 21 subsystems solution of eqn. (31) at Level 0 | 11.9 |
| N5 | two-level, 22 subsystems, solution of eqn. (31) at Level 1 | 17 |
| N6 | three-level, 22 subsystems | 18.5 |
| N7 | four-level, 131 subsystems | 56 |

The example considered shows that although decomposition-coordination techniques can be used on a set of equations without taking into account the corresponding physical system structure, the benefits may be greater if the physical structure is taken into account when partitioning the set of equations.

It should be noted that increasing the number of subsystems beyond a certain number will not necessarily lead to an improvement in the speed-up factor. This is because as the number of subsystems is increased the work load of the coordinator increases, and this leads to the coordinator taking a greater proportion of the overall time. It was observed in the example considered that increasing the number of systems beyond 22 (in the 2-level structure) did not lead to an improvement in the speed-up factor.

REFERENCES

Allidina, A.Y., ed., 1984, Development of hierarchical techniques for the simulation of large scale systems with particular application to the nuclear industry, EEC project, Phase I Report.
Allidina, A.Y., Lei, S. and Wang, L., 1985, Hierarchical simulation techniques for ODE systems, Distributed Simulation 85, San Diego, California, January 1985.
Brash, F.M. Jr., Van Ness, J.E. and Kang, S.C., 1981, Design of multi-processor structures for simulation of power-system dynamics, Report, Electric Power Research Institute, Palo Alto, California 94304.
Crorkin, W., Allidina, A.Y., Malinowski, K. and Singh, M.G., 1985, "Decomposition-coordination techniques for parallel simulation - Part 2", Large Scale Systems, North Holland.

Franklin, M.A., 1978, Parallel solution of ordinary differential
    equations, <u>IEEE Transactions on Computers</u>, vol. C-27, No. 5.
Gear, C.W., 1971, Simultaneous numerical solution of differential-
    algebraic equations, <u>IEEE Trans. on Circuit Theory</u>, vol. CT-18,
    No. 1, pp. 89-95.
Lei, S., Allidina, A.Y. and Malinowski, K., 1985, Clustering technique
    for rearranging ODE systems, Control Systems Centre,
    Report 635, UMIST.
Malinowski, K., 1983, Practical aspects of coordination processes,
    Proc. of IFAC/IFORS Symposium on LSSTA, pp. 191-196, Warsaw.
Malinowski, E., Allidina, A.Y., Singh, M.G. and Crorkin, W., 1984,
    Decomposition-Coordination techniques for parallel simulation,
    Control Systems Centre Report 599, UMIST.
Miranker, W.L. and Liniger, W., 1967, Parallel methods for the numerical
    integration of ordinary differential equations,
    <u>Math. Comput.</u>, vol. 21, pp. 303-320.
Miranker, W.L., 1981, Numerical methods for stiff equations,
    <u>Mathematics and Its Applications/5</u> D. Reidel Publishing Co.
Singh, M.G., Allidina, A.Y. and Malinowski, K., 1985,
    Parallel simulation methods for dynamical systems: Issues and
    Challenges, 2nd International Symposium on Systems Analysis
    and Simulation, Berlin, GDR, August 26-31, 1985.
Worland, P.B., 1976, Parallel methods for the numerical solution of
    ordinary differential equations, <u>IEEE Trans. on Computers</u>.

A COMPLETELY PARALLEL SCHEME FOR SIMULATION OF TRANSIENTS

IN LARGE GAS TRANSMISSION NETWORKS

Rudolf Maier and Günther Schmidt

Lehrstuhl und Laboratorium fur Steuerungs- und
Regelungstechnik, Technische Universität München
Postfach 202420, 8000 München 2, F.R.G.

ABSTRACT

A new method for digital simulation of transient pressure and flow
in large gas transmission networks is presented.    The method is also
applicable to other processes with similar system's properties.    Our
simulation method is based on a network-oriented partitioning of a large
set of nonlinear equations, which has to be solved at every time step.
By multiple utilization of certain (coupling)-equations a simple parallel
block-iteration scheme can be formulated for the solution of the nonlinear
equations.    The block-iteration is based on the Newton/Raphson method
combined with a linear block-iteration of the Jacobi type.    All blocks of
equations can be treated concurrently, while the amount of data exchange
between the blocks remains comparatively low.

The new numerical scheme is ideally suited for efficient implementation on a parallel computing facility of MIMD-type.    The assumptions that
assure its applicability are discussed in detail.    Rate of convergence
and overall expense of computing power for the solution scheme are
demonstrated by results from the simulation of a medium-sized gas transmission network.

## 1.  INTRODUCTION

Due to high and still increasing complexity of today's natural gas
transmission networks advanced automation tools are needed to meet all
technical and economical requirements of network operation.  Obviously,
efficient operation of gas transmission networks must be based on a
predictive control approach, including predictive simulation and mathematical optimization techniques.

The basic structure of control-oriented software for a modern
network supervision and control system is shown in Fig. 1.    Three subsystems are of interest in the context of our paper

- state estimation
- predictive simulation
- iterative control policy development by means of mathematical
  optimization.

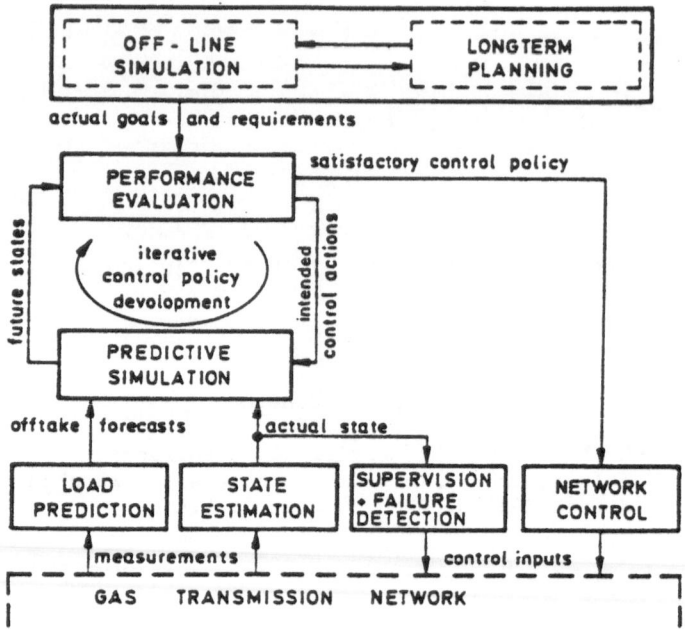

Fig. 1

All of these subsystems are required for purposes of predictive network control.

By using the small number of direct measured values of pressures and flows in a network the actual total state of a network can be estimated by means of a network state observer [4].

This actual state information provides the required initial state for a predictive simulation run, i.e. the computation of the transients of network flows and pressures for a future time interval (e.g. 4 to 8 hours). In these simulations load predictions of offtakes and intended control actions are taken into account. The calculated future network states are checked against actual goals and requirements of a safe and economic network operation. If the results are 'not yet satisfactory new predictive simulation runs will be performed with the same initial conditions but with a modified control strategy. This iterative process of control policy development may be formulated in the framework of a mathematical optimization problem.

It is obvious that many simulations will be required for this approach. A promising way to perform the time-consuming repetitive simulation runs with reasonable cost can be by means of a parallel computing facility. This is the reason for developing a new method for transient simulation of large gas transmission networks. It will prove to be ideally suited for implementation on a parallel computer system of MIMD type.

66

The main part of our paper is organized as follows. First we will summarise the mathematical model of interest and the numerical integration scheme used for digital simulation. Next, we present the new parallel simulation scheme, which is based on a block parallel method for the solution of large sets of nonlinear equations of a specific structure. We will conclude with some results from numerical experiments with a medium-sized gas transmission network.

## 2. MATHEMATICAL MODEL

A gas transmission network consists of several subsystems with similar dynamics, such as pipelegs, regulators, compressors and valves. Network dynamic behaviour depends primarily on the flow of gas through the pipelegs. Assuming a long horizontal pipeleg, with diameter d, cross section A and length L, one-dimensional flow can be modelled by a set of coupled nonlinear (hyperbolic) partial differential equations for the space and time dependent variables pressure $p(z,t)$ and massflow $q(z,t)$:

$$\frac{\partial p}{\partial t} = - \frac{c^2}{A} \frac{\partial q}{\partial z} \tag{1}$$

$$\frac{\partial q}{\partial t} = - A \frac{\partial p}{\partial z} - \frac{\lambda c^2}{2dA} \frac{|q|q}{p} \tag{2}$$

$$c^2 = Z(p,T') \cdot R \cdot T' = \frac{p}{\rho} = \frac{\partial p}{\partial \rho}\bigg|_{T=T'} = \text{const.} \tag{3}$$

c : isothermic speed of sound;
R : individual gas constant;
$\lambda$ : drag coefficient
$\rho(z,t)$ : density;
T : gas temperature.

Initial conditions at time t = 0 are given by

$$p(z,0) = p_o(z) \quad \text{and} \quad q(z,0) = q_o(z) \tag{4}$$

Boundary conditions at z = 0,L are defined by

- pressure defining functions $p_{0,L}(t)$                (5a)

- flow defining functions      $q_{0,L}(t)$                (5b)

- or coupling conditions for   $p_{0,L}(t), q_{0,L}(t)$       (5c)

Equation (3) implies isothermic gas flow. Various validation experiments have shown that this model is adequate for the simulation of underground pipelines assuming usual operating conditions. Since the dynamics of the remaining other network elements are usually negligible, they are modelled by additional algebraic equations defining boundary conditions for neighbouring pipelegs.

The total set of model equations for a network is discretized in space and time using a modified Crank-Nicholson scheme [14]. The resulting set of nonlinear equations is of the following form:

$$F_i(x_i, v) = 0 \qquad i = 1, 2, \ldots, m \tag{6}$$

$$v - g(x_1, x_2, \ldots, x_m) = 0 \tag{7}$$

with: $x_i \in R^{n_i}$ ; $v \in R^n$ ; $F_i : R^{n_i} \times R^n \to R^{n_i}$ ;

$g: R^{n_1} \times R^{n_2} \times \ldots \times R^{n_m} \to R^n$ ;

A single set of equations $F_i$ from (6) represents the model equations for one pipeleg (or one section of a pipeleg) after discretization (network-oriented partitioning). According to Fig. 2 the state variables $x_i$ are defined as

$$x_i = [q_{i,1}, p_{i,2}, \ldots, p_{i,ni-1}, q_{i,ni}]^T \quad ,$$

where $p_{i,j}$ are pressures of pipeleg sections with length $\Delta z$ and $q_{i,k}$ are flows between two pipeleg sections.

Equations (7) result from coupling conditions and algebraic equations defining boundary conditions on network nodes. The coupling variables $v$ represent the pressures on network nodes.

## 3.  BLOCK PARALLEL SOLUTION METHODOLOGY

### Modifications of the Model Equations

As a first step we will introduce new vectors $y_i$ for reduced coupling variables, i.e.

$$y_i := P_i v ; \qquad i = 1, 2, \ldots, m \tag{8}$$

$$y_i \in R^{n_{yi}} \quad .$$

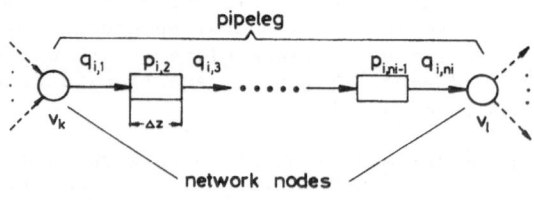

Fig. 2

$y_i$ contains the subset of elements of vector $v$ occurring in equations $F_i$. The non-quadratic matrix $P_i$ comprises certain rows of the nxn identity matrix. In other words, the application of $P_i$ reduces a vector of dimension n to a smaller format by selection of certain elements. On the other hand, the application of $P_i^T$ expands a reduced vector $y_i$ to dimension n by completion of certain entries with zeros. Moreover we can state that

$$P_i P_i^T = I \quad \text{and} \quad P_i^T P_i = \text{diag}(0 \text{ or } 1) .$$

By application of the matrices $P_i$ we can pick out of (7) a certain subset of coupling equations. We get from (6) and (7) the following m sets of subsystem equations:

$$\left\{ \begin{array}{c} y_i - h_i(x_1, x_2, \ldots, x_m) \\ \\ \\ F_i(x_i, P_i^T y_i) \end{array} \right\} = 0 ; \quad i = 1, 2, \ldots, m \qquad (9)$$

(Modification 1 of (6),(7))

with $\quad h_i := P_i g ; \quad h_i : R^{n_1} \times \ldots \times R^{n_m} \rightarrow R^{n_{yi}}$ .

In general more than one set of equations $F_i$ (pipeleg) will depend on a single component of $v$ (pressure on a network node). For this reason, certain components of $v$ will occur in more than one coupling vector $y_i$. Thus, during the solution of the total set of equations (9), coupling variables with the same physical meaning will be calculated more than once. This means that the m sets of subsystem equations (9) are in some way overlapping with respect to these coupling variables.

Let us assume that $P_i P_j^T \neq 0$, then certain components of $v$ will occur in $y_i$ as well as in $y_j$. Of course by solving (9) we will obtain equal values for these components of $y_i, y_j$, since they represent the same physical variable (node pressure). Thus we may conclude that

$$P_j^T P_j P_i^T y_i = P_i^T P_i P_j^T y_j ; \quad i = 1, 2, \ldots, m ; \quad i = j \qquad (10)$$

must hold for the solution of (9).

These constraints for the solutions of (9) with respect to $y_i$ can be formulated in a more general form as

$$R_i P_i^T y_i - \sum_{\substack{j=1 \\ j \neq i}}^{m} S_{ij} P_j^T y_j = 0 ; \quad i = 1, 2, \ldots, m \qquad (11)$$

with

$$S_{ij} := \text{diag}(s_{ij,k}) P_i^T P_i P_j^T P_j \qquad (12)$$

and

$$R_i := \sum_{\substack{j=1 \\ j \neq i}}^{m} S_{ij} \qquad (13)$$

Equations (11) can also be considered as some sort of penalty functions, penalizing deviations between components of $y_i, y_j$ $(i \neq j)$ with the same physical meaning.

By adjoining equations (11) to (9) we end up with a final set of equations which forms an adequate base for the block-iteration scheme.

$$
\left\{
\begin{array}{l}
y_i - h_i(x_1, \ldots, x_m) + P_i(R_i P_i^T y_i - \sum_{\substack{j=1 \\ j \neq i}}^{m} S_{ij} P_j^T y_j) \\[2em]
F_i(x_i, P_i^T y_i)
\end{array}
\right\}
\begin{array}{l}
= 0 ; \qquad (14) \\[2em]
i = 1, 2, \ldots m
\end{array}
$$

(Modification 2 of (6),(7))

## Block-iteration scheme

Newton's method applied to (14) results in

$$
\begin{pmatrix}
I + P_i R_i P_i^T & -\dfrac{\partial h_i}{\partial x_i} \\[1.5em]
\dfrac{\partial F_i}{\partial y_i} & \dfrac{\partial F_i}{\partial x_i}
\end{pmatrix}
\begin{pmatrix}
\Delta y_i \\[1.5em]
\Delta x_i
\end{pmatrix}
-
\begin{pmatrix}
\sum_{\substack{j=1 \\ j \neq i}}^{m} (P_i S_{ij} P_j^T \Delta y_j + \dfrac{\partial h_i}{\partial x_j} \Delta x_j) \\[2em]
0
\end{pmatrix}
$$

$$
+
\left\{
\begin{array}{l}
y_i - h_i(x_1, \ldots, x_m) + P_i(R_i P_i^T y_i - \sum_{\substack{j=1 \\ j \neq i}}^{m} S_{ij} P_j^T y_j) \\[2em]
F_i(x_i, P_i^T y_i)
\end{array}
\right\}
\begin{array}{l}
= 0 \qquad (15) \\[2em]
i = 1, 2, \ldots m
\end{array}
$$

These m sets of linear equations can be solved by means of a simple block-iteration scheme of the Jacobi type, i.e.

$$
\begin{pmatrix}
I + P_i R_i P_i^T & -\dfrac{\partial h_i}{\partial x_i} \\[1.5em]
\dfrac{\partial F_i}{\partial y_i} & \dfrac{\partial F_i}{\partial x_i}
\end{pmatrix}
\begin{pmatrix}
\Delta y_i^{k+1} \\[1.5em]
\Delta x_i^{k+1}
\end{pmatrix}
=
\begin{pmatrix}
\sum_{\substack{j=1 \\ j \neq i}}^{m} (P_i S_{ij} P_j^T \Delta y_j^k + \dfrac{\partial h_i}{\partial x_j} \Delta x_j^k) \\[2em]
0
\end{pmatrix}
+
\begin{pmatrix}
u_{yi} \\[1.5em]
u_{xi}
\end{pmatrix}
$$

$$
i = 1, 2, \ldots, m \qquad (16)
$$

with
$$
u_{yi} = -y_i + h_i(x_1, \ldots, x_m) - P_i(R_i P_i^T y_i - \sum_{\substack{j=1 \\ j \neq i}}^{m} S_{ij} P_j^T y_j) ;
$$

$$
(17)
$$

$$
u_{xi} = -F_i(x_i, P_i^T y_i) ;
$$

$$
k = 0, 1, 2, \ldots \quad \text{(iteration step)}.
$$

For carrying out one iteration step, m decoupled sets of linear equations must be solved. Thus, all computations can be performed concurrently on a maximum of $\dot{m}$ processing elements of a parallel computer system. After each iteration step a data exchange must take place between those processors handling physically interconnected subsystems.

From each subsystem $2\,n_{yi}$ variables usually have to be transferred. Assuming single pipelegs as subsystems, the $y_i$ will have dimension 2, since $y_i$ represents the node pressures of the beginning and end of a pipeleg. By arranging the equations of (16) in a suitable order, the m matrices on their left side are of tridiagonal structure. Thus, the iteration can be carried out in a very efficient way. Obviously no additional coordination step is required, because the coordination already takes place simultaneously on the subsystem level.

Although the free parameters $s_{ij,k}$ of the matrices $S_{ij}$ have no principal influence on the solution of (14) and (15), their absolute values are essential for the convergence properties of the block-iteration (16). We will show next that, under certain assumptions and by proper selection of the matrices $S_{ij}$, it will always be possible to assure convergence.

## Convergence of the Block-Iteration Scheme

Assuming that all matrices $\partial F_i / \partial x_i$ $(i = 1,2,...,m)$ are nonsingular, we can substitute $\Delta x_i$ in equation (16) by

$$\Delta x_i^{k+1} = - \left(\frac{\partial F_i}{\partial x_i}\right)^{-1} \left(\frac{\partial F_i}{\partial y_i} \Delta y_i^{k+1} + u_{xi}\right) \quad i = 1,2,...,m \tag{18}$$

and we obtain

$$(I + P_i(R_i + W_i)P_i^T)\Delta y_i^{k+1} = \sum_{\substack{j=1 \\ j \neq i}}^{m} P_i(S_{ij} - W_j)P_j^T \Delta y_j^k + .... \tag{19}$$

where the matrices $W_j$ are defined as:

$$W_j := - \frac{\partial g}{\partial x_j} \left(\frac{\partial F_j}{\partial x_j}\right)^{-1} \frac{\partial F_j}{\partial v} \tag{20}$$

From this it follows that

$$P_i W_j P_j^T = - \frac{\partial h_i}{\partial x_j} \left(\frac{\partial F_j}{\partial x_j}\right)^{-1} \frac{\partial F_j}{\partial y_j} \; . \tag{21}$$

For the matrices $W_j$ and

$$W := I + \sum_{j=1}^{m} W_j \tag{22}$$

we will make the following assumptions:

(1)   all diagonal entries of $W_j$, i.e.

$$\text{diag}(W_j) \geq 0 \quad \text{and} \quad \bar{W}_j := W_j - \text{diag}(W_j) \leq 0 \tag{23}$$

$$(2) \qquad W^{-1} \geq 0 . \tag{24}$$

These conditions imply that matrix W is a so-called M-matrix [6,12]. For the type of model equations given in (6), (7), these assumptions are always valid. More than that, all submatrices $I+P_i W_i P_i^T$ will be strictly diagonally dominant M-matrices, if we assume reasonable time ($\Delta t$) and space discretization steps ($\Delta z$).

By choosing the matrices $S_{ij}$ as

$$S_{ij} = \text{diag}(S_{ij}) = \text{diag}(W_j)P_i^T P_i \; ; \quad i,j = 1,2,\ldots,m \tag{25}$$

and consequently

$$R_i = \sum_{\substack{j=1 \\ j \neq i}}^{m} \text{diag}(W_j)P_i^T P_i \qquad\qquad i = 1,2,\ldots,m \tag{26}$$

we get from equations (19):

$$M_i \Delta y_i^{k+1} = \sum_{\substack{j=1 \\ j \neq i}}^{m} N_{ij} \Delta y_j^k +\ldots ; \qquad\qquad i = 1,2,\ldots,m \tag{27}$$

with

$$M_i := I+P_i(R_i+W_i)P_i^T = I+\sum_{j=1}^{m} P_i \text{diag}(W_j)P_i^T+P_i \bar{W}_i P_i^T \tag{28}$$

and

$$N_{ij} := P_i(S_{ij}-W_j)P_j^T = P_i \bar{W}_j P_j^T . \tag{29}$$

Because of the identity $P_i \text{diag}(W_j)P_j^T P_j = P_i \text{diag}(W_j)P_i^T P_i$ the diagonal entries of the matrices $W_j$ can be brought to the left side of equations (19). This means that there is no direct impact on $\Delta y_i^{k+1}$ from the coupling variables $P_i P_j^T \Delta y_j^k$ ($i \neq j$). $P_i P_j^T \Delta y_j^k$ contains those elements of $\Delta y_j^k$ with the same physical meaning as the corresponding elements of $\Delta y_i$. By this operation, the effects of couplings are significantly reduced. $M := \text{block-diag}(M_i)$ and also $A := M-N$ with $N := [N_{ij}]$ are obviously M-matrices. Thus, $A = M-N$ is a regular splitting of $A$ and therefore the matrix $M^{-1}N$ is convergent. The spectral radius $\rho(M^{-1}N)$ determining the convergence rate of the block iteration can thus be computed to be

$$\rho(M^{-1}N) = \frac{\rho(A^{-1}N)}{1 + \rho(A^{-1}N)} < 1 . \tag{30}$$

72

The effects of couplings can be reduced even further in order to achieve a better convergence rate. For this purpose we consider two sets of subsystem equations $i,j$ sharing only one coupling variable $v_i = y_{i,li} = y_{j,1j}$. Thus the matrices $N_{ij}$ and $N_{ji}$ are of rank 1 and they contain only one non-zero row. In this case the block-Jacobi-iteration has the following structure:

$$\vdots$$

$$M_i \Delta y_i^{k+1} = N_{ij} \Delta y_j^k + \ldots$$

$$\vdots$$

$$M_j \Delta y_j^{k+1} = N_{ji} \Delta y_i^k + \ldots$$

$$\vdots$$

or respectively

$$\Delta y_i^{k+1} = M_i^{-1} N_{ij} M_j^{-1} N_{ji} \Delta y_i^{k-1} + \ldots$$

$$= \bar{m}_{i,\cdot li} \; n_{ij,li\cdot} \; \bar{m}_{j,\cdot 1j} \; n_{ji,1j\cdot} \; \Delta y_i^{k-1} + \ldots , \tag{31}$$

with $\bar{m}_{i,\cdot li}$ : vector containing column li of $M_i^{-1}$

$n_{ij,li\cdot}$ : vector containing row li of $N_{ij}$ .

The scalar expression $n_{ij,\cdot 1j} \, \bar{m}_j \cdot 1j$ depends on the non-zero entry $s_{ij,1}$ of $S_{ij}$. There will be no direct impact on the i-th subsystem equation (after two iteration steps) from the subsystem $j$ if this scalar is equal to zero. That means that (from a local view-point),

$$n_{ij,li\cdot} \bar{m}_{j,\cdot 1j} = \sum_{\nu=1}^{n_{yj}} n_{ij,li\nu} \bar{m}_{j,\nu 1j}$$

should be equal to zero.

Since $n_{ij,li\cdot}$ represents the li-th row of $P_i (S_{ij} - W_j) P_j^T$ the equation $n_{ij,li\cdot} \bar{m}_{j,\cdot 1j} = 0$ is equivalent to

$$s_{ij,1} = w_{j,11} - \sum_{\substack{\nu=1 \\ \nu \neq 1j}}^{n_{yj}} n_{ij,li\nu} \; \bar{m}_{j,\nu 1j} / \bar{m}_{j,1j1j} \; . \tag{32}$$

Of course, we can obtain an equivalent equation for $s_{ji,1}$. With more general couplings we get from (32) a set of nonlinear equations $(i,j,1 = 1,2,\ldots,m)$, which determine all non-zero entries of the matrices $s_{ij}$. Since all submatrices $I + P_i W_i P_i^T$ are strictly diagonally dominant M-matrices, a simple iteration scheme can be used to solve these equations:

$$s_{ij,1}^{(\mu+1)} = w_{j,11} - \sum_{\substack{\nu=1 \\ \nu \neq 1j}}^{n_{yj}} n_{ij,li\nu} \; \bar{m}_{j,\nu 1j}^{(\mu)} / \bar{m}_{j,1j1j}^{(\mu)} \tag{33}$$

$$s_{ij,1}^{(0)} = w_{j,11} \; ; \quad \mu = 0,1,2,\ldots. \quad \text{(iteration step)} \; .$$

It can be shown that $s_{ij,1}^{(\mu+1)} \leq s_{ij,1}^{(\mu)}$ and that $0 \leq s_{ij,1}^{(\mu)} \leq w_{j,11}$ holds. Consequently (33) is convergent.

In the special case that a gas transmission network shows tree structure, two subsystem equations will share only one coupling variable and there will be no additional indirect couplings between these two subsystems. Therefore, all eigenvalues of $M^{-1}N$ will be zero, if we choose the matrices $S_{ij}$ according to equation (32). This means that the block-Jacobi iteration will converge in a finite number of steps. In other cases (arbitrarily interconnected networks) the convergence rate can also be significantly improved by selecting the matrices $S_{ij}$ according to (32), if compared to the choice $S_{ij} = \text{diag}(W_j)P_i^T P_i$ .

## 4. NUMERICAL RESULTS

The numerical results to be presented originate from a network consisting of 18 pipelegs, one pressure regulated compressor and three regulators for pressure (see Fig. 3). 18 sets of subsystem equations were used according to the 18 pipelegs of the network. Other partitionings, finer or coarser, would be possible but are not considered here.

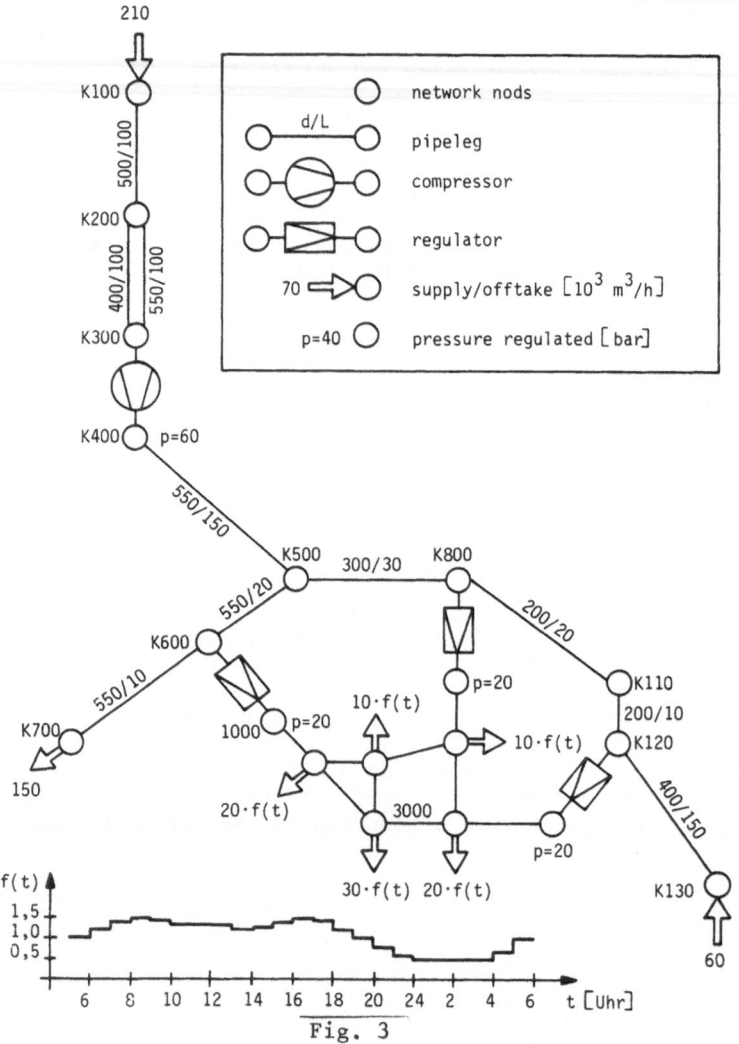

Fig. 3

The convergence rate of the block-Jacobi iteration depends on the largest eigenvalue $\lambda_{max} = max(\lambda_i)$ of $M^{-1}N$. As shown in Table 1 the value of $\lambda_{max}$ depends primarily on the type of matrices $S_{ij}$ chosen.

For these results steady state flow conditions, time step $\Delta t = 1$ h and space step $\Delta z = 5$ km were assumed.

Table 1.

| $S_{ij}$ | $S_{ij} = 0$ | $S_{ij} = diag(W_j)P_i^T P_i$ | $S_{ij}$ according to Eqn. (32) |
|---|---|---|---|
| $\lambda_{max}$ | 1.72 | 0.799 | 0.662 |

The convergence rate also depends on the ratio $\Delta t/\Delta z$: the smaller $\Delta t/\Delta z$, the better the resulting convergence rate. In spite of the rather unfavourable values of $\Delta t, \Delta z$ chosen for Table 1, we still get a satisfactory convergence rate. It may be interesting to note that simulations of various gas transmission networks have shown that the convergence rate is rather insensitive with respect to small variations of the matrices $S_{ij}$. Consequently, in most cases it suffices to carry out only one calculation of the matrices $S_{ij}$ for one simulation run.

The linear system of equations (15) is approximately solved by a fixed number (2 to 6) of block-Jacobi iteration steps. The required solutions of the subsystem equations are calculated by LU-decomposition and forward/backward substitution under consideration of the tridiagonal structure. The Newton-iteration is stopped when the vectors $\Delta y_i$ (i = 1, 2,...,m) and the deviations between overlapping calculated components (components having the same physical meaning) of $y_i, y_j$ (i ≠ j) are small enough to guarantee a certain accuracy.

Figure 4 presents the necessary number of Newton steps against the number of block-Jacobi iteration steps and Fig. 5 shows the resulting execution time against the number of block-Jacobi iteration steps per Newton step. It is evident that for our example about four block-iteration steps are nearly optimal. A greater number of iteration steps causes only small reductions in the required number of Newton steps.

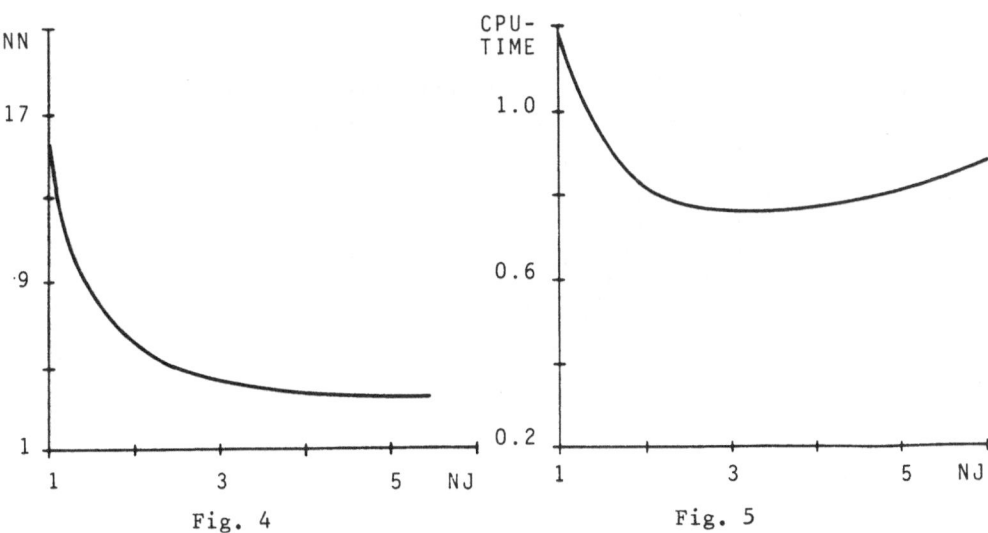

Fig. 4                    Fig. 5

Finally we compare the *total* execution time for the presented parallel method with the execution time using a sparse matrix approach [14] for the solution of the total set of linear equations. Figure 6 shows that the total execution times for both methods will be about the same (on a single processor), if a certain accuracy is assumed. This high efficiency of the new parallel method results from the fact that it fully exploits the tri-diagonal structure of the subsystem equations. As a consequence, we can expect a significant speed-up for the network simulations when implementing the parallel solution scheme on a parallel computing facility.

## 5. SUMMARY AND CONCLUSIONS

A new parallel method for the solution of large sets of nonlinear equations was presented. Such equations, as given in Section 2 of our paper, arise in connection with simulation of transient flows and pressures in gas transmission networks. A network-oriented partitioning of the network into single pipelegs or pipeleg sections is used. To reduce the strong couplings between the model equations of interconnected pipelegs, overlapping subsystem equations are included. This modification opens the possibility to solve the total set of nonlinear equations by means of a simple block parallel iteration scheme. Since all calculations can be performed concurrently, this scheme is well suited for the implementation on a parallel computing facility of the MIMD-type. However, an efficient implementation requires an adequate allocation of the workload between the processors of the parallel computer system. We are currently developing algorithms for automatic gas transmission network partitioning and task allocation under consideration of the special properties of our numerical scheme.

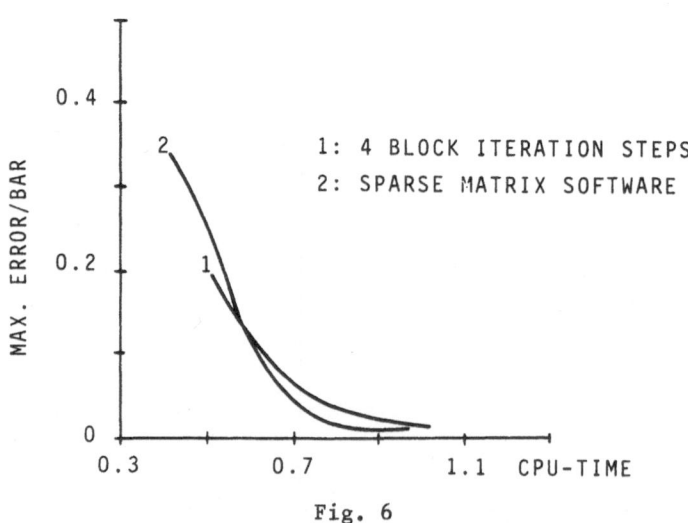

Fig. 6

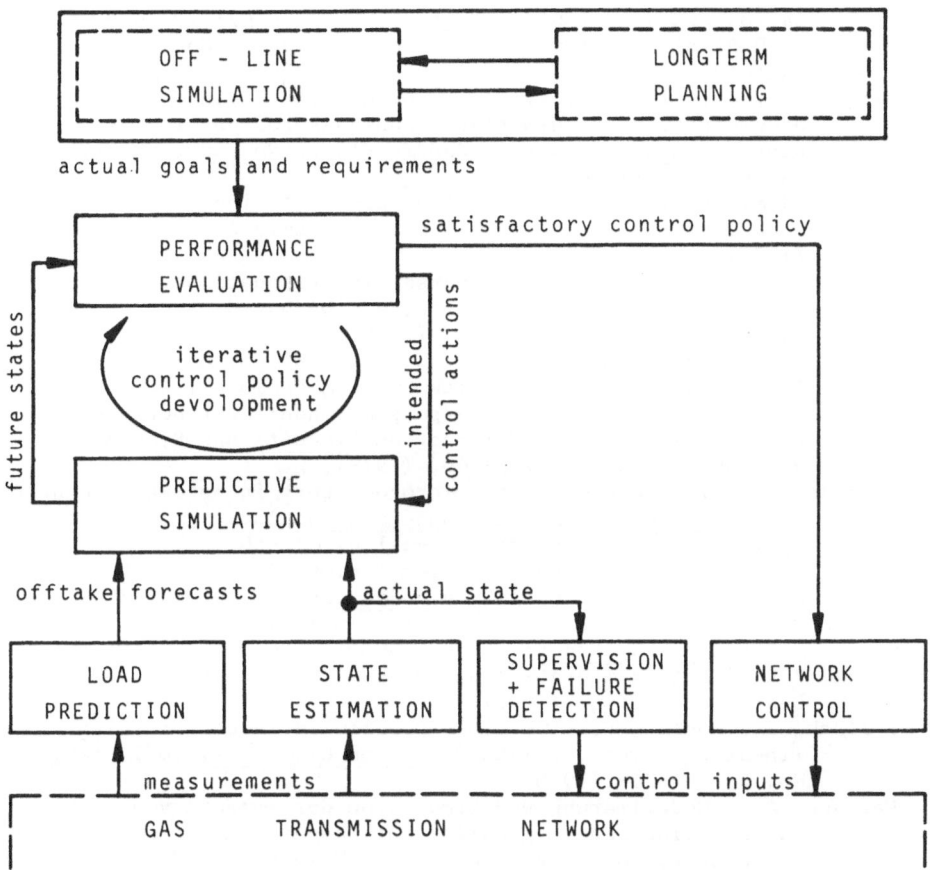

REFERENCES

[1]   Berzins, M., Buckley, T.F., Dew, P.M.   Systolic matrix iterative
      algorithms, in Parallel Computing 83, edited by M. Feilmeier
      and U. Schendel, North Holland Publications, Amsterdam-New
      York (1984).

[2]   Crorkin, W., Allidina, A.Y., Malinowski, K., Singh, M.G.
      Decomposition-coordination techniques for parallel
      simulation - Part 2, Control Systems Centre Report No. 600,
      UMIST, Manchester, (1984).

[3]   David, J.E.   Iterative methods for sparse matrices, in Sparsity
      and its Application, ed. by David, J.E., Cambridge University
      Press, (1985).

[4]   Lappus, G., Schmidt, G.   Supervision and control of gas transpor-
      tation and distribution systems, 6th IFAC/IFIP Conference
      on Digital Computer Applications to Process Control,
      Dusseldorf, (1980).

[5]   Malinowski, K., Allidina, A.Y., Singh, M.G., Crorkin, W.D.
      Decomposition-coordination techniques for parallel simulation
      part 1, Control Systems Centre Report No. 599, UMIST,
      Manchester, (1984).

[6]   Ortega, J.M., Rheinboldt, W.C.   Iterative solution of nonlinear
      equations in several variables, Academic Press, New York,
      (1970).

[7]   Peters, F.J.   Parallelism and sparse linear equations, in
      Sparsity and its Applications, ed. by David, J.E.,
      Cambridge University Press, (1985).

[8]   Rabat, N.B., Sangiovanni-Vincentelli, A.C., Hsieh, M.Y.
      A multilevel Newton-Algorithm with macromodelling and
      latency for analysis of large scale nonlinear networks in
      the time domain, IEEE Trans. on Circuits and Systems,
      vol. CAS-26, No. 9, 733-740, (1979).

[9]   Schmidt, G.   The role of multi-microcomputers in automatic control,
      Proc. of the 4th Intern. Conference on Analysis and
      Optimization of Systems, Versailles, (1980).

[10]  Siljak, D.D.   Complex dynamic systems: dimensionality, structure
      and uncertainty, Large Scale Systems, No. 4, 279-294 (1983).

[11]  Tai, H.M., Sueks, R.   Parallel system simulation, IEEE Trans. on
      Systems and Cybernetics, vol. SMC-14, No. 2, 177-183.

[12]  Varga, R.S.   Matrix iterative analysis, Prentice Hall,
      Englewood Cliffs, (1962).

[13]  Wallach, Y., Konrad, V.   On block parallel methods for solving
      linear equations, IEEE Trans. on Computers, vol. C-29,
      No. 5, 354-359,(1980).

[14]  Weimann, A.   Modellierung und Simulation der Dynamik von
      Gasverteilnetzen im Hinblick auf Gasnetzfuhrung und Gasnetzu-
      berwachung, Dissertation, Techn. Universität München, (1978).

# ON THE FACTORISATION OF CERTAIN SYMMETRIC CIRCULANT BANDED LINEAR SYSTEMS

D. J. Evans

Department of Computer Studies

Loughborough University of Technology, Loughborough, Leics

## 1. INTRODUCTION

A factorisation method is described for the fast numerical solution of certain circulant banded symmetric linear systems which occur repeatedly in the numerical solution of differential equations. It can be shown that such special banded matrices $A_r$ of semi-bandwidth $r$ can be factorised into the product of easily inverted matrices, the components of which are a cyclic matrix and its transpose and a similar circulant banded matrix $A_{r-1}$ of order 1 less. By using this factorisation, efficient algorithmic solution methods can be derived for the related linear systems ([1],[2]).

In the following we treat successively the cases r=2 (tridiagonal) and r=3 (Quindiagonal) directly. For $r \geq 4$, the resulting algorithm is too complicated for a direct solution and an iterative procedure is suggested and set up to produce a reverse recursive strategy for the solution of all such systems involving $A_r, A_{r-1}, \ldots, A_3, A_2$.

## 2. THE FACTORISATION PROCEDURE

(I)  Tridiagonal case  r=2

Consider the factorisation  $A_2 = B_2 A_1 B_2^T$

where,

and

$$A_1 \equiv \begin{bmatrix} c_1 & & & \bigcirc \\ & c_1 & & \\ & & \ddots & \\ \bigcirc & & & c_1 \end{bmatrix}$$

Then, by carrying out the required matrix multiplications and equating terms, we obtain the following relationships between the elements $a_1$, $a_2$ of $A_2$ and the elements $c_1$, $b_2$ of $A_1$ and $B_2$ respectively, i.e.

$$a_1 = c_1(1+b_2^2) \qquad (1)$$

and

$$a_2 = b_2 c_1 . \qquad (2)$$

These two equations when solved for $b_2$ give the quadratic equation

$$b_2^2 - \frac{a_1}{a_2} b_2 + 1 = 0$$

which yields the result,

$$b_2 = \frac{2}{\alpha + \sqrt{\alpha^2 - 4}} , \qquad (3)$$

where $\alpha = a_1/a_2$ and the positive sign is chosen for numerical stability. Finally, the elements of $A_1$ are determined from (2) by the relationship

$$c_1 = a_2/b_2 . \qquad (4)$$

(II)  Quindiagonal case,  r=3

A similar factorisation of the form $A_3 = B_3 A_2 B_3^T$

where

and

80

$$A_2 \equiv \begin{bmatrix} c_1 & c_2 & & & & c_2 \\ c_2 & c_1 & c_2 & & \bigcirc & \\ & & \ddots & & & \\ & \bigcirc & & & & c_2 \\ c_2 & & & c_2 & c_1 \end{bmatrix}$$

can be shown on equating terms to yield the relationships

$$a_1 = (c_1 + b_3 c_2) + (c_2 + b_3 c_1) b_3 \tag{5}$$

$$a_2 = c_2 + b_3 (c_1 + b_3 c_2) \tag{6}$$

$$a_3 = b_3 c_2 . \tag{7}$$

Now using (7) and eliminating $c_2$ from (6), we obtain

$$a_2 = a_3/b_3 + b_3 \left( \frac{(a_1 - 2a_3)}{(1 + b_3^2)} + a_3 \right)$$

which on clearing terms becomes

$$a_2 b_3 (1 + b_3^2) = a_3 (1 + b_3^2)^2 + b_3^2 (a_1 - 2a_3) \tag{8}$$

which is a quartic equation for the derivation of $b_3$ . Finally, once $b_3$ is obtained then the values of $c_2$ and $c_1$ can be easily obtained from equations (7) and (6).

(III)  Septadiagonal case,  r=4

Again, we consider the factorisation,

$$A_4 = B_4 A_3 B_4^T$$

where

$$A_4 \equiv \begin{bmatrix} a_1 & a_2 & a_3 & a_4 & & & a_4 & a_3 & a_2 \\ a_2 & a_1 & a_2 & a_3 & a_4 & & & a_4 & a_3 \\ a_3 & & & & & & \bigcirc & & a_4 \\ a_4 & & & & & & & & \\ & & & & & & & & a_4 \\ a_4 & & & & & & & & a_3 \\ a_3 & a_4 & & \bigcirc & & & & & a_2 \\ a_2 & a_3 & a_4 & & & & a_4 & a_3 & a_2 & a_1 \end{bmatrix}, \quad B_4 \equiv \begin{bmatrix} 1 & b_4 & & & & & \\ & 1 & b_4 & & & \bigcirc & \\ & & & & & & \\ & & & & & & \\ & & & & & & \\ & \bigcirc & & & & & b_4 \\ b_4 & & & & & & 1 \end{bmatrix}$$

and

$$A_3 \equiv \begin{bmatrix} c_1 & c_2 & c_3 & & & c_3 & c_2 \\ c_2 & c_1 & c_2 & c_3 & \bigcirc & & c_3 \\ c_3 & & & & & & \\ & & & & & & \\ & & & & & & c_3 \\ c_3 & & \bigcirc & & & & c_2 \\ c_2 & c_3 & & & c_3 & c_2 & c_1 \end{bmatrix}$$

Equating matrix terms again yields the relationships

$$a_1 = (c_1 + b_4 c_2) + b_4 (c_2 + b_4 c_1) \tag{9}$$

$$a_2 = (c_2 + b_4 c_1) + b_4 (c_3 + b_4 c_2) \tag{10}$$

$$a_3 = (c_3 + b_4 c_2) + b_4^2 c_3 \tag{11}$$

and

$$a_4 = b_4 c_3 . \tag{12}$$

These relationships now appear too complicated to seek an exact solution so an iterative solution is obtained as follows:

1) guess a $b_4$ value $(\le |1|)$

2) determine $c_3 = a_4/b_4$ from eqn. (12)

3) determine $c_2 = [a_3 - c_3(1 + b_4^2)]/b_4$ from (11)

4) determine $c_1 = [a_2 - b_4 c_3 - c_2(1 + b_4^2)]/b_4$ from (10)

5) finally obtain a new value of $b_4$ from (9), i.e.

$$c_1(1 + b_4^2) + 2c_2 b_4 = a_1$$

6) then, check for convergence on the value of $b_4$, otherwise return to 2.

(IV) Finally, for a symmetric matrix of semi-bandwidth $r$, we consider the following factorisation,

$$A_r \equiv B_r A_{r-1} B_r^T ,$$

where

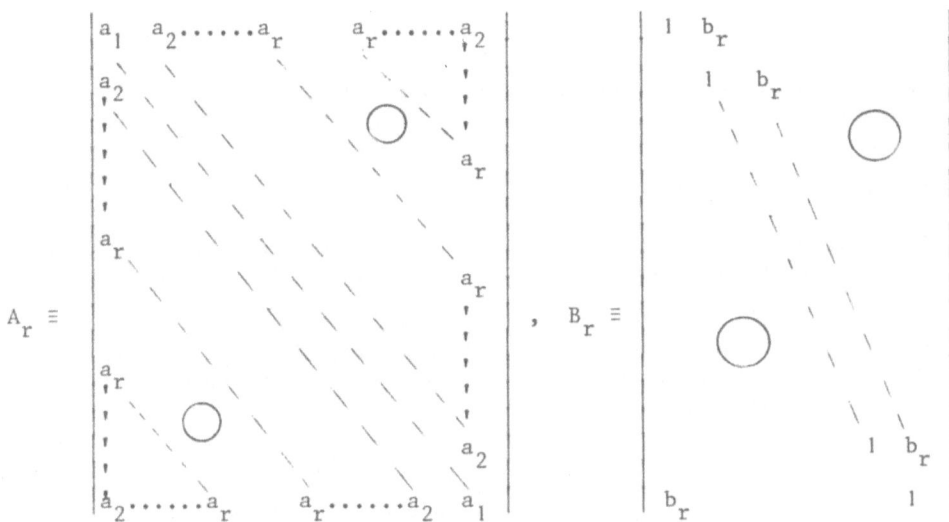

and

By equating terms in the matrix products, we can evaluate the unknown terms in an algorithmic procedure the pattern of which can be determined from the previous cases, i.e.

1) guess a $b_r$ value ($<1$)

2) determine $c_{r-1} = a_r/b_r$

3) determine $c_{r-2} = [a_{r-1} - c_{r-1}(1+b_r^2)]/b_r$

4) determine $c_{r-3} = [a_{r-2} - b_r c_{r-1} - c_{r-2}(1+b_r^2)]/b_r$

5) determine $c_{r-4} = [a_{r-3} - b_r c_{r-2} - c_{r-3}(1+b_r^2)]/b_r$ .
   . . . . . . . . . . . . . . . . . . . . . . . . . . .

6) determine $a_1 = [a_2 - b_r c_3 - c_2(1+b_r^2)]/b_r$

7) finally, obtain a new value of $b_r$ from

$$c_1(1+b_r^2) + 2c_2 b_r = a_1$$

8) then check for convergence on the value of $b_r$, otherwise return to 2).

## 3. NUMERICAL EXPERIMENTS

Finally, to compute the factors of $A$ (-2, 27, -270, 598, -270, 27, -2) we applied the iterative procedure described in III of the previous section.

The procedure appeared to have global convergence from any starting value of $b_4$ in approximately 40 iterations. The final results obtained for the factorisation were as follows:

$$b_4 = -2.169$$

and $c_3 = 0.922$, $c_2 = -10.021$ and $c_1 = 97.180$.

Further experiments are currently underway to further our knowledge of this algorithmic technique.

## ACKNOWLEDGMENT

The author is indebted to Mr. G. Samra for programming assistance.

## REFERENCES

1.  D.J. Evans and A. Hadjidimos, (1979). On the factorisation of special symmetric periodic and non-periodic quindiagonal matrices. Computing 21, 259-266.
2.  D.J. Evans, (1980). On the solution of certain Toeplitz tridiagonal linear systems. SINUM 17, 675-680.

AN ABS METHOD FOR SOLVING SUITABLY STRUCTURED LINEAR SYSTEMS

IS SUPPORTED BY PARALLEL ARCHITECTURES

V. Fragnelli  and  Giovanni Resta

Mathematical Institute,  University of Genova

SUMMARY

We present an algorithm based on a method belonging to the class of Abaffy-Broyden-Spedicato [3].  The linear system to be solved can be sparse and in this situation must have a row-polychromatic structure.

Suggestions are given in order to exploit the high degree of freedom of the class.  A suitable data structure reduces computation and memory requirements.  The computations specified by the algorithm have a high degree of parallelism.

The computational complexity of the algorithm is discussed. Parallel architectures supporting the algorithm are examined.

## 1.  ROW-MONOCHROMATIC STRUCTURE  $\mu$  FOR A MATRIX

In [1] were introduced Row-Monochromatic Structure  $\mu$  and Row-Polychromatic Structure  $\pi$  for a matrix.  Here we shortly recall the given definitions.

Let  $m$  be a positive integer.  If  $s$  is a positive integer with $s \leq m$, an s-disaggregator is a vector  $e$  of positive integers with $e(s) = m$  and  $e(i) > e(i-1)$, $i = 2,\ldots,s$ .

An $s \times m$ matrix  $B$  is  $\mu$  if it exists an s-disaggregator  $e$  for which:

$$B(i,j) = 0 \quad \begin{array}{ll} i = 1,\ldots,s-1 & j = e(i)+1,\ldots,m \\ i = 2,\ldots,s & j = 1,\ldots,e(i-1) \end{array}$$

Such a matrix has at most one non-zero element in each column;  it can be compressed in a row vector  $b := c(B,e)$.  The disaggregator  $e$  allows for each component of  $b$  to identify the row position in the matrix  $B$, or, in other words, to decompress a row vector  $b$  into a $\mu$ matrix $B := d(b,e)$  (Fig. 1).

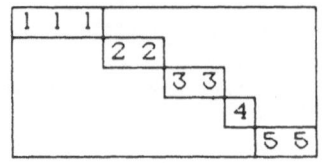

e = ( 3, 5, 7, 8, 10 )
disaggregator

$c(A,e) = $  | 1 | 1 | 1 | 2 | 2 | 3 | 3 | 4 | 5 | 5 |

μ matrix A                                   compressed form

Fig. 1

We also define the block diagonal structure  β  for a square matrix
G.   An m×m matrix  G  is  β  if there exists an s-disaggregator  e   for
which:

$$G(i,j) = 0 \quad h = 1,\ldots,s-1 \quad i = 1,\ldots,e(h) \quad j = e(h)+1,\ldots,e(h+1)$$
$$h = 1,\ldots,s-1 \quad j = 1,\ldots,e(h) \quad i = e(h)+1,\ldots,e(h+1)$$

The computational complexity in terms of space for storing a μ matrix
B is of m real words, since we can store  c(B,e).   We investigate the
computational complexity in terms of time for linear algebra operations
on μ matrices.

If   e  is an s-disaggregator,
     f  is a t-disaggregator,

 there exists an integer vector  j  for which  e(j(i)) = f(i),  i = 1,t

     S and R are μ  with respect to e,
     A and W are μ  with respect to f,
     x  is an m-dimensional vector,
     y  is an s-dimensional vector,

     $G := S^T R$, $V := SR^T$, $F^T := S^T RA^T$, $B := WS^T R$, $v := Sx$ and $w^T := y^T S$

then the following results hold:

Lemma 1    G is β with respect to e.

Lemma 2    V is an sxs diagonal matrix.

Lemma 3    F and B are μ with respect to f.

Lemma 4    The computational complexity in terms of time for computing G,
           V, F, B, v or w is of order m floating point operations.

Lemma 5    On a super computer or on a suitable parallel architecture
           each of the computations for obtaining G, V, F, B, v or w can
           be efficiently implemented reducing the corresponding
           computational complexity.

## 2. ROW-POLYCHROMATIC STRUCTURE $\pi$ FOR A MATRIX

Let $n$ be a positive integer with $n \leq m$. If $k$ is a positive integer a k-disaggregating tree is a collection $E = [e_h, h = 1,k]$ of $k$ $n_h$-disaggregators $e_h$, for which

$$\Sigma_h \, n_h = n \qquad \text{for} \quad h = 1,\ldots,1-1$$

for $h = 1,\ldots,k-1$ there exists an integer vector $j_j$ for which

$$e_h(j_h(i)) = e_{h+1}(i), \qquad i = 1,n_{h+1} \, .$$

An nxn matrix $A$ is $\pi$ if there exists a k-disaggregating tree $E$ for which $A = \text{rows}(A_h, h = 1,k)$ is the rowise aggregate of k matrices $A_h$ where the matrix $A_h$ is an $n_h$xm matrix $\mu$ with respect to $e_h$, $h = 1,k$. The matrix $A$ can be compressed into a kxm matrix $Q = C(A,E)$, whose h-th row is $c(A_h,e_h)$, i.e. the compression of the $\mu$ matrix $A_h$. Matrix $Q$ can be decompressed into the matrix $A = D(Q,E)$ utilizing the k-disaggregating tree $E = [e_h, h = 1,k]$, (Fig. 2).

We now define $\sigma_h$ and $\beta_h$ structures, $h = 1,k$. If $E = [e_h, h = 1,k]$ is a k-disaggregating tree, we define an mxm matrix $G$ to be $\beta_h$ if it is block diagonal with respect to $e_h$, $h = 1,k$. We define an $n_h$x$n_h$ matrix $V$ to be $\sigma_h$ if it is diagonal, $h = 1,k$.

As a consequence of lemma 1, if $S$ and $R$ are $\mu$ with respect to $e_h$ then $S^T R$ is $\beta_h$, $h = 1,k$.

As a consequence of lemma 2, if $S$ and $R$ are $\mu$ with respect to $e_h$ then $SR^T$ is $\sigma_h$, $h = 1,k$.

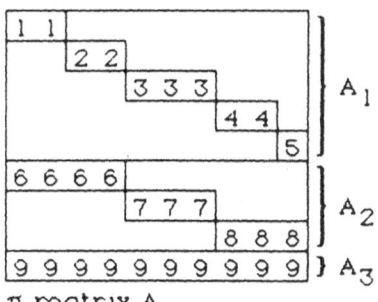

$$e_1 = (2, 4, 7, 9, 10)$$
$$e_2 = (4, 7, 10)$$
$$e_3 = (10)$$
$$E = \{e_1, e_2, e_3\}$$

disaggregating tree E

$\pi$ matrix A

compressed form $Q = C(A,E)$

Fig. 2

## 3. OPERATIONS ON ROW-POLYCHROMATIC MATRICES

Let $A = \text{rows}(A_h, h = 1,k)$, $W = \text{rows}(W_h, h = 1,k)$ and $P = \text{rows}(P_h, h = 1,k)$ be three matrices $\pi$ with respect to the disaggregating tree $E = [e_h, h = 1,k]$, for which $V_h := P_h A_h^T$ is nonsingular, $h = 1,\ldots,k$. Let

$$S_1 := A_1$$

$$R_1 := W_1$$

$$S_{h+1} := (I - \sum_{j=1}^{h} S_j^T R_j)A_{h+1} \qquad R_{h+1}^T := W_{h+1}^T(I - \sum_{j=1}^{h} S_j^T R_j)$$

$$h = 1,\ldots,k-1 \; ,$$

then the following results hold:

Lemma 6    $S = \text{rows}(S_h, h = 1,k)$ is $\pi$ with respect to E.

Lemma 7    $R = \text{rows}(R_h, h = 1,k)$ is $\pi$ with respect to E.

Lemma 8    $V_h$ is $\sigma_h$, $h = 1,\ldots,k$.

Lemma 9    $S_h^T R_h$ is $\beta_h$ , $h = 1,\ldots,k$.

Lemma 10   The computational complexity in terms of time for computing $V_h$ is of order m floating point operations, $h = 1,\ldots,k$.

Lemma 11   The computational complexity in terms of time for computing $S_h$ or $R_h$ is of order $h \cdot m$ floating point operations, $h = 1,\ldots,k$.

Lemma 12   On a super computer or on a suitable parallel architecture each of the computations for obtaining $V_h$, $S_h$ or $R_h$ can be efficiently implemented reducing the corresponding computational complexity.

Lemma 13   The computational complexity in terms of space for storing A, W, P, S and R is $5 \cdot h \cdot m$ real words.

Let

$$G_1 = I$$

$$G_{h+1} = (I - \sum_{j=1}^{h} S_j^T R_j), \quad h = 1,\ldots,k$$

$$s(1) = 1$$

$$s(2) := 1+n_1$$

$$s(h+1) = 1+n_1+\ldots+n_h \, , \quad h = 2,\ldots,k \; .$$

## 4. MULTISTAGE PROCESSES

The $\pi$ matrices are typical in multistage production processes in which some constraints apply to the variables of each stage and other constraints relate an increasing number of different stages. The resulting structure may be called block-diagonal constraints with recursively coupling constraints [2]. A π matrix results by suitably permuting the constraints (Fig. 3).

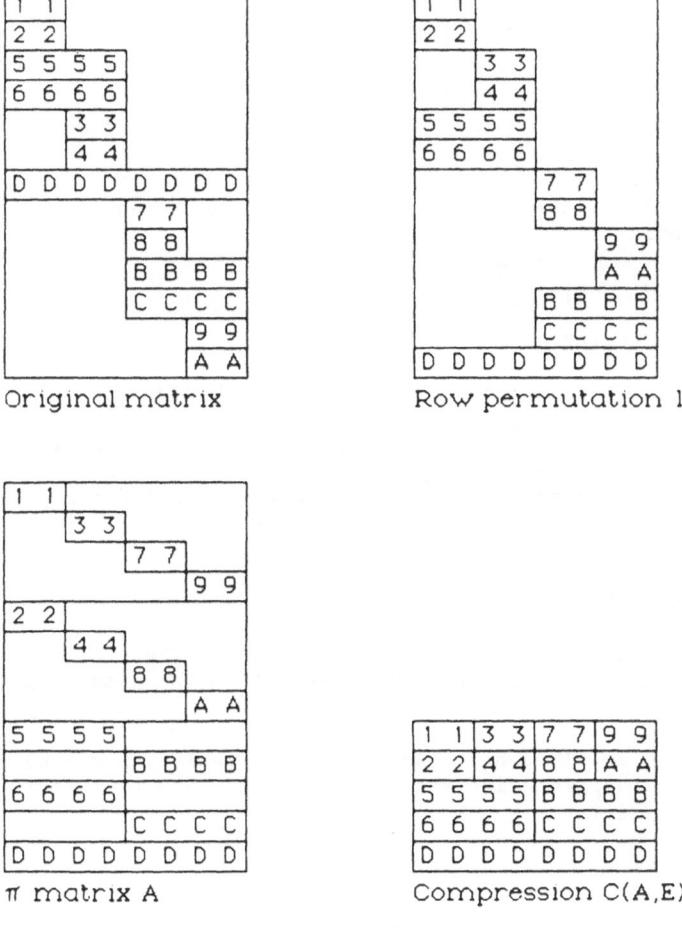

Fig. 3

## 5. ABS CLASS OF METHODS

A generalized class of iterative methods for solving a linear system $Ax = b$ is presented in [3]. This class has some interesting properties, in our case the most important are those related to the high degree of freedom in choosing some parameters. Before analyzing the possible choices of the parameters we recall the generalized algorithm of Abaffy-Broyden-Spedicato (ABS).

Let $Ax = b$ be the given system, where A is an $n \times m$ matrix ($n \leq m$). Starting from a point $x_1$, the algorithm generates a sequence of points $x_i$, where $x_{n+1}$ is a solution of the system of equations, according to the following formula

$$x_{i+1} = x_i - \delta_i p_i$$

where $p_i$ is a search direction and $\delta_i$ is the stepsize.

The search direction $p_i$ is defined by

$$p_i = H_i^T z_i$$

where $H_i$ is the deflection matrix and $z_i$ is a vector of parameters. The stepsize $\delta_i$ is defined by

$$\delta_i = (a_i^T x_i - b_i)/p_i^T a_i$$

where $a_i^T$ is the i-th row of the matrix A and $p_i^T a_i \neq 0$ (well-definiteness condition).

The deflection matrix $H_i$ is updated at each step according to the following formula

$$H_{i+1} = H_i - H_i a_i w_i^T H_i$$

where $w_i$ is a vector of parameters.

The vectors $z_i$ and $w_i$ may be arbitrarily chosen respecting the well-definiteness condition $a_i^T H_i^T z_i \neq 0$, and the projection condition $a_i^T H_i^T w_i = 1$. The first deflection matrix $H_1$ is an arbitrary positive definite $m \times m$ matrix. The starting solution $x_1$ is an arbitrary vector. An high degree of freedom is a consequence of the selection of $z_i$, $w_i$, $H_1$ and of $x_1$.

## 6. GENERALIZED ABS ALGORITHMS FOR A $\pi$ MATRIX

In this section we investigate the situation where A is a $\pi$ matrix. Let $A = \text{rows}(A_h, h = 1,k)$ be $\pi$ with respect to the disaggregating tree $E = [e_h, h = 1,k]$.

Select $Z = \text{rows}(Z_h, h = 1,k)$ and $W = \text{rows}(W_h, h = 1,k)$ $\pi$ with respect to the disaggregating tree $E$ in such a way to satisfy well-definiteness condition

$$A_h G_h^T Z_h^T \neq 0 , \qquad h = 1,\ldots,k$$

and projection condition

$$A_h G_h^T W_h^T = 1 \qquad h = 1,\ldots,k .$$

90

One possible choice to meet all these requirements is the following:

$$Z_h := A_h \qquad\qquad h = 1,\ldots,k$$

$$W_h = (A_h G_h{}^T A_h{}^T)^{-1} A_h \qquad h = 1,\ldots,k .$$

We denote by $b_h$ the $n_h$-dimensional vector defined by the relation

$$b_h(i) := b(i+s(h)-1) \qquad i = 1,n_h , \qquad h = 1,k$$

and introduce the following algorithm which, starting from an arbitrary point $y_1$, terminates after $k$ iterations to a solution $y_{k+1}$ of the system of equations $Ax = b$ :

| | |
|---|---|
| Step 1 | $(Z_h := A_h, \quad h = 1,\ldots,k)$ |
| Step 2 | $G_1 := I$ |
| Step 3 | $h := 1$ |
| Step 4 | $W_h = (A_h G_h{}^T A_h{}^T)^{-1} A_h$ |
| Step 5 | $P_h{}^T := G_h{}^T Z_h{}^T$ |
| Step 6 | $V_h := P_h A_h{}^T$ |
| Step 7 | $d_h := V_h{}^{-1}(A_h y_h - b_h)$ |
| Step 8 | $y_{h+1} := y_h - P_h{}^T d_h$ |
| Step 9 | $S_h{}^T := G_h A_h{}^T$ |
| Step 10 | $R_h := W_h G_h$ |
| Step 11 | $G_{h+1} := G_h - S_h{}^T R_h$ |
| Step 12 | $h := h+1$ |
| Step 13 | if $h < k+1$ go to Step 4. |

Observe that Step 11 is never performed and that the products involving $G_h$ in Steps 5, 9 and 10 are performed utilizing the relation

$$G_1 = I$$

$$G_{h+1} = (I - \sum_{j=1}^{h} S_j{}^T R_j) , \qquad h = 1,\ldots,k .$$

Steps 1, 2 and 4 can be modified, obtaining different algorithms belonging to the ABS class, provided that $G_1$ is a nonsingular $\beta_1$ matrix and that the well-definiteness and projection conditions are satisfied.

Observe that the $h$-th iteration of the algorithm has a computational complexity of $h\cdot m$ in terms of floating point operations. The computational complexity of the algorithm is therefore of order $k^2 m$ floating point operations.

We consider the method presented in Section 5 corresponding to the choices $H_1 = I$ and $y_1 = x_1$ . We select as $z_i$ the $i$-th row of $Z$

and as $w_i$ the i-th row of W, $i = 1,n$ . Utilizing the notations introduced in Section 2 we observe that

Lemma 15     $H_{i+1}$ is $\beta_h$           $i = s(h),\ldots,s(h+1)-1$   $h = 1,\ldots,k$

Lemma 16     $p_i = H_i^T z_i = H_{s(h)}^T z_i$   $i = s(h),\ldots,s(h+1)-1$   $h = 1,\ldots,k$

Lemma 17     $H_{s(h)} = G_h$                            $h = 1,\ldots,k$

Since        $P_h^T = G_h^T Z_h^T$                           $h = 1,\ldots,k$

and          $P = \text{rows}(P_h, h = 1,k)$             $h = 1,\ldots,k$

because of lemma 16 and lemma 17 $p_i^T$ is the i-th row of the matrix P, $i = 1,n$, and therefore

$$y_h = x_{s(h)} \qquad\qquad h = 1,\ldots,k$$

Thus the algorithm introduced in this section starting from an arbitrary point $y_1$ terminates after k iterations to a solution $y_{k+1} = x_{n+1}$ of the system of equations $Ax = b$.


# 7. A DATA STRUCTURE FOR IMPLEMENTING THE ALGORITHM

We now define an m-dimensional vector u and k $n_h$-dimensional vectors $u_h$, $h = 1,k$ posing $u_k(i) := 1$, $i = 1,n_h$, $h = 1,k$ and $u(i) := 1$, $i = 1,m$. We define k $\mu$-matrices $U_h$, $h = 1,k$ posing $U_h := D(u,e_h)$, $h = 1,k$. If $V_h$ and $b_h$ are the ones introduced in Section 6, $h = 1,k$, let

$$r_h := U_h^T b_h, \; v_h := V_h u_h, \; s_h := A_h y_h, \; z_h := (U_h^T s_h) - r_h$$

and     $w_h := U_h^T v_h$ .

The following result holds:

Lemma 18     $d_h(i) = z_h(i)/w_h(i)$, $i = 1,m$, $h = 1,k$ .

Therefore we can substitute to the corresponding steps 6 and 7 of the algorithm introduced in Section 6 the following steps:

Step 6*     $z_h := (U_h^T A_h y_h) - r_h$,   $w_h := U_h^T V_h u_h$

Step 7*     $d_h := z_h/w_h$

where / denotes the componentwise division of m-dimensional vectors. In this way each of the steps of the algorithm operates on m-dimensional vectors.

We suggest, for an efficient implementation, a data structure rowise storing each of the matrices C(A,E), C(W,E), C(Z,E), C(P,E), C(S,E) and C(R,E). The h-th iteration in fact will utilize only the h-th row of each of these matrices.

An efficient data structure for storing a disaggregator e is an m-dimensional vector of type Boolean whose components are false in correspondence to the components of e and true otherwise. For instance, if m = 9 we store the disaggregator e = (3,7,9) as the Boolean vector (TTFTTTFTF).

An efficient data structure for storing a k-disaggregating tree E is
the rowise storage of a kxm matrix of type Boolean whose h-th row is the
m-dimensional vector of type Boolean corresponding to the disaggregator
$e_h$, h = 1,k .

## 8. PARALLEL ARCHITECTURES

In [4] some particular architectures are introduced. These
architectures support a very efficient implementation if the disaggregating tree E = [$e_h$, h = 1,k] satisfies some restrictions. Namely, m must
be a power of two and there must exist an integer k-dimensional vector  i
for which

$$i(h) \geq i(h-1) \quad , \quad h = 2,k$$

$$n_h = m/2^{i(h)} \quad , \quad h = 1,k$$

$$e_h(j) = j \cdot i(h) \quad , \quad j = 1,n_h \quad , \quad h = 1,k$$

One of these architectures is the $\log_2(m)$-dimensional binary cube [5]. In
this situation the computational complexity of the algorithm introduced
in Section 6 reduces to $k^2 \cdot \log_2(m)$.

## REFERENCES

[1]     G. Resta,  An algorithm for the least squares solution of
             suitably structured linear systems.   To appear.
[2]     P.E. Gill, W. Murray,  Numerical methods for constrained
             optimization.  Academic Press, (1974).
[3]     J. Abaffy, S. Broyden, E. Spedicato,  A class of methods for
             linear equations II: Non-singular representation, parameter
             solution and other tales.  Report IAM1.
[4]     L. Galli, G. Resta,  Implementation of a data-flow algorithm for
             linear programming.  Parallel Computing 83,  Feilmeier,
             Joubert and Schendel, editors, Elsevier (1984).
[5]     G. Resta,  The binary cube: a superstructure for super computers,
             Numerical Optimisation Centre T.R. 144, Hatfield Pol technic
             (U.K.), (1984).

FREQUENCY-DOMAIN SEPARABLE DECOMPOSITION OF 2-DIMENSIONAL SYSTEMS

N.M. Mitrou[*], G.I. Stassinopoulos[**]  and
E.N. Protonotarious[**]

[*]Department of Electronics, Nuclear Research Center
"Democritos", Aghia Paraskevi, Attiki, Greece
[**]Division of Computer Science, National Technical
University of Athens, Greece

ABSTRACT

    Fan filters in seismic data processing, directional filters in image
processing and 2-D equalizers are but a few examples of 2-Dimensional
systems described explicitly in the frequency domain.  Current methods
for handling such systems in a discrete form require an NxN sampling grid
resulting in considerable design complexity and a high implementation
cost.

    The approach presented here consists of approximating a desired 2-
Dimensional response function through decomposing it into a sum of  L
separable terms.  The advantages gained by such a decomposition are the
following:

1.  The analysis and design problem is reduced to a 1-D one.
2.  Application of parallel-processing techniques is allowed.
3.  The number of elements required for the implementation is now 2xLxN;
    in the case where L << N, appreciable simplification and economy is
    obtained.

    Applications are given through design examples of 2-D low-pass, fan
and Wiener filtering.

1.  INTRODUCTION

    Parallelism of computational means is very often used in signal
processing or other system implementation/simulation to gain speed and
structure modularity.  In this context many architectural styles have
appeared offering a high degree of parallelism, like pipelining arrangements of multiple functional units, etc. [1,2].  Situations demanding
high speed, on the other hand, arise in many modern and important
applications, especially with real-time implementations, like image
processing, control, robotics, etc.

    To get a favourable parallel structure a suitable decomposition of
the system to be simulated or implemented must be performed.   Such a

decomposition depends of course on the system description (the model) and it must conform to other constraints of the individual application. In other words, the adopted architecture must solve a space-time tradeoff in the environment of the particular task. Nevertheless, some general features which are always desirable can be addressed. Apart from the obvious requirements for simplicity and modularity, some further pursuits are the following:

- Independence of the subsystems connected in parallel. This allows each subsystem to be handled (i.e. designed and implemented) independently.
- Optimal utilization of the computational resources available.
- Fourier coefficient-like orthogonality of the subsystems. This allows a finer decomposition to be obtained by just adding some more parallel units without affecting the previous structure.

This paper examines a decomposition method exhibiting some of the advantages mentioned above for 2-Dimensional (2-D) systems. Because of their high complexity 2-D filters require a considerable amount of storage and a great deal of computation time for implementation/simulation [3]. Hence, even when no parallelism aspects are in the foreground, there is an urgent need for simplification through suitable decomposition. Treitel and Shanks [4] have proposed a method based on the well-known Singular Value Decomposition (SVD) technique for such a decomposition to a sum of separable subsystems. According to their method the impulse response matrix $h(.,.)$ of the system (original domain) is approximated by a sum of separable terms thus allowing an implementation through a parallel connection of products of 1-D subsystems. If a small number of such terms, compared to the size of the matrix, is sufficient for an acceptable approximation, as is often the case, a substantial simplification is obtained.

There are several motivations for extending Treitel and Shanks' method in the frequency (transform) domain. An obvious reason is that in many cases the system description is readily obtained or given in terms of its frequency rather than its impulse response which may be of infinite extent. Characteristic examples of 2-D systems described explicitly in the transform domain are given in Section II. Another reason is that of overcoming some restrictions imposed by the application of the method in the original domain, like the requirement for same impulse-response length of the resultant 1-D subsystems [3, 3.5.2] .

As a natural generalization we examine here the corresponding problem in the continuous frequency domain. The decomposition leads now to a sum of products of 1-D functions, instead of vectors, and the approximation error is expressed by the integral of the squared difference. The associated theory is well-known from the field of linear operators [5] and integral equations [6]. Some relevant results are stated very briefly in Section III. To perform the decomposition in the general case, with $H(.,.)$ any square integrable response, we must discretize the problem in order to carry out the necessary numerical integrations. In that case the procedure does not differentiate itself from that outlined in [4], apart from the mathematical and physical insight. However, the problem can be simplified in some special cases, as with some real, binary frequency responses. More significantly, completely analytical results are derived for linear effective-domain boundary, as for the fan-shaped response. Systems with such a response appear in direction-finding or beamforming applications as in seismic data processing, image processing, wide-band array processing, etc., as explained in Section II. All these are applications with high demand in computational power and critical speed requirements, where gains due to parallelism are highly appreciated.

96

The paper is organized as follows. Some characteristic 2-D systems described explicitly in the frequency domain are addressed in Section II, which supplements this introduction. Section III presents some background theoretical results from the linear-operator field. Section IV deals with a decomposition algorithm, while Section V considers some implementation and error aspects. Section VI demonstrates the applicability of the method by means of some examples, including the fan-filter analytical design.

## II. 2-D SYSTEMS WITH FREQUENCY DOMAIN SPECIFICATIONS

The Fourier transform, despite its 150 years of history, remains ever new and active in many branches of physics and engineering. Apart from being a standard engineering tool it always poses interesting and challenging new problems and methods in a variety of frameworks. In this section we address some important applications involving 2-D systems which are explicitly described in the transform domain. We skip the somewhat exhausted cases of low-pass, high-pass and band-pass filters. In all cases listed below, our design through separable decomposition in the continuous frequency domain is particularly attractive.

### 1. Deconvolution and Wiener Filtering in Image Processing

Every image-forming system is characterised by aberrations due to diffraction or other imperfections, thus the resultant image is blurred. Sometimes this blurring can be described in the transform domain by a frequency response function $H(u,v)$. Then a deconvolution process is necessary to correct the image (deblurring) which in that case is modelled by the frequency response function $H_I(.,.) = 1/H(.,.)$ .

The inverse filtering just described does not work well, however, in the presence of noise, especially in regions of the frequency plane where $H(.,.)$ exhibits small magnitudes. In such a case a more suitable filter is suggested by the Wiener theory in direct analogy to 1-D case. Suppose that the blurred and noise-corrupted image is expressed in the original domain as

$$\tilde{s}(x,y) = s(x,y) \circledast h(x,y) + n(x,y) \tag{1}$$

where $s(.,.)$ is the correct image with a power density spectrum $\phi_s(u,v)$, $h(.,.)$ the impulse response of the blurring system and $n(.,.)$ an additive noise with a power density spectrum $\phi_n(u,v)$. Then the system with a frequency response

$$H_w(u,v) = F\{h_w(x,y)\} = \frac{H^*(u,v)}{|H(u,v)|^2 + \phi_n(u,v)/\phi_s(u,v)} \tag{2}$$

where

$$H(u,v) = F\{h(x,y)\} \tag{3}$$

is the optimal deblurring system in the Mean-Square sense, i.e.

$$E\{[s(x,y) - \tilde{s}(x,y) \circledast h_w(x,y)]^2\} = \min \tag{4}$$

### 2. Wide-band Array Processing - Fan Filtering

Fourier transform methods are quite advantageous in analyzing signals carried by propagating waves. This is so because specifications concerning velocity and direction of propagation are set directly in the multidimensional frequency domain [3,6]. This can be easily understood by considering the elementary signal

$$e^{-j(\omega t - \bar{k}^T \bar{x})},$$

with $\bar{k}$ the wavenumber vector, $\omega$ the cyclic temporal frequency and $\bar{x}$ the position vector; it represents a plane wave travelling towards $\bar{x}_0 = \bar{k}$ with a velocity $\nu = \omega/|\bar{k}|$. Hence, each point of the 4-D transform space $\omega - \bar{k}$ represents a plane wave travelling on a certain direction with a certain velocity.

Many applications in a variety of technological fields, like geophysical data processing, radar and sonar, handle information concerning direction of propagation. If we confine ourselves in the 2-D space $\omega - k_x$, i.e. of one spatial and one temporal dimension, it is readily seen that all signals travelling with a certain velocity $\nu$ on directions confined by certain angles $\gamma_1, \gamma_2$ with respect to the axis $x$, are represented by an angular sector $\Theta = \Theta_2 - \Theta_1$ in the $\omega - k_x$ plane. The angles $\Theta_1$ and $\Theta_2$ are defined by the $k_x$ axis and the lines (see Fig.1)

$$\frac{\omega}{k_x} = \frac{\nu}{\cos\gamma_1} \quad . \quad \text{resp.} \quad \frac{\omega}{k_x} = \frac{\nu}{\cos\gamma_2} \tag{5}$$

Obviously, the "visible region", i.e. the region of the $\omega - k_x$ plane corresponding to existing signals, is confined by the lines

$$\frac{\omega}{k_x} = -\nu \quad \text{and} \quad \frac{\omega}{k_x} = \nu \tag{6}$$

Filters with a frequency response

$$H(k_x,\omega) = \begin{cases} 1, & \text{within } \Theta \\ 0, & \text{elsewhere} \end{cases} \tag{7}$$

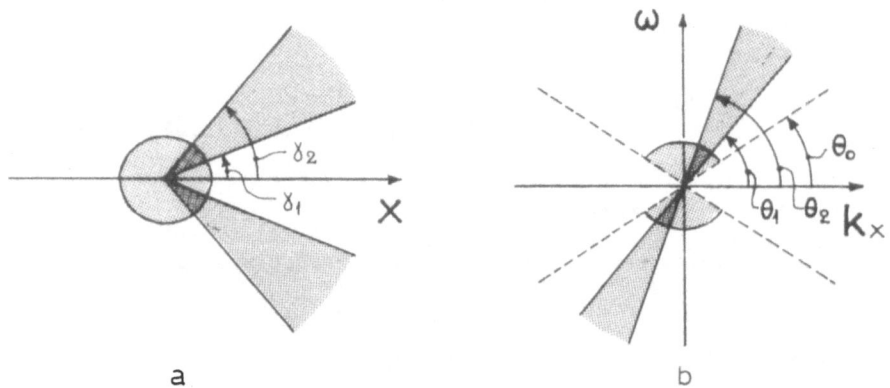

Fig. 1    Direction and velocity specifications
(a) spatial domain,   (b) transform domain.

are widely used in seismic data processing, named fan or pie-slice or velocity filters. They could also be used to other wide-band array processing applications like in sonar. In such applications, if no such a 2-D (or multidimensional) space-time filtering is performed, we have the well-known problem of the different radiation patterns at different frequencies.

3.  Image Coding Through Directional Decomposition

Modern image-coding techniques, often called "Second-generation image-coding techniques" [9], perform some preprocessing before applying the classical coding theory to achieve high compression. These usually exploit the properties of the human visual system to filter out redundant information. One of them proceeds through a decomposition in the frequency domain by a low-pass and a set of directional filters [9], as in Fig. 2. The resultant subimages at the output of the individual filters are then encoded, thus achieving a very high compression.

A directional filter, with a pie-slice-like frequency response, symmetric about the origin, has the property of preserving only edges on a certain direction, i.e. perpendicular to its axis of symmetry. This can be readily understood by considering the 2-D Fourier transform of a line which is also a line in the frequency plane on a perpendicular direction. Apparently, these filters resemble the pie-slice filters for processing signals carried by propagating waves, discussed in the previous paragraph. The only difference is that they concern two spatial directions instead of one spatial and one temporal.

III. THEORETICAL BACKGROUND

In the previous section there have been discussed some signal-processing systems meeting certain transform-domain specifications. Of course, any other linear or linearized system, 2-D or multidimensional, can be modelled in terms of its frequency response through appropriate experimental measurements.

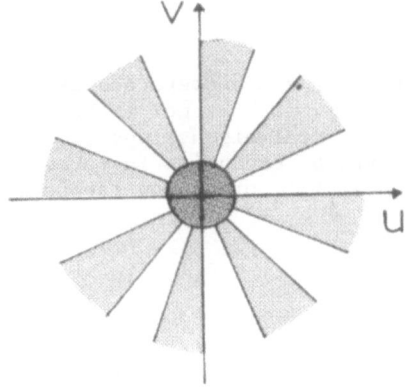

Fig. 2    Transform-domain directional decomposition

Suppose the desired (experimentally derived) frequency response of a 2-D system to be implemented (simulated) is given by the complex-valued function $H(u,v)$, defined on the rectangle $J_1 x J_2 = [u_o^-, u_o^+] x [v_o^-, v_o^+]$. We take $H(u,v)$ in the space $L^2(J_1 x J_2)$ of square integrable functions with the $L^2$-norm. In the following it will be proved that $H(u,v)$ admits an expansion of the form

$$H(u,v) = \sum_{i=1}^{\infty} \lambda_i q_i(u) r_i(v) \tag{8}$$

where $q_i(.)$ and $r_i(.)$, $i = 1,2,\ldots$, are the orthonormal eigenfunctions of the integral operators with kernels

$$Q(u',u) = \int_{J_2} H(u',v) H^*(u,v) dv \quad \text{on } J_1 x J_1 , \tag{9a}$$

$$R(v',v) = \int_{J_1} H(u,v') H^*(u,v) du \quad \text{on } J_2 x J_2 , \tag{9b}$$

possessing the same eigenvalues $\lambda_i^2$ (the star means complex conjugate). Moreover, it will be shown that $q_i(.)$ and $r_i(.)$ are related with each other through

$$r_i(v) = \frac{1}{\lambda_i} \int_{J_1} H(u,v) q_i^*(u) du , \quad v \in J_2 \tag{10a}$$

$$q_i(u) = \frac{1}{\lambda_i} \int_{J_2} H(u,v) r_i^*(v) dv , \quad u \in J_1 \tag{10b}$$

In analogy to [4, Appendix I], let us define $W(.,.)$ on

$$I_1 x I_2 = [u_o^- - (v_o^+ - v_o^-), u_o^+] x [v_o^- - (u_o^+ - u_o^-), v_o^+]$$

by

$$W(\omega,\nu) = \begin{cases} H(v_o^- + u_o^- - \nu, u_o^- + v_o^- - \omega), & (\omega,\nu) \in R_4 \triangleq [u_o^- - (v_o^+ - v_o^-), u_o^+] x [v_o^- - (u_o^+ - u_o^-), v_o^-] \\ H^*(\omega,\nu), & (\omega,\nu) \in R_3 \triangleq [u_o^-, u_o^+] x [v_o^-, v_o^+] \\ 0, & (\omega,\nu) \in R_2 \triangleq (u_o^-, u_o^+] x [v_o^- - (u_o^+ - u_o^-), v_o^-) \\ 0, & (\omega,\nu) \in R_1 \triangleq [u_o^- - (v_o^+ - v_o^-), u_o^-) x (v_o^-, v_o^+] \end{cases} \tag{11}$$

(see Fig. 3).

Since $W(.,.) \in L^2 (I_1 x I_2)$ it can be regarded as the kernel of a Hilbert-Schmidt operator [5, XI.8.44]. The kernel is Hermitian and the operator compact so it has countable, real eigenvalues [5, X5,X4.2]. Moreover, by [5, XI.8.56] $W(.,.)$ admits an expansion in terms of the normalized orthogonal eigenfunctions $\phi_i(.)$ of the operator, i.e.

$$W(\omega,\nu) = \sum_i \lambda_i \phi_i(\omega) \phi_i^*(\nu) , \quad \omega \in I_1, \nu \in I_2 \tag{12}$$

where

$$\int_{I_2} W(\omega,\nu) \phi_i(\nu) d\nu = \lambda_i \phi_i(\omega) , \quad \omega \in I_1 \tag{13}$$

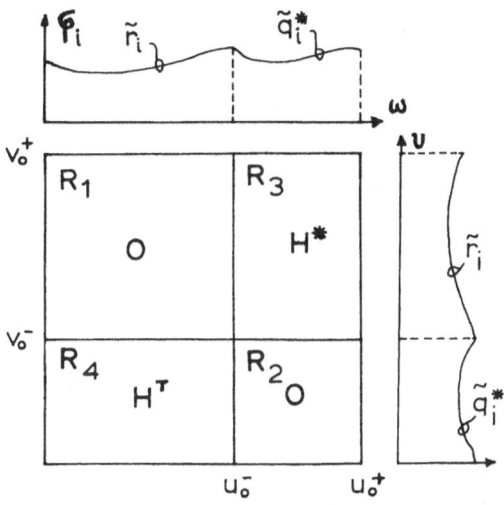

Fig. 3 Region of support for the Hermitian
kernel $W(.,.)$ and partition of the
associated eigenfunctions $\phi_i(.)$ .

$$\int_{I_1} \phi_i(\omega)\phi_j^*(\omega)d\omega = \delta_{ij} \tag{14}$$

By partitioning $\phi_i(.)$ as (see Fig. 3)

$$\phi_i(\omega) = \begin{cases} \tilde{r}_i(u_o^- + v_o^- - \omega), & \omega \in [u_o^- - (v_o^+ - v_o^-), u_o^-] \\ \tilde{q}_i^*(\omega), & \omega \in (u_o^-, u_o^+) \end{cases} \tag{15}$$

the following properties are easily derived:

(i) If $\lambda_i$ is an eigenvalue of the operator with a corresponding eigenfunction as in (15), then $\lambda_{-i} \triangleq -\lambda_i$ is also an eigenvalue for an eigenfunction given by

$$\phi_{-i}(\omega) = \begin{cases} \phi_i(\omega), & \omega \in [u_o^- - (v_o^+ - v_o^-), u_o^-] \\ -\phi_i(\omega), & \omega \in (u_o^-, u_o^+] \end{cases} \tag{16}$$

(ii) The orthonormality condition (14) for $\phi_i(.)$ and $\phi_{-i}(.)$ yields

$$\int_{J_1} \tilde{q}_i(u)\tilde{q}_i^*(u)du = \int_{J_2} \tilde{r}_i(v)\tilde{r}_i^*(v)dv = \frac{1}{2} \tag{17}$$

(iii) The orthonormality condition (14) for $\phi_i(.), \phi_j(.)$ and $\phi_i(.),$
$\phi_{-j}(.)$ with $i \neq j$ yields

101

$$\int_{J_1} \tilde{q}_i(u)\tilde{q}_j^*(u)\,du \;=\; \int_{J_2} \tilde{r}_i(v)\tilde{r}_j^*(v)\,dv \;=\; 0 \tag{18}$$

Combination of (17) and (18) establishes that for

$$q_i(u) \;=\; \sqrt{2}\,\tilde{q}_i(u)\;,\qquad r_i(v) \;=\; \sqrt{2}\,\tilde{r}_i(v) \tag{19}$$

$\{q_i(.)\}_{i=1}^{\infty}$ and $\{r_i(.)\}_{i=1}^{\infty}$ are orthonormal sets.

We then examine (12) on the region $R_3$ or $R_4$ of the domain $(I_1 x I_2)$ in the definition (11) of $W(.,.)$ and obtain

$$H(u,v) \;=\; \sum_i \lambda_i \tilde{q}_i(u)\tilde{r}_i(v) \qquad\qquad u \in J_1, \; v \in J_2 \tag{20}$$

By enumerating only the positive eigenvalues, the desired expansion (8) results. Equations (10) readily follow from (13) and (14). Finally, substitution of (10a) to (10b) and (10b) to (10a) shows that $q_i(.)$ and $r_i(.)$ are eigenfunctions of the operators with kernel given by (9a) and (9b) corresponding to the eigenvalues $\lambda_i^2$ . Notice further that if $\lambda_i$, $r_i(.)$, $q_i(.)$ appear in (8) and satisfy (10), so do $-\lambda_i$, $-r_i(.)$, $q_i(.)$ or $-\lambda_i$, $r_i(.)$, $-q_i(.)$ . Hence in (8) we take all $\lambda_i$'s positive by changing sign of $r_i(.)$ or $q_i(.)$ if needed.

At this point some salient features of the expansion (8) are highlighted. Suppose that the terms in (8) have been arranged in descenting magnitude order of $\lambda_i$'s. In full analogy to [4, Appendix II], it can be proved that each term $\lambda_k q_k(u) r_k(v)$ is the best approximation of the error function from the last-term truncation, given by

$$e_{k-1}(u,v) \;=\; H(u,v) \;-\; \sum_{i=1}^{k-1} \lambda_i q_i(u) r_i(v) \tag{21}$$

in the Mean-Square (M-S) sense (i.e. with respect to the $L^2$-norm). Moreover, the M-S error at each stage of truncation is given by the sum of the squares of the remaining $\lambda_i$'s, i.e.

$$||e_k(.,.)||_2^2 \;=\; \sum_{i=k+1}^{\infty} \lambda_i^2 \;=\; ||H(.,.)||_2^2 \;-\; \sum_{i=1}^{k} \lambda_i^2 \;, \tag{22}$$

while

$$\lim_{k\to\infty} ||e_k(.,.)|| \;=\; 0 \;, \tag{23}$$

therefore

$$||H(.,.)||_2^2 \;=\; \sum_{i=1}^{\infty} \lambda_i^2 \;. \tag{24}$$

Thus, expansion (8) possesses the basic properties of the Fourier series expansion of functions. It is unique, except for isomorphisms within each subspace corresponding to a multiple eigenvalue. By means of (22) the designer is provided with an accurate assessment of the error at each stage of truncation.

We now present some modifications of (10a), (10b) applying when the frequency response is constrained to be a real, binary function, i.e.

for $\underline{0} \subset J_1 x J_2$, we have

$$H(u,v) = \begin{cases} 1 & \text{on} \quad \underline{0} \\ 0 & \text{on} \quad J_1 x J_2 - \underline{0} \end{cases} \tag{25}$$

Suppose further that the effective domain $\underline{0}$ is a compact set with

(i)   piecewise continuously differentiable boundary

(ii)  for every $\bar{u} \in J_1$, $\bar{v} \in J_2$ the sets

$$\{v \in J_2 / (\bar{u},v) \in \underline{0}\} \quad \text{and} \quad \{u \in J_1 / (u,\bar{v}) \in \underline{0}\}$$

are simple intervals.

As illustrated in Fig. 4, we describe the boundary $\Gamma$ of $\underline{0}$ by continuously differentiable functions $\gamma_1(.)$, $\gamma_2(.)$ of $u \in [u^-,u^+]$ and $\delta_1(.)$, $\delta_2(.)$ of $v \in [v^-,v^+]$. Any continuously differentiable function $F(.,.)$ on $\underline{0}$ satisfies then the Leibnitz formula

$$\frac{d}{du} \int_{\gamma_1(u)}^{\gamma_2(u)} F(u,v)dv = \int_{\gamma_1(u)}^{\gamma_2(u)} \frac{\partial F}{\partial u} dv + F(u,\gamma_2) \frac{d\gamma_2}{du} - F(u,\gamma_1) \frac{d\gamma_1}{du} \tag{26}$$

In particular (10a) and (10b) become

$$q_i(u) = \frac{1}{\lambda_i} \int_{\gamma_1(u)}^{\gamma_2(u)} r_i(v)dv , \qquad u \in J_1 , \tag{27a}$$

$$r_i(v) = \frac{1}{\lambda_i} \int_{\delta_1(v)}^{\delta_2(v)} q_i(u)du , \qquad v \in J_2 , \tag{27b}$$

which through (26) yield

$$\dot{q}_i(u) = \frac{1}{\lambda_i} [r_i(\gamma_2(u))\dot{\gamma}_2(u) - r_i(\gamma_1(u))\dot{\gamma}_1(u)] \tag{28a}$$

$$\dot{r}_i(v) = \frac{1}{\lambda_i}[q_i(\delta_2(v))\dot{\delta}_2(v) - q_i(\delta_1(v))\dot{\delta}_1(v)] \tag{28b}$$

with boundary conditions

$$q_i(u^-) = q_i(u^+) = 0 \tag{29a}$$

$$r_i(v^-) = r_i(v^+) = 0 \tag{29b}$$

We observe that nontrivial solutions to (28), (29) can exist only for appropriate $\lambda_i$-values, hence the problem of finding the expansion (8) by solving (13) manifests itself here through the eigenvalue problem (28), (29).

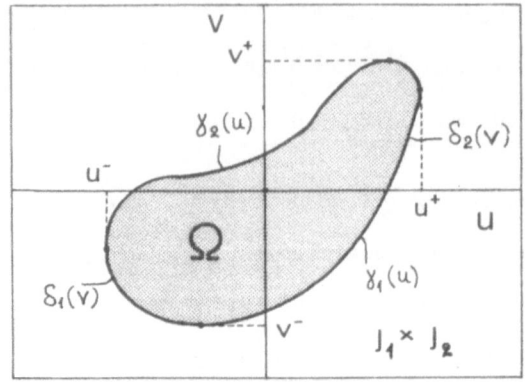

Fig. 4    Effective domain and its boundary.

## IV.  A DECOMPOSITION ALGORITHM

In this section an iterative algorithm for the separable expansion (8) in the general case (i.e. without the additional assumption (25)) is presented, in analogy to the power method [10, 5.9] for the finite dimensional case.

Consider the iterative relations

$$z_1^{(k+1)}(u) = \int_{J_2} H(u,v)w_1^{(k)*}(v)\,dv / ||w_1^{(k)}(.)||_2^2 \qquad (30a)$$

$$w_1^{(k+1)}(v) = \int_{J_1} H(u,v)z_1^{(k+1)*}(u)\,du / ||z_1^{(k+1)}(.)||_2^2 \qquad (30b)$$

$$k = 0,1,2,\dots.$$

on an arbitrary initial estimate $w_1^{(o)}(.)$ .

### Assertion:

The $L^2$-limits $w_1(.)$ and $z_1(.)$ of the sequences defined in (30) will satisfy the relation

$$w_1(v).z_1(u) = \lambda_1 r_1(v)q_1(u) , \qquad (31)$$

where $\lambda_1$ is the largest eigenvalue in (8).

The proof which follows is in an intuitive rather than a very strict mathematical basis.

### Proof:

Substitution of (30a) to (30b) gives:

104

$$w_1^{(k+1)}(v) = \int_{J_2} R(v,v')w_1^{(k)}(v')dv' \,/B_k , \qquad (32)$$

where

$$B_k = ||\int_{J_2} H(u,v)w_1^{(k)*}(v)dv||_2^2 \,/||w_1^{(k)}(.)||_2^2 , \qquad (33)$$

and $R(.,.)$ as in (9b).

Equation (32) expresses an iterative application of the operator

$$\int_{J_2} R(v,v').dv'$$

on the initial estimate $w_1^{(o)}(.)$ . Then, through [6,5.13-14], it can be written in terms of the eigenvalues $\lambda_i^2$ and the eigenfunctions $r_i(.)$ as

$$w_1^{(k)}(v) = [c_1(\lambda_1^2)^k r_1(v) + c_2(\lambda_2^2)^k r_2(v) + \ldots]/(B_0 B_1 \ldots B_{k-1}) \qquad (34)$$

with

$$c_i = \int_{J_2} r_i^*(v)w_1^{(o)}(v)dv , \qquad i = 1,2,\ldots . \qquad (35)$$

As seen from (34) the w-sequence tends to a function which belongs to the subspace $S_1$ , spanned by the eigenfunctions corresponding to the largest eigenvalue $\lambda_1$, under the assumption that the initial guess $w_1^{(o)}(.)$ is not orthogonal to $S_1$ .

Let us assume for simplicity that $\lambda_1 > \lambda_2 \geq \lambda_3 \ldots$ . Then

$$\lim_{k \to \infty} \frac{w_1^{(k)}(v)}{||w_1^{(k)}(.)||_2} = r_1(v) \qquad (36)$$

and, by (30a) and (36) we get

$$\lim_{k \to \infty} z_1^{(k+1)}(u)w_1^{(k+1)}(v) = \lim_{k \to \infty} \int_{J_2} H(u,v)r_1(v)dv.w_1^{(k+1)}(v)/||w_1^{(k)}(.)||_2 \qquad (37)$$

which by (10a) and

$$\lim_{k \to \infty} w_1^{(k+1)}(v)/||w_1^{(k)}(.)||_2 = \lambda_1^2 r_1(v)$$

reduces to (31).

Successive application of the iterative algorithm for the functions $H_n(.,.) = H_{n-1}(.,.) - \lambda_n q_n(.)r_n(.)$, $n = 1,2,\ldots$, with $H_o(.,.) = H(.,.)$, gives the desired expansion (8).

To implement the algorithm described above one has to discretize the functions involved and invoke a numerical integration method to perform (30a) and (30b). The simplest such method consists in sampling the functions and calculating the sums. In that case we fall into the matrix decomposition problem as in [4]. In contrast to [4], however, we have the freedom to perform a very fine sampling or use a more sophisticated integration method to acquire any accuracy we want.

Finally, it is noteworthy to consider the special case of a binary function as defined in the previous section through referring to Fig. 4. To compute the n-th term of the expansion (8), (30b) has to be written as

$$w_n^{(k)}(v) = \int_{J_1} (H(u,v) - \sum_{i=1}^{n-1} z_i(u).w_i(v)) \frac{z_n^{(k)}(u)}{||z_n^{(k)}(.)||_2^2} du, \quad v \in J_2 \qquad (38)$$

or, by (25),

$$w_n^{(k)}(v) = \int_{\delta_1(v)}^{\delta_2(v)} \frac{z_n^{(k)}(u)}{||z_n^{(k)}(.)||_2^2} du - \sum_{i=1}^{n} w_i(v) \int_{J_1} \frac{z_i(u) z_n^{(k)}(u)}{||z_n^{(k)}(.)||_2^2} du, \qquad (39)$$

$$v \in J_2 ,$$

with $w_i(.), z_i(.)$, $i = 1,2,\ldots,n-1$, determined at the previous stages.

Equation (39), along with its counterpart for $z_n^{(k)}(.)$, may lead to simplified implementations of (30a), (30b), for they involve only 1-D functions.

## V.  IMPLEMENTATION AND ERROR ASPECTS

Except for very special cases, expansion (8) extends to an infinite number of terms. The contribution of each term is measured by the square of the corresponding eigenvalue, as seen from (22), (24), and it tends to zero as the corresponding index tends to infinity. In practice we must retain only a finite number of terms, depending on the desired accuracy. The error of a K-term approximation is given by (22). The efficiency of the decomposition approach depends on the separability of the response function $H(.,.)$, i.e. the degree of convergence of the $\lambda_i$'s to zero.

A second thing we have to decide on is how to realize the 1-D subsystems which are combined in pairs for each term of the expansion. That also depends on the particular application and we have to choose a synthesis method which complies with the application constraints. For example, if a linear-phase 2-D discrete filter is required (as in many image processing applications), then special nonrecursive implementations have to be used. The inverse transforms of the 1-D frequency response functions $q_i(.)$ and $r_i(.)$ must be sampled, truncated and possibly weighted to counter the Gibb's phenomenon at discontinuities of $H(.,.)$. At this stage of realization a second approximation usually comes into foreground, e.g. the truncation of the 1-D impulse responses, and we have to account for that error too. The triangle inequality may be used to assess an upper bound of the overall error due to both approximations, i.e. the truncation of (8) and the 1-D response approximation.

Notice finally that the requirement for real 1-D impulse responses imposes a restriction of symmetry w.r.t. both axes for the function $H(.,.)$.

## VI.  APPLICATION EXAMPLES

Three examples are considered in this section, demonstrating the applicability of the proposed method. The first concerns a fan-filter design, where the decomposition is given through analytical formulae. In the same example a directional filter is constructed as a difference of two fan-filters. A circular low-pass filter is then considered in

Example II, while Example III concerns a Wiener-filter design.

Example I

The efficiency of the decomposition approach for fan filters was discussed in [4]. Application of our decomposition method is considered here, fully justifying itself because it leads to analytical results. Consider Fig. 5 showing the first-quadrant region of support for the fan-shaped response. In accordance with Fig. 4 and the discussion of Section III, the boundary of the effective support is described by the linear equations

$$\gamma_1(u) = 0 , \quad \gamma_2(u) = \gamma(u) = \frac{v_o}{u_o} u , \quad u \in [0, u_o] \tag{40a}$$

$$\delta_1(v) = \gamma^{-1}(v) = \frac{u_o}{v_o} v , \quad \delta_2(v) = u_o , \quad v \in [0, v_o] \tag{40b}$$

The above equations, (40), enable the decoupling of the eigenvalue problem (28), (29) into two independent eigenvalue subproblems, one on each direction, as follows. By subsituting (40) into (28) we get

$$\dot{q}_i(u) = \frac{1}{\lambda_i} r_i(\gamma(u)) \dot{\gamma}(u) \tag{41a}$$

$$\dot{r}_i(v) = \frac{1}{\lambda_i} q_i(\gamma^{-1}(v)) \dot{\gamma}^{-1}(v) \tag{41b}$$

Differentiation of (41b) with respect to $v$ and substitution of (41a) gives

$$\lambda_i^2 \ddot{r}_i(v) + r_i(v) \frac{u_o}{v_o} = 0 \tag{42a}$$

resp.

$$\lambda_i^2 \ddot{q}_i(u) + q_i(u) \frac{v_o}{u_o} = 0 \tag{42b}$$

which are simple, harmonic relations for $r_i(.)$'s and $q_i(.)$'s. The boundary conditions

$$\dot{r}_i(0) = r_i(v_o) = 0 \tag{43a}$$

$$\dot{q}_i(u_o) = q_i(0) = 0 \tag{43b}$$

readily follow from the corresponding integral equations (27). The solution of (42), (43) then follows:

$$q_i(u) = \frac{1}{\sqrt{u_o}} \sin(\frac{|u|}{u_o} \pi \frac{2i-1}{2}) , \quad |u| \le u_o , \quad i = 1, 2, \ldots \tag{44a}$$

$$r_i(v) = \frac{1}{\sqrt{v_o}} \cos(\frac{v}{v_o} \pi \frac{2i-1}{2}) , \quad |v| \le v_o , \quad i = 1, 2, \ldots \tag{44b}$$

$$\lambda_i = 4(u_o v_o)^{\frac{1}{2}} / (2i-1)\pi \tag{45}$$

By taking into account the symmetry of the response with respect to both axes, the inverse transforms of (44) are easily found by direct application of its definition:

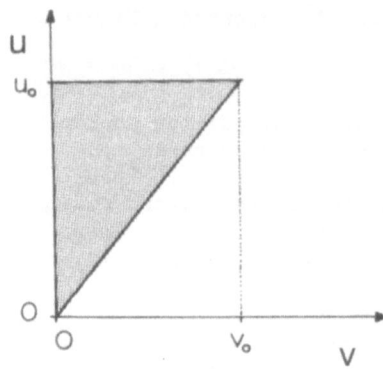

Fig. 5 Effective domain for fan filter.

$$Q_i(t) = \sqrt{u_o}\{c_a[2\pi(\frac{2i-1}{4} + tu_o)] + c_a[2\pi(\frac{2i-1}{4} - tu_o)]\}, \quad i = 1,2,\ldots \quad (46a)$$

$$R_i(s) = \sqrt{v_o}\{s_a[2\pi(\frac{2i-1}{4} + sv_o)] + s_a[2\pi(\frac{2i-1}{4} - sv_o)]\}, \quad i = 1,2,\ldots \quad (46b)$$

with

$$c_a(z) = \frac{1-\cos(z)}{z}, \quad s_a(z) = \frac{\sin(z)}{z} \quad (46c)$$

The difference of two fan-shaped functions, e.g. with effective domains spanned by the points $(0,0)$, $(0,v_o)$, $(u_o,0)$ and $(0,0)$, $(0,v_1)$, $(u_o,0)$ correspondingly, gives the response of a directional filter as

$$H_w(u,v) = \sum_{i=1}^{\infty} \frac{4}{(2i-1)\pi} \sin(\frac{|u|}{u_o} \pi \frac{2i-1}{2})[\cos(\frac{v}{v_o} \pi \frac{2i-1}{2})w_o(v)$$

$$- \cos(\frac{v}{v_1} \pi \frac{2i-1}{2})w_1(v)] \quad (47)$$

with

$$w_j(v) = \begin{cases} 1, & |v| \leq v_j \\ 0, & \text{elsewhere} \end{cases} \quad j = 0,1 \quad (48)$$

Apparently, (47) is not the expansion developed in Section III, because the v-functions do not constitute an orthogonal set. However, as a direct consequence of that expansion, it has the advantage of supplying analytical formulae for the implementation.

### Example II

Consider Fig. 6(a) showing the first-quadrant region of support of a circular low-pass filter. By applying the iterative procedure suggested in Section IV, we obtain the first four eigenfunctions as they appear in Fig. 6(b), normalized over [0,.5]. The corresponding eigenvalues are $\lambda_1 = .4195$, $\lambda_2 = .1010$, $\lambda_3 = .0576$ and $\lambda_4 = .0404$. Figures 6(c), (d) show the contours and the perspective plot of the corresponding 4-term approximate response. The normalized M-S error for that

108

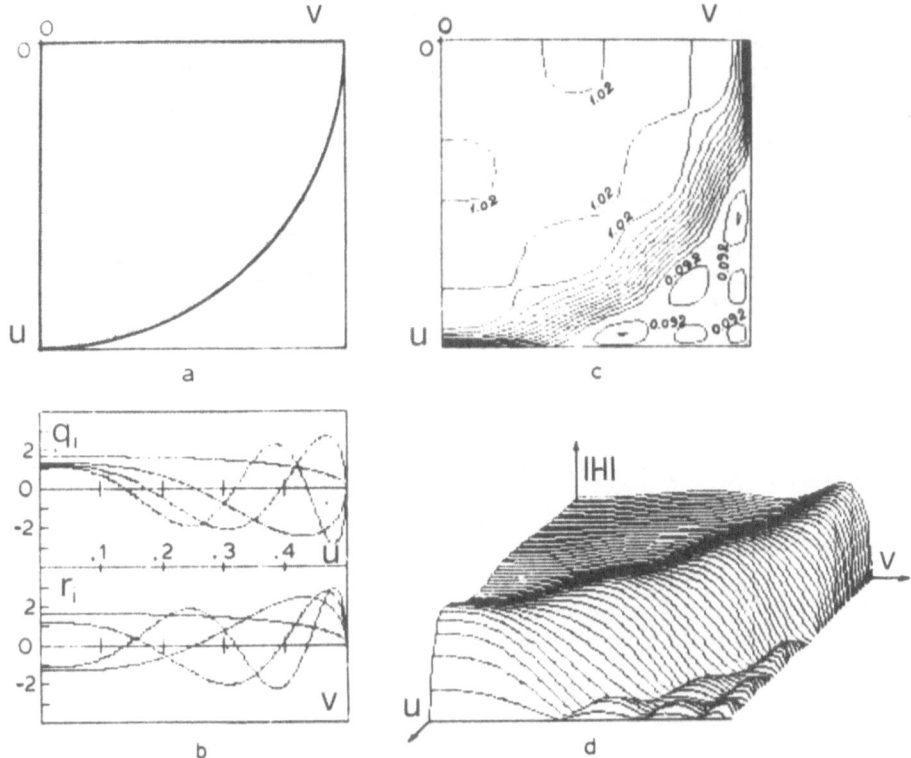

Fig. 6  Circular low-pass filter design
        (a) effective domain,  (b) eigenfunctions,
        (c) contour plot of 4-term approximation,
        (d) perspective plot of 4-term approximation.

approximation is found through (22), i.e.

$$||e_4(.,.)||_2^2 \ / \ ||H(.,.)||_2^2 \ = \ 0.0266 \tag{49}$$

Example III

   This concerns a Wiener filter, as described in II.1 . The frequency response of the blurring system is taken Gaussian on $J_1 \times J_2 = [-.5,.5] \times [-.5,.5]$, given by

$$H(u,v) \ = \ e^{-\dfrac{u^2+v^2}{0.20}} \tag{50}$$

while the power density spectra of the signals and the additive noise are taken white on $J_1 \times J_2$, with a Signal-to-Noise ratio equal to 0.1 . The perspective plots and the contours of the functions $H(.,.)$, as in (50), and $H_w(.,.)$, given by (2) with $\Phi_n(.,.)/\Phi_s(.,.) = 0.1$, are shown in Fig. 7. Application of the expansion algorithm of Section IV for the function $H_w(.,.)$ gives the eigenfunctions (first two) shown in Fig. 8(a). The squares of the corresponding eigenvalues, normalized over $||H_w(.,.)||_2^2$ are

$$\tilde{\lambda}_1^2 = 0.9839 \quad \text{and} \quad \tilde{\lambda}_2 = 0.0159 .$$

The perspective plot and the contours for the 2-term approximate response are shown in Figs. 8(b),(c). Notice that the functions $H(.,.)$ and $1/H(.,.)$ are separable, while $H_w(.,.)$ is not. However, because it slightly differs from $1/H(.,.)$ it is almost separable. So, the 2-term approximation error is of the order of 0.001.

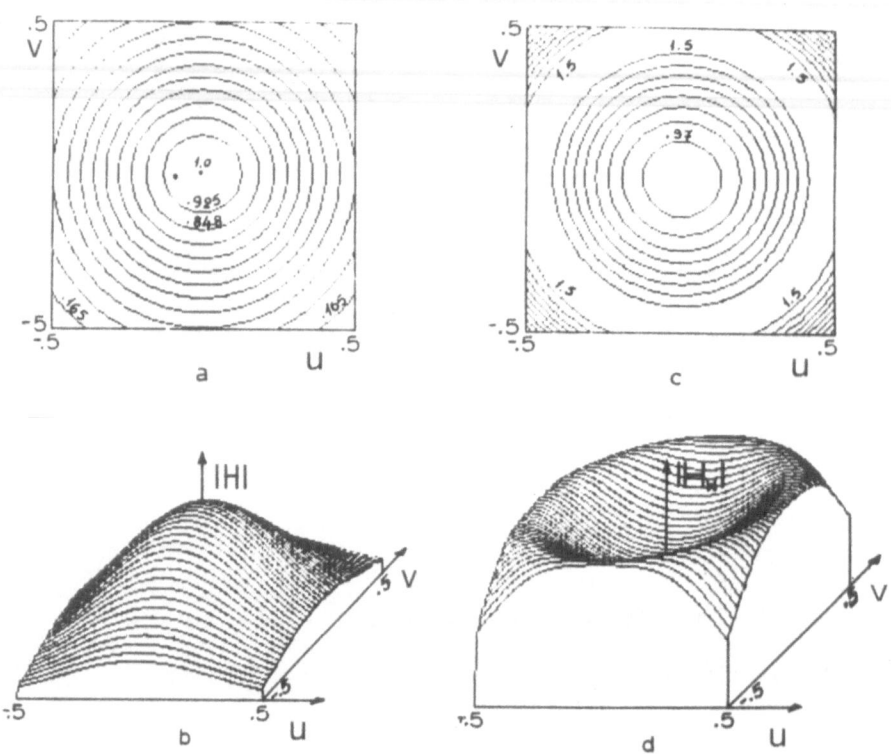

Fig. 7  Gaussian and Wiener filter frequency response
(a),(c)  contour plots,
(b),(d)  perspective plots.

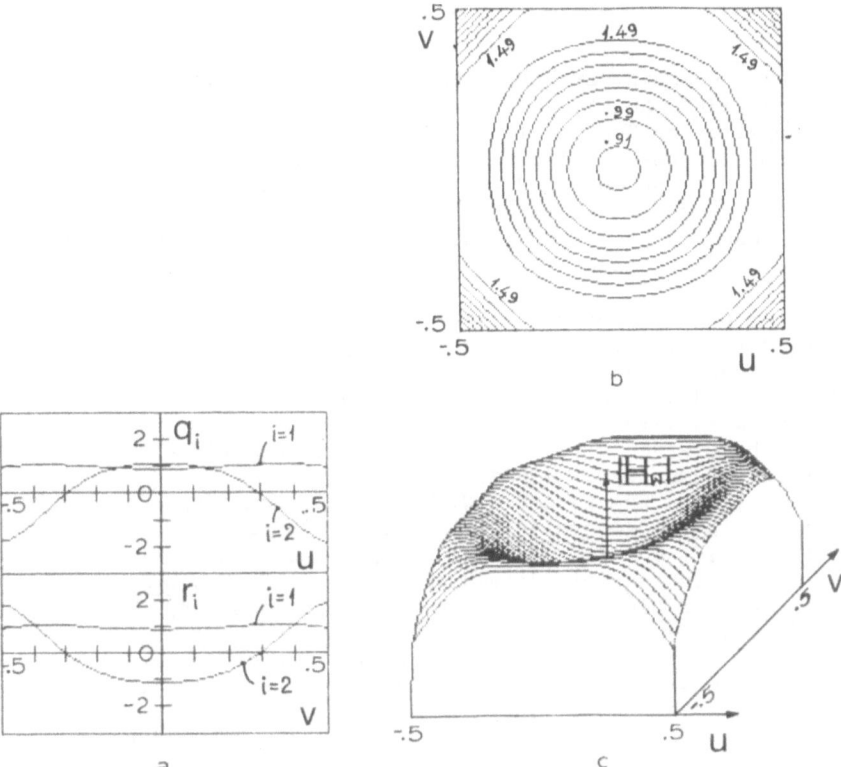

Fig. 8  Wiener filter design
        (a)   eigenfunctions,
        (b)   contour plot of 2-term approximation,
        (c)   perspective plot of 2-term approximation.

VII. CONCLUSION

We have shown that the Singular Value Decomposition (SVD) method
used to decompose a 2-D system in the discrete, original domain has a
theoretically interesting and practically useful counterpart in the continuous frequency domain.  An iterative algorithm for solving the
decomposition problem in the general case has been presented and some
simplifications concerning the special case of binary frequency responses
have been considered.  The applicability of the method has been demonstrated by means of characteristic examples.  Completely analytical
results have been derived for the fan filter design.

REFERENCES

[1]   J. Allen,   Computer architecture for digital signal processing,
      Proc. IEEE, No. 5, vol. 73, pp. 852-73, (May, 1985).
[2]   P.R. Cappello and K. Steiglitz,   Completely-pipelining
      architectures for digital signal processing,   IEEE Trans.
      Accoust., Speech and Signal Processing, No. 4, vol. 31,
      pp. 1016-23, (Aug. 1983).
[3]   D.E. Dudgeon and R.M. Mersereau,   Multidimensional digital signal
      processing,   Prentice-Hall, Englewood Cliffs, New Jersey,
      (1984).
[4]   S. Treitel and J.L. Shanks,   The design of multistage separable
      planar filters,   IEEE Trans. Geos. Electron., No. 1, vol. 9,
      pp. 10-27, (Jan. 1971).
[5]   N. Dunford and J.T. Schwartz,   Linear operators,   John Wiley,
      New York, (1963).
[6]   G.F. Roach,   Green's functions,   2nd Ed., Cambridge University
      Press, Cambridge, (1982).
[7]   S. Treitel, J.L. Shanks and C.W. Frasier,   Some aspects on fan
      filtering,   Geophysics, vol. 32, pp. 789-800, (Oct. 1967).
[8]   K.L. Peacock,   On the practical design of discrete velocity
      filters for seismic data processing,   IEEE Trans Accoust.,
      Speech and Signal Processing, No. 1, vol. 30, pp. 52-60,
      (Feb. 1982).
[9]   M. Kunt, A. Ikonomopoulos and M. Kocher,   Second-generation
      image-coding techniques,   Proc. IEEE, No. 4, vol. 73,
      pp. 549-574, (Apr. 1985).
[10]  G. Strang,   Linear algebra and its applications,   2nd Ed.,
      Academic Press, New York, (1980).

# A PARALLEL ESTIMATION ALGORITHM

M. Hodzic   and   D. D. Siljak

School of Engineering, EECS Department
Santa Clara University
Santa Clara, California 95053

ABSTRACT

A parallel estimation algorithm is proposed for large scale dynamic systems. The algorithm is bi-modal in that the local estimates are computed in parallel at high sampling rates considering the subsystems as decoupled from each other. At the same time, the outputs of the local estimators are used as initial data for the overall estimator which takes the interaction effects into account in a block-iterative fashion. Each individual locally optimal estimator can have a different sampling rate to match the subsystem dynamics.

## 1. INTRODUCTION

In this paper, we initiate a study of parallel processing in estimation of large-scale systems, which is oriented toward multirate operation using multiprocessor schemes with all their inherent advantages over the single large computing machine. The approach is a natural continuation of the hierarchical estimation scheme (Hodzic and Siljak, 1985) developed for large sparse systems. The essential feature introduced in this paper, is the parallel operation of subsystem estimators at high sampling rates which can be chosen freely to match the dynamics of each individual subsystem.

Fundamental to the parallel processing schemes involving large scale systems, is the convergence of the resulting solution schemes. Without satisfactory convergence properties, the gain in speed and accuracy provided by the parallelism in computation, is inevitably lost or problem dependent, to say the least. To ensure a satisfactory convergence of our parallel multirate estimator, we decompose the system into a lower block-triangular (LBT) structure (Pichai, Sezer and Siljak, 1983), which is an hierarchical ordering of subsystems. The resulting parallel overall estimator, which is obtained by building locally optimal estimators for each individual subsystem, going from top to the bottom of the hierarchy, is suboptimal and stable.

## 2. HIERARCHICAL (LBT) ESTIMATOR

The task of building an estimator for a large system is broken down into a number of smaller problems of building estimators for each individual subsystem. First, the subsystems are considered as decoupled and the standard locally optimal estimators are obtained in a highly parallel fashion. A variety of sampling rates can be used to fit the dynamics of each individual subsystem. The locally computed estimates for the subsystems considered as decoupled, serve as a starting data for the calculation of the estimates for the interconnected subsystems.

The calculations, which include the interconnection effects in another type of locally optimal procedure, are performed at a slow sampling rate with sampling period which is an integer multiple of each subsystem sampling period. The operation of the slow-rate estimator, at each sampling instant, is block-iterative (subsystem-by-subsystem) from top to bottom of the overall system. The output of the block-iterative estimator serves as the input to the decoupled subsystem estimators, and so on.

Let us first describe the design of the slow-rate estimator, which includes the interconnection effects (Hodzic and Siljak, 1985). For this purpose, we assume that a given system

$$S: \quad x(t+1) \quad = \quad A\ x(t) + w(t)$$

$$y(t) \quad = \quad C\ x(t) + v(t) \tag{2.1}$$

is decomposed as

$$S: \quad x_i(t+1) \quad = \quad A_{ii}x_i(t) + w_i(t) + \sum_{j=1}^{i} A_{ij}x_j(t)$$

$$y_i(t) \quad = \quad C_{ii}x_i(t) + v_i(t) + \sum_{j=1}^{i} C_{ij}x_j(t), \quad i = 1,2,\ldots,N. \tag{2.2}$$

In this way, the system $S$ is represented as an interconnection of $N$ subsystems

$$S_i: \quad x_i(t+1) \quad = \quad A_{ii}x_i(t) + w_i(t)$$

$$y_i(t) \quad = \quad C_{ii}x_i(t) + v_i(t), \quad i = 1,2,\ldots,N \tag{2.3}$$

where $x_i(t) \in R^{n_i}$, $y_i(t) \in R^{\ell_i}$ are the state and output of the subsystem $S_i$ at time $t = 0,1,2,\ldots$; $w_i(t) \in R^{n_i}$ and $v_i(t) \in R^{\ell_i}$ are independent Gaussian processes with zero mean and covariances $R_w^{ii}$ and $R_v^{ii}$, respectively, such that $x(t) = [x_1^T(t), x_2^T(t), \ldots, x_N^T(t)]^T$, $w(t) = [w_1^T(t), w_2^T(t), \ldots, w_N^T(t)]^T$, $y(t) = [y_1^T(t), y_2^T(t), \ldots, y_N^T(t)]^T$, $v(t) = [v_1^T(t), v_2^T(t), \ldots, v_N^T(t)]^T$. The constant matrices $A = (A_{ij})$ and $C = (C_{ij})$ of the system $S$ are regarded as block matrices with blocks $A_{ij}$ and $C_{ij}$ having appropriate dimensions. Similarly, the covariance matrices $R_w = (R_w^{ij})$ and $R_v = (R_v^{ij})$ of the state noise $w(t)$ and the measurement noise $v(t)$, have blocks compatible with the dimensions of the subsystems. Finally, we note that the sampling period $T$, which is time between the two consecutive time instants $t$ and $t+1$, is considered as the unit of time.

We follow the development of Hodzic and Siljak (1985), and assume that the updated estimates are already computed for the first $i-1$ subsystems. Then, on the ith subsystem level, we introduce the adjusted measurement as

$$y_i^*(t) = y_i(t) - \sum_{j=1}^{i-1} C_{ij} \hat{x}_j(t|t) \tag{2.4}$$

or

$$y_i^*(t) = C_{ii} x_i(t) + v_i^*(t) \tag{2.5}$$

where $v_i^*(t)$ represents the adjusted zero-mean measurement noise at the level of the ith subsystem

$$v_i^*(t) = v_i(t) + \sum_{j=1}^{i-1} C_{ij} \tilde{x}_j(t|t) \tag{2.6}$$

with $\tilde{x}_j(t|t) = x_j(t) - \hat{x}_j(t|t)$ representing updated estimation error. By $Y_i^*(t)$ we denote the space of all adjusted measurements

$$Y_i^*(t) = \{y_i^*(0), y_i^*(1), \ldots, y_i^*(t)\} . \tag{2.7}$$

Then we obtain the updated estimate $\hat{x}_i(t|t)$ by projecting the state $x_i(t)$ onto the space $Y_i^*(t)$, that is,

$$\hat{x}_i(t|t) = E[x_i(t)|Y_i^*(t)], \tag{2.8}$$

where $E$ is the expectation operator, so that

$$\hat{x}_i(t|t) = \hat{x}_i(t|t-1) + K_{ii}^*(t) \, \bar{\bar{y}}(t) . \tag{2.9}$$

In (2.9), $\bar{\bar{y}}_i(t)$ is the innovation sequence given by

$$\bar{\bar{y}}_i(t) = y_i^*(t) - E[y_i^*(t)|Y_i^*(t-1)], \tag{2.10}$$

which can also be written as a function of $v_i^*(t)$,

$$\bar{\bar{y}}_i(t) = C_{ii} \tilde{x}_i(t|t-1) + v_i^*(t) . \tag{2.11}$$

The vector $\hat{x}_i(t|t-1)$ is the predicted estimate

$$\hat{x}_i(t+1|t) = \sum_{j=1}^{i} A_{ij}\hat{x}_j(t|t), \tag{2.12}$$

and $K_{ii}^*(t)$ is the gain matrix defined as

$$K_{ii}^*(t) = \text{COV}[x_i(t),\bar{\bar{y}}_i(t)]\ \text{COV}^{-1}[\bar{\bar{y}}_i(t),\bar{\bar{y}}_i(t)], \tag{2.13}$$

where COV stands for covariance. The overall LBT estimator can now be represented in a compact form

$$\hat{S}^{\oplus}: \quad \hat{x}(t|t) = \hat{x}(t|t-1) + K_D^*(t)\bar{\bar{y}}(t), \tag{2.14}$$

where $K_D^*(t) = \text{diag}[K_{11}^*(t),K_{22}^*(t),\ldots,K_{NN}^*(t)],$ and

$$\bar{\bar{y}}(t) = y(t) - C_L\hat{x}(t|t) - C_{DR}\hat{x}(t|t-1) \tag{2.15}$$

is a quasi-innovation vector (Hodzic and Siljak, 1985), with

$$C_L = \begin{bmatrix} 0 & & & \\ C_{21} & 0 & & \bigcirc \\ \cdot & \cdot & \cdots & \cdot \\ C_{N1} & C_{N2} & \cdots & C_{N,N-1} & 0 \end{bmatrix} \tag{2.16}$$

and $C = C_L + C_{DR}$.

In order to calculate gain matrix $K_{ii}^*(t)$ in (2.13) we note that (2.9) can be written in an equivalent form

$$\hat{x}_i(t|t) = \hat{x}_i(t|t-1) + \sum_{j=1}^{i} K_{ij}^*(t)\bar{y}_j(t), \tag{2.17}$$

where

$$\bar{y}_i(t) = y_i(t) - \sum_{j=1}^{i} C_{ij}\hat{x}_j(t|t-1), \tag{2.18}$$

and $K_{ij}^*(t)$, $i > j$, is defined by

$$K_{ij}^*(t) = - K_{ii}^*(t) \sum_{k=j}^{i-1} C_{ik}K_{kj}^*(t). \tag{2.19}$$

By forming the gain matrix $K^*(t)$ in the lower block triangular form

$$K^*(t) = \begin{bmatrix} K_{11}^*(t) & & & \\ K_{21}^*(t) & K_{22}^*(t) & \bigcirc & \\ \cdot & \cdot & \cdots & \cdot \\ K_{N1}^*(t) & K_{N2}^*(t) & \cdots & K_{NN}^*(t) \end{bmatrix} \tag{2.20}$$

we obtain the overall LBT estimator in the following form

$$\mathcal{S}^{\theta}: \quad \hat{x}(t|t) = \hat{x}(t|t-1) + K^*(t)\bar{y}(t) . \qquad (2.21)$$

Another equivalent form of the estimator is obtained by using predicted rather than updated estimate

$$\hat{x}_i(t+1|t) = \sum_{j=1}^{i} [A_{ij}\hat{x}_j(t|t-1) + K_{ij}^+(t)\bar{y}_j(t)] \qquad (2.22)$$

where

$$K_{ij}^+(t) = \sum_{\ell=j}^{i} A_{i\ell} K_{\ell j}^*(t) , \qquad (2.23)$$

so that $K^+(t) = [K_{ij}^+(t)]$ is equal to

$$K^+(t) = A K^*(t), \qquad (2.24)$$

and

$$\hat{\mathcal{S}}^{\theta}: \quad \hat{x}(t+1|t) = A \hat{x}(t|t-1) + K^+(t)\bar{y}(t). \qquad (2.25)$$

The updated and predicted estimation errors, $\tilde{x}(t|t)$ and $\tilde{x}(t|t-1)$, are defined according to

$$\tilde{x}(t|t) = [I - K^*(t)C]\tilde{x}(t|t-1) - K^*(t)v(t) \qquad (2.26)$$

$$\tilde{x}(t+1|t) = A \tilde{x}(t|t) + w(t) , \qquad (2.27)$$

respectively. Then the corresponding error covariances satisfy the equations

$$P(t|t) = \bar{C}(t)P(t|t-1)\bar{C}^T(t) + K^*(t)R_v K^{*T}(t) \qquad (2.28)$$

$$P(t|t-1) = A P(t-1|t-1) A^T + R_w(t-1), \qquad (2.29)$$

and the matrix $\bar{C}(t) = [I - K^*(t)C] = [\bar{C}_{ij}]$ is defined by

$$\bar{C}_{ij}(t) = I_i - \sum_{\ell=1}^{i} K_{i\ell}^*(t)C_{\ell i} , \quad i = j$$

$$= - \sum_{\ell=1}^{i} K_{i\ell}^*(t)C_{\ell j} , \quad i \neq j . \qquad (2.30)$$

From the equations (2.28)-(2.30), we write the submatrices $P_{ij}(t|t)$ and $P_{ij}(t|t-1)$ as follows

$$P_{ij}(t|t) = \sum_{k=1}^{i} \sum_{\ell=1}^{i} [\bar{C}_{ik}(t)P_{k\ell}(t|t-1)\bar{C}_{j\ell}^T(t)]$$

$$+ \sum_{k=1}^{i} \sum_{\ell=1}^{i} [K_{ik}^*(t)R_v^{k\ell} K_{j\ell}^{*\,T}(t)] \qquad (2.31)$$

$$P_{ij}(t|t-1) = \sum_{k=1}^{i} \sum_{\ell=1}^{i} [A_{ik}P_{k\ell}(t-1|t-1)A_{j\ell}^T] + R_w^{ij} \qquad (2.32)$$

for $i = 1,2,\ldots,N$, $i < j$, and $P_{ij}(.|.) = P_{ij}^T(.|.)$.

Let us now use the standard performance index $J_i = \text{tr } P_{ii}(t|t)$, and minimize it at the ith subsystem level

$$J_i^* = \min_{K_{ii}^*(t)} \text{tr } P_{ii}(t|t), \qquad (2.33)$$

where $P_{ii}(t|t)$ is given by (2.31), for $i = j$.

The locally optimal gain matrix $K_{ii}^*(t)$ is obtained as

$$K_{ii}^*(t) = [\sum_{j=1}^{i} P_{ij}(t|t-1)C_{ij}^{*T}(t)]$$

$$\{\sum_{j=1}^{i} C_{ij}^*(t) \sum_{k=1}^{i} P_{jk}(t|t-1)C_{ik}^{*T}(t)$$

$$+ [\sum_{j=1}^{i} D_{ij}^*(t)]R_v[\sum_{j=1}^{i} D_{ij}^*(t)]^T\}^{-1} \qquad (2.34)$$

where the matrices $C_{ij}^*(t)$ and $D_{ij}^*(t)$ are given by

$$C_{ij}^*(t) = C_{ii}, \qquad i = j$$

$$= -\sum_{k=j}^{i-1} \sum_{\ell=k}^{i-1} C_{i\ell}K_{\ell k}^*(t)C_{kj} + C_{ij}, \qquad i \neq j$$

$$D_{ij}^*(t) = D_{ii}, \qquad i = j$$

$$= -\sum_{k=j}^{i-1} C_{ik}K_{kj}^*(t)D_{jj}, \qquad i > j \qquad (2.35)$$

and $\ell_j \times \ell$ matrix $D_{jj}$ has the form $D_{jj} = [0 \ I_{\ell j} \ 0]$ where the $\ell_j \times \ell_j$ identity matrix $I_{\ell j}$ is positioned to correspond to the subsystem $S_j$.

A LAN (Local Area Network) implementation of the described estimator is discussed in (Hodzic and AlKhatib, 1983).

The block diagram of the iterative LBT estimator is shown in Fig. 1, where the blocks denoted by $S_i^\oplus$ correspond to the computations described above. These computations occur at a slow rate, and constitute a sub-optimal estimator with properties described by Hodzic and Siljak (1985).

As soon as the iteration drops to the (i+1)st level of the hierarchy, the ith estimator continues to operate as decoupled at a fast rate until it is revisited again by the iteration procedure of the LBT estimator. In this way, all subsystem estimators work in parallel at different rates, until they are periodically picked-up by the iterative estimator operating in a pipeline regime.

To describe the parallel operation of the decoupled subsystem estimators, we rewrite (2.3) as

$$S_i: \quad x_i[(k+1)T_i] = A_{ii}x_i(kT_i) + w_i(kT_i)$$

$$y(kT_i) = C_{ii}x_i(kT_i) + v_i(kT_i) \qquad (2.36)$$

where $k = 0,1,2,\ldots,$ and $T_i$ is the sampling period of $S_i$. For each $S_i$, we build the optimal Kalman filter,

$$\hat{S}_i^o: \quad \hat{x}_i(kT_i|kT_i) = \hat{x}_i[kT_i|(k-1)T_i] + K_i^o(kT_i)\bar{y}_i(kT_i)$$

$$\bar{y}_i(kT_i) = y_i(kT_i) - C_{ii}\hat{x}_i[kT_i|(k-1)T_i]$$

$$i = 1,2,\ldots,N \qquad (2.37)$$

by standard methods (e.g. Anderson and Moore, 1979). Of course, the only restriction imposed on $T_i$ is that $m_i T_i = T$, where $m_i$ is an integer chosen for the corresponding subsystem $S_i$.


3. HIERARCHICAL (LBT) DECOMPOSITION

The purpose of this section is to describe graph-theoretic approach to decomposition of large dynamic systems, suitable for parallel implementation of estimation algorithms described in Section 2. When a directed graph (digraph) is associated with the system, the nodes of which represent the input, state, and output variables, whereas the lines represent their relations, the digraph can be relabelled to obtain a hierarchical ordering of subgraphs with hierarchically independent inputs and outputs (Pichai, Sezer and Siljak, 1983).

The hierarchical structure of the digraph corresponds to a lower-block triangular (LBT) form of the system, with each block being a subsystem with its own inputs and outputs. For the hierarchical design strategy (estimation, and control) to work, we need controllability and observability of each subsystem. When decomposing the overall system, it is difficult to assure these properties of the subsystems. In the graph-theoretic setting, a natural way to approximate controllability and observability is to use the concept of input and output reachability, which can be suitably included in the digraph decomposition framework. Once the input and output reachability of the subsystems are established, structural controllability and observability of the hierarchically ordered subsystems follow, provided their associated digraphs contain no dilation. In all but pathological cases (ideal matching of parameters), structural controllability and observability imply standard controllability and observability.

We now outline a scheme for acyclic decomposition of dynamic systems into input/output reachable subsystems (Pichai, Sezer and Siljak, 1983).

Let us consider discrete time dynamical system $S$

$$S: \quad x(t+1) = A\,x(t) + B\,u(t)$$

$$y(t) = C\,x(t) \tag{3.1}$$

which can be decomposed as

$$S: \quad x_i(t+1) = \sum_{j=1}^{i} A_{ij} x_j(t) + \sum_{j=1}^{i} B_{ij} u_j(t)$$

$$y_i(t) = \sum_{j=1}^{i} C_{ij} x_j(t), \quad i = 1,2,\ldots,N. \tag{3.2}$$

Equation (3.2) represents an interconnection of $N$ hierarchically ordered subsystems where $x_i(t) \in R^{n_i}$, $u_i(t) \in R^{m_i}$, and $y_i(t) \in R^{l_i}$ are the states, inputs, and outputs of the subsystems $S_i$, and the subsystem dimensions are such that $n = n_1+n_2+\ldots+n_N$, $m = m_1+m_2+\ldots+m_N$, $l = l_1+l_2+\ldots+l_N$ .

The decomposition (3.2) is obtained by a permutation of the variables of $S$, which brings the original matrices A, B, and C of (3.1) into a compatible lower block-triangular form. The corresponding interconnection matrix M (Siljak, 1978) is of the form

$$\tag{3.3}$$

The above decomposition of $S$ corresponds to a partitioning of the associated digraph $D = (U \cup X \cup Y, E)$ into subgraphs $D_i = (U_i \cup X_i \cup Y_i, E_i)$ corresponding to the decoupled subsystems $S_i$ of (3.2), where $U$, $X$, $Y$, $U_i$, $X_i$, and $Y_i$ are corresponding sets of input, state and output vertices.

At this point we note that input/output reachability of the digraph $D_i$ is the minimal requirement for structural controllability and observability of the subsystem $S_i$ . An additional structural condition, that of dilation, is checked post factum.

When the hierarchical estimator of Section 2 is built around hierarchically decomposed system, all the gain matrices eventually become constant and the estimator reaches steady state. This comes as a result of the special lower-block triangular structure of the system and the fact that the described graph-theoretic algorithm of Pichai, Sezer and Siljak (1983) guarantees structural controllability and observability of subsystems (Hodzic and Siljak, 1985).

## CONCLUSION

In this paper we have presented a parallel estimation algorithm for large-scale dynamic systems, based on the decomposition of the original system into a set of subsystems. The estimation algorithm consists of the fast multirate subsystem estimators operating independently (in parallel), and an interconnected (slow) hierarchical estimator which is turned on from time to time. The algorithm is suitable for implementation on a multiprocessor computer system. To ensure convergence of the proposed algorithm, a hierarchical (LBT) graph-theoretic decomposition technique has also been discussed. The decomposition guarantees convergence in both decoupled and interconnected cases.

## REFERENCES

B.D.O. Anderson and J.B. Moore (1979). Optimal Filtering, Prentice Hall, Englewood Cliffs, New Jersey.

M. Hodzic and H. AlKhatib (1983). Distributed architecture for implementation of an algorithm for estimation of sparse large scale dynamic systems, ISMM Conference MIMI'83, San Antonio, Texas.

M. Hodzic and D.D. Siljak (1985). Estimation and control of large sparse systems, Automatica, 21, pp. 277-292.

Pichai, V., M.E. Sezer and D.D. Siljak (1983). A graph-theoretic Algorithm for hierarchical decomposition of dynamic systems with application to estimation and control, IEEE Transactions, SMC-13, pp. 197-207.

Siljak, D.D. (1978). Large-Scale Dynamic Systems : Stability and Structure, North Holland, New York.

DESIGN OF PARALLEL NUMERICAL ALGORITHMS

D. J. Evans

Department of Computer Studies, Loughborough University

of Technology, Loughborough, Leicestershire, U.K.

ABSTRACT

In this paper some techniques for exposing parallelism in a problem
are surveyed and some new parallel numerical algorithms for the  direct
solution of  linear systems presented and compared with the existing
sequential methods.

Further, a new explicit iterative scheme is presented for the
numerical solution of 2 point boundary value problems.  From the usual
central difference approximations to the differential operator, a tridiagonal system of equations has to be solved.  In this new approach,
the matrix is split into component matrices $G_1, G_2$  and an iterative
method formulated which is easily expressed in explicit form.  By alternating this strategy on the grid points results in the Alternating Group
Explicit (AGE) method which is analogous to the ADI method.

Finally, a new explicit method for the finite difference solution of
parabolic partial differential equations is derived.  The new method uses
stable asymmetric approximations to the partial differential equation
which when coupled in groups of 2 adjacent points (4 points for 2
dimensions) on the grid result in implicit equations which can be easily
converted to explicit form and offer many advantages especially for use
on parallel computers.  A judicious use of alternating this strategy on
the grid points of the domain results in new explicit parallel algorithms
which possess unconditional stability.

1.  INTRODUCTION

Parallelism can arise at many different levels within a computational problem, which if exposed can be efficiently exploited by parallel
computers.  Some well known techniques of doing this are:

1.  *Vectorising existing software*.  This is often achieved by changing the
    order in the evaluation of terms in a complicated expression so that a
    vector or matrix of components can be handled in one operation.

2.  To decompose the problem into a number of independent sub-problems,
    all of which can proceed independently.  The solutions of these sub-

problems are then combined in some way to yield the answer of the original problem. This technique is usually known as a *Divide and Conquer* strategy or partitioning , e.g. for the evaluation of $\sum_{i=1}^{k} a_i$ it is possible to decompose the problem in the following manner

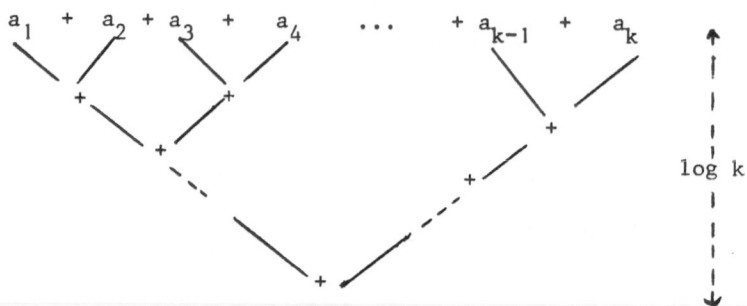

Thus, if $t$ is the time unit for the addition operation, then the total times for the sequential and parallel computations are

$$T_{sequential} = (k-1)t, \quad T_k = (\log_2 k)t ,$$

while the speed-up of the computation due to parallel evaluation is determined as

$$S_k = (k-1)/(\log_2 k) = O(k/\log_2 k) .$$

This result is true if we neglect:

i) the interconnection cost for S.I.M.D. computers;
ii) the synchronisation and shared memory conflicts for M.I.M.D. computers.

3. By the discovery of independent sub-expressions in the calculation which can proceed in parallel. Often this is termed *Implicit Parallelism* and examples such as *recursive decoupling and cyclic reduction* are such that the extraction of these sub-expressions can lead to a more balanced decomposition for parallel evaluation.

4. By developing new parallel methods such as the *Quadrant Interlocking Methods* for Computational Linear Algebra which will be described in Section 2.

5. Another technique of achieving parallelism in a numerical algorithm is by the use of *Explicit methods*. Usually, such algorithms are the oldest methods for the solution of many problems. Unfortunately, they suffer from major defects such as poor stability and convergence characteristics and require unacceptable large solution times. Undoubtedly the more recent *Implicit Methods* are better but often we are not able to exploit to the full any Implicit Parallelism within the algorithm. Thus, the discovery of new *Explicit methods* of solution as given in Sections 3 and 5 is important for the development of parallel algorithms.

# 2. DIRECT METHODS FOR THE SOLUTION OF LINEAR SYSTEMS

The usual approach for solving linear systems is by Gaussian Elimination or triangular decomposition.

Given the matrix, A, i.e.

$$A = \begin{bmatrix} a_{11} & a_{12} & a_{13} & a_{14} \\ a_{21} & a_{22} & a_{23} & a_{24} \\ a_{31} & a_{32} & a_{33} & a_{34} \\ a_{41} & a_{42} & a_{43} & a_{44} \end{bmatrix}, \text{ where det A } \neq 0 \text{ so that A is nonsingular.} \tag{2.1}$$

We now attempt to find the matrix factors L and U of the form:

$$L = \begin{bmatrix} 1 & & & \\ \ell_{21} & 1 & \bigcirc & \\ \ell_{31} & \ell_{32} & 1 & \\ \ell_{41} & \ell_{42} & \ell_{43} & 1 \end{bmatrix} \text{ and } U = \begin{bmatrix} u_{11} & u_{12} & u_{13} & u_{14} \\ & u_{22} & u_{23} & u_{24} \\ \bigcirc & & u_{33} & u_{34} \\ & & & u_{44} \end{bmatrix}$$

$$\tag{2.2}$$

such that, $\qquad A \equiv LU$ . $\tag{2.3}$

By equating the coefficients in the matrix product (2.3), the following relations can be obtained to determine the coefficients of L and U. These are for rows 1, 2 and 3, i.e.

$$u_{11} = a_{11}, \quad u_{12} = a_{12}, \quad u_{13} = a_{13}, \quad u_{14} = a_{14}$$

$$\ell_{21}u_{11} = a_{21}, \quad \ell_{21}u_{12}+u_{22} = a_{22}, \quad \ell_{21}u_{13}+u_{23} = a_{23}, \quad \ell_{21}u_{14}+u_{24} = a_{24} \tag{2.4}$$

$$\ell_{31}u_{11} = a_{31}, \quad \ell_{31}u_{12}+\ell_{32}u_{22} = a_{32}, \quad \ell_{31}u_{13}+\ell_{32}u_{23}+u_{33} = a_{33}, \text{ etc.}$$

with similar results for the last row.

These equations are essentially all *sequential* relations, since each of the unknowns $\ell_{i,j}$ and $u_{i,j}$ are brought into the above relations one at a time recursively and then determined in a similar manner.

The reason why such a factorisation is sought specifically in L.U. form is that the matrix factors L and U are known as easily inverted matrix forms and so the solution of the linear system

$$A\underline{x} = \underline{b} , \tag{2.5}$$

can be obtained by making use of the substitution $A = LU$ to reduce the problem to the solution of the coupled systems,

$$L\underline{y} = \underline{b} \tag{2.6}$$

and $\qquad U\underline{x} = \underline{y} ,$ $\tag{2.7}$

where $\underline{y}$ is an intermediate vector.

The linear systems (2.6) and (2.7) are easily solvable systems and can be solved by well known forward or backward substitution processes, i.e.

$$L\underline{y} = \underline{b},$$

$$
\begin{bmatrix}
1 & & & \\
\ell_{21} & 1 & & \bigcirc \\
\ell_{31} & \ell_{32} & 1 & \\
\ell_{41} & \ell_{42} & \ell_{43} & 1
\end{bmatrix}
\begin{bmatrix}
y_1 \\ y_2 \\ y_3 \\ y_4
\end{bmatrix}
=
\begin{bmatrix}
b_1 \\ b_2 \\ b_3 \\ b_4
\end{bmatrix},
\qquad (2.8)
$$

can be solved as follows:

$$
\begin{aligned}
y_1 &= b_1 & &\rightarrow y_1 = b_1 \\
\ell_{21}y_1 + y_2 &= b_2 & &\rightarrow y_2 = b_2 - \ell_{21}y_1 \\
\ell_{31}y_1 + \ell_{32}y_2 + y_3 &= b_3 & &\rightarrow y_3 = b_3 - \ell_{31}y_1 - \ell_{32}y_2 \\
\ell_{41}y_1 + \ell_{42}y_2 + \ell_{43}y_3 + y_4 &= b_4 & &\rightarrow y_4 = b_4 - \ell_{41}y_1 - \ell_{42}y_2 - \ell_{43}y_3 .
\end{aligned}
\qquad (2.9)
$$

Similarly, for the system $U\underline{x} = \underline{y}$ .

These relations are again all *sequential processes*.

The question now is can we find a matrix factorisation that is more suitable for parallel computation?

Consider then a factorization of the matrix A of the form

$$A = WZ , \qquad (2.10)$$

where

$$
W =
\begin{bmatrix}
1 & & & 0 \\
w_{21} & 1 & 0 & w_{24} \\
w_{31} & 0 & 1 & w_{34} \\
0 & & & 1
\end{bmatrix},
\quad \text{and} \quad
Z =
\begin{bmatrix}
z_{11} & z_{12} & z_{13} & z_{14} \\
& z_{22} & z_{23} & \\
\bigcirc & z_{32} & z_{33} & \bigcirc \\
z_{41} & z_{42} & z_{43} & z_{44}
\end{bmatrix}
$$

In general, the matrices  W  and  Z  will have the forms  (2.11)

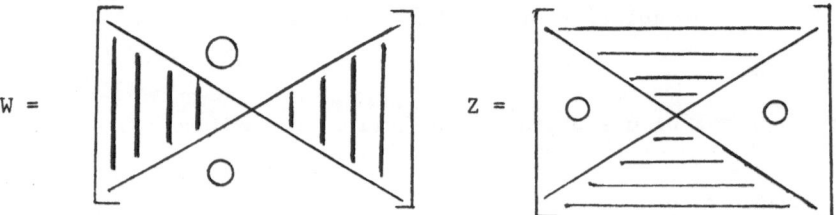

and are termed the quadrant interlocking factors (Q.I.F.) of  A.  It can be noticed that they have a butterfly shape (Evans and Hatzopoulos, 1979).

To determine the coefficients of W and Z we equate the coefficients of A and WZ in (2.10). Thus, for rows I and IV we have,

I $\quad z_{11} = a_{11}, \quad z_{12} = a_{12}, \quad z_{13} = a_{13}, \quad z_{14} = a_{14},$

IV $\quad z_{41} = a_{41}, \quad z_{42} = a_{42}, \quad z_{43} = a_{43}, \quad z_{44} = a_{44} .$
$$\tag{2.12}$$

Whilst for row 2 we have the equations,

II $\quad w_{21}z_{11} + w_{24}z_{41} = a_{21}, \quad w_{21}z_{12} + z_{22} + w_{24}z_{42} = a_{22} ,$

$\quad w_{21}z_{13} + z_{23} + w_{24}z_{43} = a_{23}; \quad w_{21}z_{14} + w_{24}z_{44} = a_{24} .$
$$\tag{2.13}$$

From the first and last equations we obtain $w_{21}$ and $w_{24}$ and by substitution in the 2nd and 3rd equations we obtain $z_{22}$ and $z_{23}$ .

Similarly for row 3, we have the equations

III $\quad w_{31}z_{11} + w_{34}z_{41} = a_{31}, \quad w_{31}z_{12} + z_{32} + w_{34}z_{42} = a_{32} ,$

$\quad w_{31}z_{13} + z_{33} + w_{34}z_{43} = a_{33}, \quad w_{31}z_{14} + w_{34}z_{44} = a_{34} .$
$$\tag{2.14}$$

As before, we obtain from the first and last equations the values of $w_{31}$ and $w_{34}$ and by substituting in the 2nd and 3rd equations, we obtain $z_{32}$ and $z_{33}$ .

Thus, we can see that the first and last rows of Z are given immediately. Then (2x2) sets of linear equations are solved to obtain $w_{i,1}$ and $w_{i,4}$ for i = 2,3 and $z_{2,j}$ and $z_{3,j}$ for j = 2,3 .

Thus, the calculation proceeds as follows,

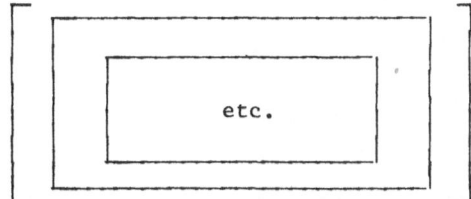

where the outermost peripheral elements of the matrices W and Z are obtained. Then, the calculation proceeds to the innermost next layer of elements. Thus, only (n-1)/2 stages are required to compute all the elements of W and Z.

In comparison, the determination of the coefficients in the LU decomposition is given as

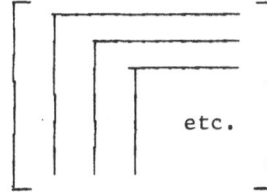

## Solution of the Linear Systems

Using the relationship $A = WZ$, then the linear system $A\underline{x} = \underline{b}$, can now be reformulated as the solution of 2 related linear systems,

$$W\underline{y} = \underline{b}, \text{ and } Z\underline{x} = \underline{y} \ .$$

To solve $W\underline{y} = \underline{b}$ we proceed as follows:

$$\begin{bmatrix} 1 & & & 0 \\ w_{21} & 1 & 0 & w_{24} \\ w_{31} & 0 & 1 & w_{34} \\ 0 & & & 1 \end{bmatrix} \begin{bmatrix} y_1 \\ y_2 \\ y_3 \\ y_4 \end{bmatrix} = \begin{bmatrix} b_1 \\ b_2 \\ b_3 \\ b_4 \end{bmatrix}$$

We see immediately that

$$y_1 = b_1 \text{ and } y_4 = b_4 \ ,$$

$$w_{21}y_1 + y_2 + w_{24}y_4 = b_2 \text{ and } w_{31}y_1 + y_3 + w_{34}y_4 = b_3 \ ,$$

or

$$y_2 = \tilde{b}_2 = (b_2 - w_{21}y_1 - w_{24}y_4) \ ,$$

and

$$y_3 = \tilde{b}_3 = (b_3 - w_{31}y_1 - w_{34}y_4) \ .$$

The solutions for $\underline{y}$ are obtained in pairs working from the top and bottom components of the vector.

Once the vector $\underline{y}$ has been determined then to solve the system $Z\underline{x} = \underline{y}$ we proceed as follows,

$$\begin{bmatrix} z_{11} & z_{12} & z_{13} & z_{14} \\ & z_{22} & z_{23} & \\ & z_{32} & z_{33} & \\ z_{41} & z_{42} & z_{43} & z_{44} \end{bmatrix} \begin{bmatrix} x_1 \\ x_2 \\ x_3 \\ x_4 \end{bmatrix} = \begin{bmatrix} y_1 \\ y_2 \\ y_3 \\ y_4 \end{bmatrix}$$

Starting at the centre we solve the (2x2) linear system,

$$z_{22}x_2 + z_{23}x_3 = y_2 \ ,$$

$$z_{32}x_2 + z_{33}x_3 = y_3 \ ,$$

to evaluate $x_2$ and $x_3$.

Then, we proceed outwards and solve the (2x2) linear system,

$$z_{11}x_1 + z_{14}x_4 = \tilde{y}_1 = (y_1 - z_{12}x_2 - x_{13}x_3)$$

$$z_{41}x_1 + z_{44}x_4 = \tilde{y}_4 = (y_4 - z_{42}x_2 - z_{43}x_3)$$

to evaluate $x_1$ and $x_4$.

Similarly, the system $Z\underline{x} = y$ can be treated in a similar manner.

For parallel computers with $O(n^2)$ processors - this is an $O(n)$ method.

Finally, it can be shown that by suitably chosen permutation matrices the method is identical to a (2x2) block Gaussian Elimination technique.

## 3. THE ALTERNATING GROUP EXPLICIT (AGE) METHOD

Consider the solution of the two-point boundary value problem:

$$-\frac{d^2y}{dx^2}(x) + \rho y(x) = f(x), \quad 0 < x < 1, \quad (3.1)$$

with

$$y(0) = \alpha, \; y(1) = \beta. \quad (3.2)$$

We assume that $\alpha, \beta$ and $\rho$ are given constants with $\rho \geq 0$, and $f(x)$ is a given function such that $d^4y/dx^4$ exists in $0 \leq x \leq 1$ and

$$\left| \frac{d^4y}{dx^4}(x) \right| \leq M, \quad 0 \leq x \leq 1. \quad (3.3)$$

Using Taylor's Theorem we can express $- d^2y/dx^2$ in terms of a three-point central difference approximation plus a truncation error:

$$-\frac{d^2y}{dx^2}(x_i) = [2y(x_i)-(y(x_i+h)+y(x_i-h))]/h^2 + h^2 \frac{d^4y}{dx^4}(x_i+\theta_i h)/12, \quad (3.4)$$

where $|\theta_i| < 1$, $x_i \equiv ih$, $1 \leq i \leq m$ and $h \equiv 1/(m+1)$.

With $y(x_i) \equiv y_i$, the differential equation (3.1) can be written for the nodal values $x_i$, $1 \leq i \leq m$, in matrix form as

$$Ay = k, \quad (3.5)$$

where $A$ is an $m \times m$ real matrix, and $y$ and $k$ are vectors, given explicitly by

$$A = \frac{1}{h^2}\begin{bmatrix} 2+\rho h^2 & & -1 & & & \\ -1 & 2+\rho h^2 & & & \bigcirc & \\ & & & & & \\ & & & & & \\ & \bigcirc & & & & -1 \\ & & & & -1 & 2+\rho h^2 \end{bmatrix} \; ; \; y = \begin{bmatrix} y_1 \\ y_2 \\ , \\ , \\ , \\ , \\ y_m \end{bmatrix} \; ; \; k = \begin{bmatrix} f_1+\alpha/h^2 \\ f_2 \\ , \\ , \\ , \\ , \\ f_m+\beta/h^2 \end{bmatrix}$$

$$(3.6)$$

Thus, the problem has been reduced to the matrix equation

$$Az = k ,\qquad\qquad (3.7)$$

whose solution $z$ is defined to be the discrete approximation to the solution $y(x)$ of (3.1)-(3.2).

We now consider a class of methods for solving the system (3.6) which is based on the "splitting" of the matrix A into the sum of three matrices,

$$A = G_1 + G_2 + \Sigma \qquad\qquad (3.8)$$

where $\Sigma$ is a non-negative diagonal matrix and where $G_1, G_2$ and $\Sigma$ satisfy the following conditions:

(a) $G_1+\theta\Sigma+rI$ and $G_2+\theta\Sigma+rI$ are non-singular for any $\theta \geq 0$, $r > 0$ ;

(b) for any vectors c and d and for any constants $\theta \geq 0$ and $r > 0$ it is "convenient" to solve the systems explicitly, i.e.

$$x = G_1^{-1}c \quad \text{and} \quad y = G_2^{-1}d$$

for x and y respectively.

We shall be concerned here with the situation where $G_1$ and $G_2$ are either small (2x2) block systems or can be made so by a suitable permutation of their rows and corresponding columns (Evans, 1985). This procedure is "convenient" in the sense that the work required is much less than would be required to solve the original system (3.6) directly.

Now choose $h^{-1} = 5$, then for the given boundary conditions (3.2) we have

$$A \equiv \begin{bmatrix} 2+\rho h^2 & -1 & & \\ -1 & 2+\rho h^2 & -1 & \\ & -1 & 2+\rho h^2 & -1 \\ & & -1 & 2+\rho h^2 \end{bmatrix} \equiv G_1 + G_2 + \Sigma , \qquad (3.10)$$

where

$$G_1 = \begin{bmatrix} 1 & -1 & & \\ -1 & 1 & & \\ & & 1 & -1 \\ & & -1 & 1 \end{bmatrix}, \quad G_2 = \begin{bmatrix} 1 & & & \\ & 1 & -1 & \\ & -1 & 1 & \\ & & & 1 \end{bmatrix}, \quad \Sigma \equiv h^2 \begin{bmatrix} \rho & & & \\ & \rho & & \\ & & \rho & \\ & & & \rho \end{bmatrix}.$$

(3.11)

Let us write (3.6) in the form

$$(G_1 + G_2 + \Sigma)y = b,$$  (3.12)

and let us consider two equivalent forms

$$
\left.
\begin{aligned}
(G_1 + \theta\Sigma + rI)y &= b - (G_2 + (1-\theta)\Sigma - rI)y , \\
(G_2 + \theta\Sigma + r'I)y &= b - (G_1 + (1-0)\Sigma - r'I)y.
\end{aligned}
\right\}
$$  (3.13)

Analogous to the Peaceman-Rachford Method (1955) one selects positive iteration parameters $r$ and $r'$ and determines $y^{(n+\frac{1}{2})}$ by

$$(G_1 + \Sigma\theta + rI)y^{(n+\frac{1}{2})} = b - (G_2 + (1-\theta)\Sigma - rI)y^{(n)} .$$  (3.14)

Then one determines $y^{(n+1)}$ by

$$(G_2 + \hat{\theta}\Sigma + r'I)y^{(n+1)} = b - (G_1 + (1-\hat{\theta})\Sigma - r'I)y^{(n+\frac{1}{2})} .$$  (3.15)

For simplicity, we shall consider here the special case where

$$\theta = \hat{\theta} = \frac{1}{2} , \quad r = r' ,$$

and we let

$$\bar{G}_1 = G_1 + \frac{1}{2}\Sigma , \quad \bar{G}_2 = G_2 + \frac{1}{2}\Sigma .$$

Evidently $\bar{G}_1$ and $\bar{G}_2$ satisfy the following conditions:

(a)  $\bar{G}_1 + rI$ and $\bar{G}_2 + rI$ are non-singular for any $\rho > 0$,

(b)  for any vectors c and d and for any $r > 0$ it is practical to solve the systems

$$(\bar{G}_1 + rI)\tilde{x} = c, \quad (\bar{G}_2 + rI)\tilde{y} = d$$

in explicit form since they consist of only (2x2) subsystems.

Thus (3.12) becomes

$$(\bar{G}_1 + \bar{G}_2)y = b ,$$  (3.16)

and (3.14)-(3.15) become respectively

$$(\bar{G}_1 + rI)y^{(n+\frac{1}{2})} = b - (\bar{G}_2 - rI)y^{(n)}$$  (3.17)

$$(\bar{G}_2 + rI)y^{(n+1)} = b - (\bar{G}_1 - rI)y^{1n+\frac{1}{2})} .$$  (3.18)

## 4. NUMERICAL RESULTS

We now consider the linear problem

$$U_1' = U_2 , \tag{4.1}$$

$$U_2' = 400(U_1 + \cos^2(\pi x)) + 2\pi^2 \cos(2\pi x) , \tag{4.2}$$

subject to the boundary conditions

$$U_1(0) = U_1(1) = 0 . \tag{4.3}$$

The exact solution for this problem is given by

$$\left. \begin{aligned} U_1(x) &= \frac{e^{-20}}{1+e^{-20}} \cdot e^{20x} + \frac{1}{1+e^{-20}} \cdot e^{-20x} - \cos^2(\pi x) \\[2ex] U_2(x) &= \frac{20e^{-20}e^{20x}}{1+e^{-20}} - \frac{20}{1+e^{-20}} e^{-20x} + \pi \sin(2\pi x) \end{aligned} \right\} \tag{4.4}$$

From (4.1) and (4.2) we have

$$U_1'' = 400(U_1 + \cos^2(\pi x)) + 2\pi^2 \cos(2\pi x) . \tag{4.5}$$

By following the usual finite difference procedure, equation (4.5) can be approximated to obtain the linear difference equation (assuming that $u = U_1$)

$$\frac{u_{i-1} - 2u_i + u_{i+1}}{h^2} = 400[u_i + \cos^2(\pi x_i)] + 2\pi^2 \cos(2\pi x_i),$$
$$i = 1, 2, \ldots, m .$$

This equation can be simplified to the form

$$-u_{i-1} + (2 + 400h^2)u_i - u_{i+1} = -2h^2[200\cos^2(\pi x_i) + \pi^2 \cos(2\pi x_i)]$$
$$i = 1, 2, \ldots, m . \tag{4.6}$$

The boundary conditions are replaced by

$$u_0 = 0 \quad \text{and} \quad u_{m+1} = 0 \tag{4.7}$$

where $\quad h = \dfrac{1}{m+1} .$

The linear system (4.7) can be represented in matrix notation as

$$A\underline{u} = (\bar{G}_1 + \bar{G}_2)\underline{u} = \underline{b} . \tag{4.8}$$

The vector $\underline{u}$ is defined in the usual way, and $\underline{b}$ is given by

$$\underline{b} = (c_1, c_2, \ldots, c_{m-1}, c_m)^T , \tag{4.9}$$

where

$$c_i = -2h^2[200\cos^2(\pi x_i) + \pi^2 \cos(2\pi x_i)], \; i = 1, 2, \ldots, m \tag{4.10}$$

$\bar{G}_1$ and $\bar{G}_2$ are given by

$$\bar{G}_1 = \begin{pmatrix} g & -1 & & & & \\ -1 & g & & & & \\ & & g & -1 & & \\ & & -1 & g & & \\ & & & & g & -1 \\ & & & & -1 & g \end{pmatrix} \quad \text{and} \quad \bar{G}_2 = \begin{pmatrix} g & & & & \\ & g & -1 & & \\ & -1 & g & & \\ & & & g & -1 \\ & & & -1 & g \\ & & & & g \end{pmatrix}$$

$$(4.11)$$

with $g = 1+200h^2$ .

Hence, by applying the A.G.E. method of (3.13), we can determine $\underline{u}^{(n+\frac{1}{2})}$ and $\underline{u}^{(n+1)}$ successively from equations (3.17)-(3.18). It is obvious that the (2x2) submatrices of $(\bar{G}_1+rI)$ and $(\bar{G}_2+rI)$ are of the form

$$\hat{G} = \begin{pmatrix} \alpha & -1 \\ -1 & \alpha \end{pmatrix} \quad ,$$

where $\alpha = g+r$, and the inverse of $\hat{G}$ is given by

$$\hat{G}^{-1} = d\begin{pmatrix} \alpha & 1 \\ 1 & \alpha \end{pmatrix} \quad , \quad \text{where} \quad d = \frac{1}{\alpha^2-1} \quad . \quad (4.12)$$

Hence the vector $\underline{u}^{(n+1)}$ can be determined from $\underline{u}^{(n)}$ in two steps, we first determine $\underline{u}^{(n+\frac{1}{2})}$ as follows

$$\begin{pmatrix} u_1 \\ u_2 \\ u_3 \\ u_4 \\ \vdots \\ u_{m-1} \\ u_m \end{pmatrix}^{(n+\frac{1}{2})} = d \begin{pmatrix} \alpha & 1 & & & & \\ 1 & \alpha & & & & \\ & & \alpha & 1 & & \\ & & 1 & \alpha & & \\ & & & & \alpha & 1 \\ & & & & 1 & \alpha \end{pmatrix} \begin{pmatrix} c_1-\beta u_1 \\ c_2-\beta u_2+u_3 \\ c_3+u_2-\beta u_3 \\ c_4-\beta u_4+u_5 \\ \vdots \\ c_{m-1}+u_{m-2}-\beta u_{m-1} \\ c_m-\beta u_m \end{pmatrix}^{(n)}$$

$$(4.13)$$

and by using the values of $\underline{u}^{(n+\frac{1}{2})}$ we determine $\underline{u}^{(n+1)}$,

133

$$
\begin{pmatrix} u_1 \\ u_2 \\ u_3 \\ \vdots \\ \\ \\ \\ u_{m-2} \\ u_{m-1} \\ u_m \end{pmatrix}^{(n-1)}
=
\begin{pmatrix}
1/\alpha & & & & & \\
& \alpha d & d & & & \bigcirc \\
& d & \alpha d & & & \\
& & & \ddots & & \\
& & & & & \\
& \bigcirc & & & \alpha d & d \\
& & & & d & \alpha d \\
& & & & & 1/\alpha
\end{pmatrix}
\begin{pmatrix}
c_1 - \beta u_1 + u_2 \\
c_2 + u_1 - \beta u_2 \\
c_3 - \beta u_3 + u_4 \\
\vdots \\
\\
\\
c_{m-2} + u_{m-3} - \beta u_{m-2} \\
c_{m-1} - \beta u_{m-1} + u_m \\
c_m + u_{m-1} - \beta u_m
\end{pmatrix}^{(n+\frac{1}{2})}
\qquad (4.14)
$$

where $\beta = g - r$ .

The accompanying tables show the results obtained by solving the given problem by the A.G.E. method and the S.O.R. method. The convergence test used was the average test

$$
||u_i^{(n+1)} - u_i^{(n)}|| / (1 + ||u_i^{(n)}||) < \varepsilon \qquad (4.15)
$$

with $\varepsilon = 10^{-5}$. For each method, the logarithm of the minimum number of iterations was plotted against $\log(h^{-1})$, the graph which supports the A.G.E. theory is shown in Fig. 4.1 .

Table 4.1

| | S.O.R. | | A.G.E. | |
|---|---|---|---|---|
| $h^{-1}$ | $\omega_b$ | n | r | n |
| 13 | 1.075-1.077 | 9 | 1.75-1.9 | 5 |
| 25 | 1.206-1.238 | 16 | 0.58-0.98 | 9 |
| 37 | 1.325-1.374 | 23 | 0.53-0.59 | 12 |
| 49 | 1.42 -1.47 | 30 | 0.25-0.465 | 16 |

For the linear problem $\underline{u}^{(n+\frac{1}{2})}$ and $\underline{u}^{(n+1)}$ can be obtained from (4.13) and (4.14) respectively. Hence, we can show that the number of operations required to solve this problem by the A.G.E. iterative method is

$$(6m-7) \text{ multiplications} + (4m - \frac{11}{2}) \text{ additions} + \text{R.H.S. unit} \qquad (4.16)$$

per iteration.

On the other hand, to solve this problem by the S.O.R. iterative method we require per iteration

$$2m \text{ multiplications} + (4m-2) \text{ additions} + \text{R.H.S. unit} \qquad (4.17)$$

Hence, by combining the results shown in Table 1 with the corresponding number of operations per iteration required to solve the problem by the S.O.R. and A.G.E. methods, we can obtain the total number of arithmetic operations required.

Table 4.2

| Method $h^{-1}$ | S.O.R. | | A.G.E. | |
|---|---|---|---|---|
| | M | A | M | A |
| 13 | 18m | 36m-18 | 30m-35 | 20m-27.5 |
| 25 | 32m | 64m-32 | 54m-63 | 36m-49.5 |
| 37 | 46m | 92m-46 | 72m-84 | 48m-66 |
| 49 | 60m | 120m-60 | 96m-112 | 64m-88 |

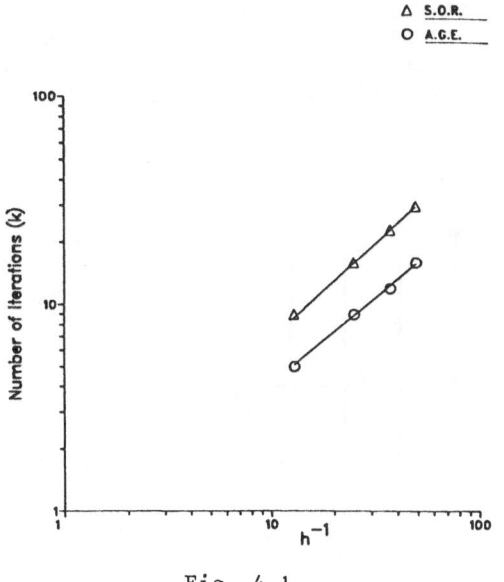

Fig. 4.1

Further, it can be seen from Table 4.2 that the A.G.E. algorithm is more efficient than the S.O.R. iterative approach to solve the linear problem.


## 5. CONVERSION OF IMPLICIT METHODS TO EXPLICIT FORM

Another technique of achieving parallelism in a numerical algorithm is by the use of explicit methods.

However, these methods are the oldest methods and suffer from poor stability and convergence characteristics that require unacceptable computer solution times.

The newer implicit methods are better but often we are not able to exploit to the full the implicit parallelism in the solution algorithm.

Hence we must find new explicit methods with improved stability and convergence characteristics.

Consider the simple heat-conduction problem, (Fig. 5.1),

$$\frac{\partial u}{\partial t} = \frac{\partial^2 u}{\partial x^2} , \quad 0 \le x \le 1, \quad t > 0 \tag{5.1}$$

with initial conditions, $u(x,0) = f(x), \quad 0 \le x \le 1,$

and boundary conditions, $u(0,t) = g_0(t), \quad 0 < t \le T,$

$$u(1,t) = g_1(t), \quad 0 < t \le T .$$

The simplest explicit method uses a forward difference operator approximation to $\partial u/\partial t$ and a central difference operator approximation to $\partial^2 u/\partial x^2$ . The formula

$$u_{i,j+1} = r u_{i-1,j} + (1-2r) u_{i,j} + r u_{i+1,j} + 0(\Delta t + \Delta x^2) \tag{5.2}$$

is well known (Fig. 5.2) but is unstable for values of $r = \Delta t/\Delta x^2 > \frac{1}{2}$ . However, the algorithm is ideal for parallel application since every point on the grid can be evaluated at the same time. The method requires long solution times due to the small time step of integration.

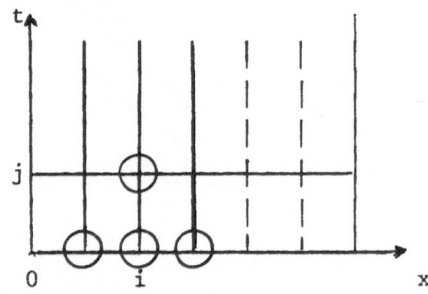

Fig. 5.1

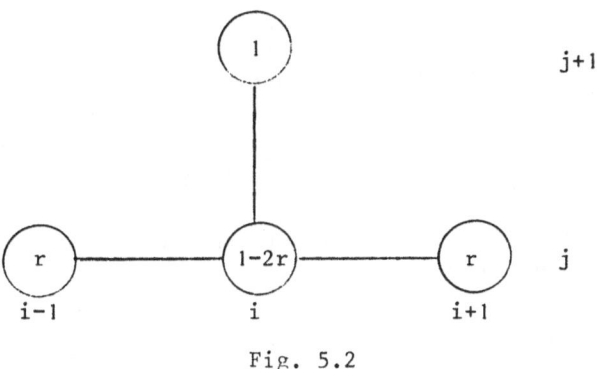

Fig. 5.2

An implicit method uses a backward difference operator approximation to $\partial u/\partial t$ and a central difference operator approximation to $\partial^2 u/\partial x^2$. The equation

$$-ru_{i-1,j+1} + (1+2r)u_{i,j+1} - ru_{i+1,j+1} = u_{i,j} \qquad (5.3)$$

is also well known and is stable for all values of r (Fig. 5.3). However, the algorithm requires the solution of a system of 3 term finite difference equations at every time step in which we are not able to exploit the parallelism to the full.

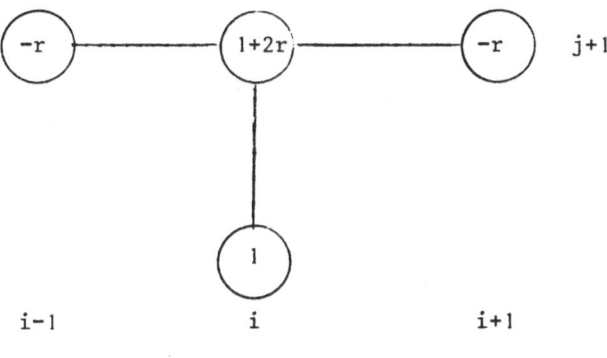

Fig. 5.3

137

In order to facilitate the solution of these implicit equations, asymmetric techniques due to Saul'yev (1964) have been used, i.e. the computational molecule Fig. 5.4 representing the equation

$$-ru_{i-1,j+1}+(1+r)u_{i,j+1} = (1-r)u_{i,j}+ru_{i+1,j}+0(\Delta t+\Delta x^2 + \frac{\Delta t}{\Delta x}) \qquad (5.4)$$

is explicit if solved from left to right and the computational molecule Fig. 5.5 representing the equation

$$-ru_{i+1,j+1}+(1+r)u_{i,j+1} = (1-r)u_{i,j}+ru_{i-1,j}+0(\Delta t+\Delta x^2 - \frac{\Delta t}{\Delta x}) \qquad (5.5)$$

is explicit if solved from right to left.

These two schemes are often referred to as semi-explicit formulae.

### A New Group Explicit Method

If we now couple the use of the asymmetric equations (5.4) and (5.5) at 2 adjacent points (Fig. 5.6), then they result in a (2x2) set of implicit difference equations.

For the group of two points, i.e. $\{i\Delta x,(j+\frac{1}{2})\Delta t\}$ and $\{(i+1)\Delta x,(j+\frac{1}{2})\Delta t\}$ in which equations (5.5) and (5.4) are used simultaneously to calculate the values of u at these points respectively. Therefore, at point $\{i\Delta x,(j+\frac{1}{2})\Delta t\}$ the solution is approximated by

$$-ru_{i+1,j+1} + (1+r)u_{i,j+1} \simeq ru_{i-1,j} + (1-r)u_{i,j} \qquad (5.4a)$$

whilst at point $\{(i+1)\Delta x,(j+\frac{1}{2})\Delta t\}$, the solution is approximated by

$$-ru_{i,j+1} + (1+r)u_{i+1,j+1} = (1-r)u_{i+1,j} + ru_{i+2,j} \qquad (5.5a)$$

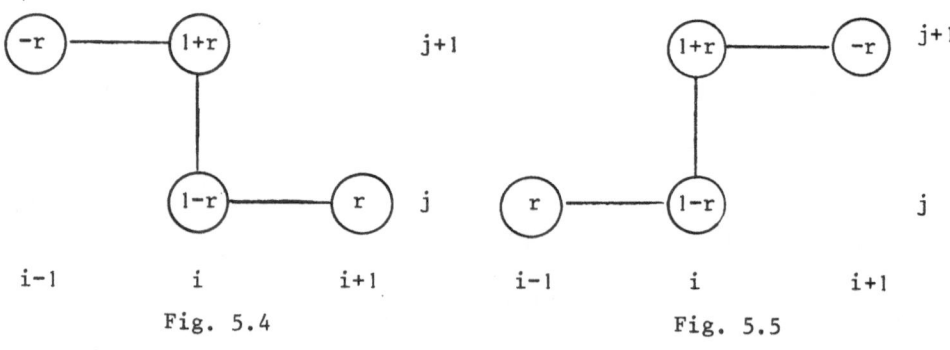

Fig. 5.4                    Fig. 5.5

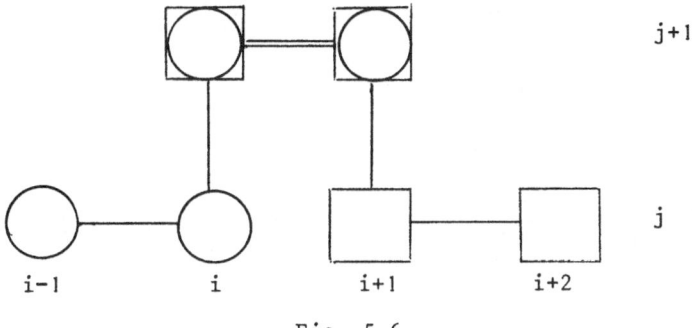

j+1

j

i-1       i       i+1       i+2

Fig. 5.6

If we now rewrite equations (5.4) and (5.5) in matrix form,

$$\begin{bmatrix} 1+r & -r \\ -r & 1+r \end{bmatrix} \begin{bmatrix} u_{i,j+1} \\ u_{i+1,j+1} \end{bmatrix} = \begin{bmatrix} 1-r & 0 \\ 0 & 1-r \end{bmatrix} \begin{bmatrix} u_{i,j} \\ u_{i+1,j} \end{bmatrix} + \begin{bmatrix} ru_{i-1,j} \\ ru_{i+2,j} \end{bmatrix} \quad (5.6)$$

in which the (2x2) matrix of coefficients can easily be inverted so that the equation can be written in explicit form as

$$\begin{bmatrix} u_{i,j+1} \\ u_{i+1,j+1} \end{bmatrix} = \frac{1}{|A|} \begin{bmatrix} 1+r & r \\ r & 1+r \end{bmatrix} \left\{ \begin{bmatrix} 1-r & 0 \\ 0 & 1-r \end{bmatrix} \begin{bmatrix} u_{i,j} \\ u_{i+1,j} \end{bmatrix} + \begin{bmatrix} ru_{i-1,j} \\ ru_{i+2,j} \end{bmatrix} \right\}$$

$$(5.7)$$

where $|A| = 1+2r$. This simplifies to

$$\begin{bmatrix} u_{i,j+1} \\ u_{i+1,j+1} \end{bmatrix} = \frac{1}{|A|} \begin{bmatrix} r(1+r)u_{i-1,j}+(1-r^2)u_{i,j}+r(1-r)u_{i+1,j}+r^2u_{i+2,j} \\ r^2u_{i-1,j}+r(1-r)u_{i,j}+(1-r^2)u_{i+1,j}+r(1+r)u_{i+2,j} \end{bmatrix}$$

$$(5.8)$$

For any ungrouped (single) points near the right and left boundaries equations (5.4) and (5.5) can be used respectively, i.e. for the right boundary,

$$u_{m-1,j+1} = \frac{1}{(1+r)} (ru_{m,j+1}+ru_{m-2,j}+(1-r)u_{m-1,j}) , \quad (5.9)$$

and for the left boundary

$$u_{1,j+1} = \frac{1}{(1+r)} (ru_{0,j+1}+ru_{2,j}+(1-r)u_{1,j}) . \quad (5.10)$$

Finally, equation (5.6) can be easily converted to explicit form

139

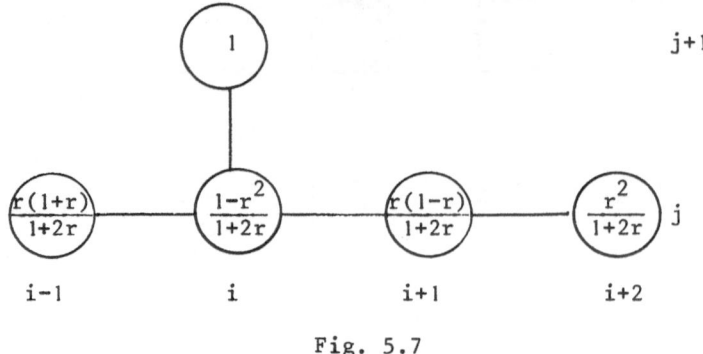

Fig. 5.7

resulting in the computational molecule (Fig. 5.7), representing the equation

$$u_{i,j+1} = \frac{1}{(1+2r)} [r(1+r)u_{i-1,j}+(1-r^2)u_{i,j}+r(1+r)u_{i+1,j}+r^2u_{i+2,j}]$$

(5.11)

and the molecule (Fig. 5.8), representing

$$u_{i+1,j+1} = \frac{1}{(1+2r)}[r^2u_{i-1,j}+r(1-r)u_{i,j}+(1-r^2)u_{i+1,j}+r(1+r)u_{i+2,j}],$$

(5.12)

which when used in the alternating group explicit (AGE) method results in a stable explicit algorithm which is ideal for parallel application (Evans and Abdullah, 1983).

The given problem (5.1) was solved using the AGE algorithm on the NEPTUNE 4 processor parallel MIMD system at Loughborough University and the results obtained when compared with the standard explicit method confirm its suitability for parallel implementation.

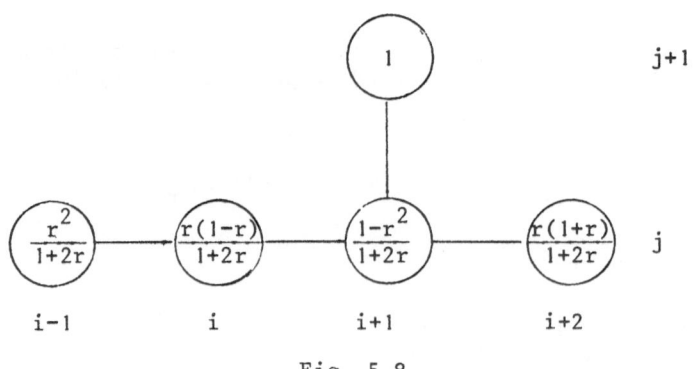

Fig. 5.8

Table 5.1

| No. of points | No. of processors | The Explicit Method | | The Group Explicit Method | |
|---|---|---|---|---|---|
| | | Speed-up | Efficiency | Speed-up | Efficiency |
| 1920 | 0.1 | 1.93 | 0.9650 | 1.98 | 0.9900 |
| | 0,1,2 | 2.85 | 0.9500 | 2.95 | 0.9833 |
| | 0,1,2,3 | 3.77 | 0.9425 | 3.91 | 0.9775 |
| The relative speed up = $\dfrac{\text{explicit}}{\text{Group explicit}}$ = 1.1619 | | | | | |

Finally, a fast algorithmic solution can be obtained for the special value of r = 1.  For this case equation (5.8) becomes

$$\begin{bmatrix} u_{i,j+1} \\ u_{i+1,j+1} \end{bmatrix} = \frac{1}{3} \begin{bmatrix} 2u_{i-1,j} + u_{i+2,j} \\ u_{i-1,j} + 2u_{i+2,j} \end{bmatrix}$$

with the corresponding computational molecule given by Fig. 5.9 .

The  amount of computational work required/point is just 2 additions per
and 1 multiplication which is less than the standard explicit method
(which would be unstable at r = 1 anyway) together with the advantages of
parallelisation which is derived from the explicitness.

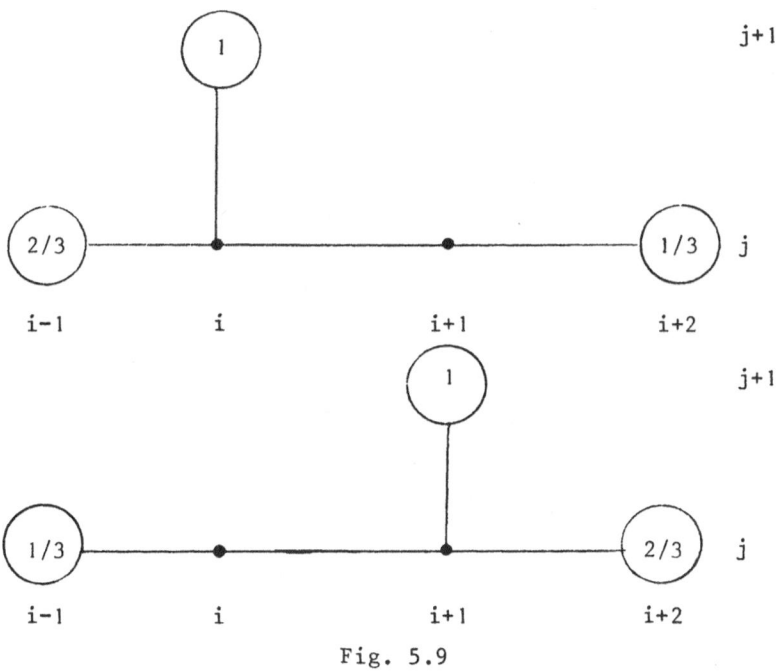

Fig. 5.9

Finally, it can be noticed that for the Fast Group Explicit method it is possible to obtain the solution at alternate sets of nodes (either ○ or ●) as shown in Figure 5.9 . Thus, the method (2 line Hopscotch) can advance the solution over the whole domain leaving half of the grid points uncalculated. This in effect reduces the cost of the computation by half.

REFERENCES

1    Evans, D.J. and Hatzopoulos, M., 1979, A parallel linear system solver, Int. J. Comp. Math. 7, 227-238.
2    Varga, R.S., 1963, "Matrix Iterative Analysis", Prentice Hall.
3    Evans, D.J., 1985, Group explicit iterative methods for solving large linear systems, Int. Journ. Comp. Math., 17, 81-108.
4    Saul'yev, V.K., 1964, "Integration of Equations of Parabolic Type by the Method of Nets", Macmillan, New York.
5    Evans, D.J. and Abdullah, A.R.B., 1983, A new explicit method for the diffusion equation, pp. 330-347, in "Numerical Methods in Thermal Problems III", eds. R.W. Lewis et al, Pineridge Press.

DETERMINATION AND SEPARATION OF DYNAMICS FOR MULTI-TIME SCALE

BILINEAR SYSTEMS

F. Rotella, G. Dauphin-Tanguy, and P. Borne

Laboratoire d'Automatique et d'Informatique Industrielle
(L.A.I.I.)
Institut Industriel Du Nord (I.D.N.) - B.P. 48
59651 Villeneuve d'Ascq Cedex - France

ABSTRACT

This paper deals with multi-model representation and parallel simulation of bilinear continuous systems.

Two points of view are discussed here for these systems: the quasi-diagonalisation of original ill-conditioned models and the application of singular perturbations technique.

We propose a method to build bilinear reduced order models and to determine their validity domain in regard to the inputs.

I. INTRODUCTION

The class of continuous bilinear systems is described by the following mathematical model:

$$\dot{x}(t) = \sum_{i=1}^{n} u_i(t) [A_i x(t) + B_i] = A(u)x + Bu$$

$$y = Cx(t) \qquad\qquad (1)$$

where $\forall~t \in \mathbb{R}^q$ is the state-vector

$u^T = [u_1(t),\ldots,u_n(t)] \in \mathbb{R}^n$ is the input vector

$y(t) \in \mathbb{R}^m$ is the output vector

and $C, A_i, B_i, i \in [1,\ldots,n]$ are $\mathbb{R}^{m\times q}, \mathbb{R}^{q\times q}, \mathbb{R}^q$ constant matrices.

$A(u)$ will be called in the following the system-matrix [BRA, 81]. It is obvious that this class includes linear systems; their interest is in a more exact modelling of non-linear systems than a linear model [MOH, 73]. In recent years, an extensive literature [MOH, 80] has been concerned with the study of several particular problems for these bilinear systems like controllability [RIN, 68], observers [DER, 79],

optimal control [DER, 82; CEB, 84], identification [KAR, 78], stabilisa-
bility [TZA, 84].   For large-scale bilinear systems, the problem of order
reduction by the use of singular perturbations technique has been
studied in [GUI, 80].   They obtain slow and fast reduced order models,
but with strong conditions on the matrices $A_i$,  which does not allow to
apply their method in the general case.

Two points of view are exposed in our paper.  The singular perturbations technique is linked to the existence of small parameters [KOK, 76]
in (1), so , first, we propose a reconditioning of (1).  In a second part,
after the use of the singular perturbations method, we impose constraints
on the input-vector to get a bilinear slow-reduced model.


## II.   PSEUDO-DIAGONALISATION OF THE SYSTEM MATRIX

The system (1) can be considered as a linear non-stationary system,
then following [O'MA, 82] we have:

Definition:   The system

$$\dot{x} = A(t)x + B(t)u(t) \quad 0 \le t \le T \tag{2}$$

is said two-time-scale on [0, T] if the spectrum of  $A(t)$
can be partitioned into two sets  $S(t)$  and  $F(t)$   such that:

$$\forall\ t \in [0, T] \quad \max_{s_i(t) \in S(t)} |s_i(t)| \ << \ \min_{f_j(t) \in F(t)} |f_j(t)|$$

By use of a non-constant transformation:

$$y = T(t)\ x \tag{3}$$

where

$$T(t) = \begin{bmatrix} I+K(t)L(t) & K(t) \\ L(t) & I \end{bmatrix}$$

and  $K(t)$ and $L(t)$  are solutions of the differential Riccati equations:

$$\left. \begin{aligned} \dot{L} &= A_{22}L - LA_{11} + LA_{12}L - A_{21} \ ; \ L(0) = 0 \\ \dot{K} &= (A_{11} - A_{12}L)K - K(A_{22} + LA_{12}) - A_{12} \ ; \ K(T) = 0 \end{aligned} \right\} \tag{4}$$

the system can be block-diagonalised as:

$$\dot{y} = \begin{bmatrix} \tilde{A}_{11}(t) & 0 \\ 0 & \tilde{A}_{22}(t) \end{bmatrix} y + \begin{bmatrix} \tilde{B}_1(r) \\ \tilde{B}_2(r) \end{bmatrix} u(r) \tag{5}$$

In the case of bilinear systems, the transformation will depend on
inputs which is not a practical way.   We prefer, in order to solve the
problem, to use the block-diagonalising transformation proposed in
[CHA, 72] and [CHO, 76] for linear systems.

## II.1 Pseudo-triangularisation

Let us consider, first, the state basis change defined by the matrix $T_1$ and the decomposition of the system-matrix:

$$T_1 = \begin{bmatrix} I & 0 \\ L & I \end{bmatrix}, \quad \forall\ i\ \epsilon\ [1,\ldots,n] \qquad A_i = \begin{bmatrix} A_{11}^i & A_{12}^i \\ A_{21}^i & A_{22}^i \end{bmatrix} \tag{6}$$

thus, if we note $y = T_1 x$, it comes:

$$\dot{y} = \tilde{A}(u)y + \tilde{B}u$$

where

$$\tilde{A}(u) = \sum_{i=1}^{n} \tilde{A}_i u_i\ ; \qquad \tilde{B} = T_1 B$$

and

$$\forall\ i\ \epsilon\ [1,\ldots,n]$$

$$\tilde{A}_i = \begin{bmatrix} A_{11}^i - A_{12}^i L & A_{12}^i \\ \\ R_i(L) & A_2^i + LA_{12}^i \end{bmatrix} = \begin{bmatrix} \tilde{A}_{11}^i & A_{12}^i \\ \\ R_i(L) & \tilde{A}_{22}^i \end{bmatrix} \tag{7}$$

with

$$R_i(L) = A_{21}^i + LA_{11}^i - A_{22}^i L - LA_{12}^i L$$

The system-matrix is block-triangular if the following condition is satisfied:

$$\forall\ i \quad [1,\ldots,n] \qquad R_i(L) = 0 \tag{8}$$

This condition cannot be realized for each $i$, so if we introduce the notation:

$$\forall\ (k,1)\ \epsilon\ [1,\ 2] \qquad A_{k1} = \begin{bmatrix} A_{k1}^1 \\ \hline A_{k1}^2 \\ \vdots \\ \hline A_{k1}^n \end{bmatrix} \tag{9}$$

$L$ is the "solution" of the expanded algebraic Riccati equation:

$$A_{21} + L \otimes [A_{11} - A_{12} L] = A_{22} L \tag{10}$$

where $\otimes$ is the Kronecker product of partitioned matrices. To solve equation (10) in the least squares sense, we can use the classical algorithm [KOK, 75] defined by:

$$k \in \mathbb{N}$$

$$L_{k+1} = A_{22}^+ \left[ \bar{A}_{21} + L_k \, \Theta \, [A_{11} - A_{12} \, L_k] \right]$$

with $\quad L_o = A_{22}^+ \, A_{21}$

where $\quad A_{22}^+ = (A_{22}^T \, A_{22})^{-1} \, A_{22}^T \quad$ is the left pseudo-inverse

of the matrix $\quad A_{22} \quad$ [KOR, 67].

$$\left. \right\} \quad (11)$$

To ensure existence and uniqueness of a real root of (10), we can enounce the following lemma, the proof of which is analogue to the proof of Lemma 1 in [KOK, 75].

<u>Lemma 1</u> :    If $A_{22}$ is full of rank and if:

$$||A_{22}^+|| \leq \frac{1}{3}(||A_o|| + ||A_{12}|| \; ||L_o||)^{-1} \qquad (12)$$

with $\quad A_o = A_{11} - A_{12} \, L_o$

then a unique real root of (2-9) exists satisfying:

$$L = L_o + D \qquad \text{with}$$

$$0 \leq ||D|| \leq \frac{2 \; ||A_o|| \; ||L_o||}{||A_o|| + ||A_{12}|| \; ||L_o||}$$

As comments we can say that as in [KOK, 75], this sufficient condition is not necessary and the convergence of the algorithm is efficient without verifying the condition of the Lemma.

II.2  Pseudo-diagonalisation

In a second step, we can apply on $\tilde{A}(u)$ the basis change defined by the matrix $T_2$ :

$$T_2 = \begin{bmatrix} I & -M \\ 0 & I \end{bmatrix} \qquad (13)$$

thus, if we note $z = T_2 \, y$, it comes:

$$\dot{z} = \tilde{\tilde{A}}(u)z + \tilde{\tilde{B}}u \qquad (14)$$

where

$$\tilde{A}(u) = \sum_{i=1}^{n} \tilde{A}_i u_i \; ; \; \tilde{B} = T_2 \tilde{B}$$

and $\quad \forall \; i \in [1,\ldots,n]$

$$\tilde{A}_i = \begin{bmatrix} \tilde{A}_{11}^i - M\,R_i(L) & R_i(M) \\ \\ R_i(L) & \tilde{A}_{22}^i + R_i(L)\,M \end{bmatrix}$$

with $\quad R_i(M) = A_{12}^i - M\,\tilde{A}_{22}^i + \tilde{A}_{11}^i\,M - M\,R_i(L)\,M$

$\tilde{A}(u)$ is a block-diagonalised system-matrix if:

$$\forall \; i \in [1,\ldots,n] \qquad R_i(M) = 0 \tag{15}$$

So, if we transpose these relations, we obtain:

$$\forall \; i \in [1,\ldots,n] \quad A_{12}^{iT} - \tilde{A}_{22}^{iT} M^T + M^T \tilde{A}_{11}^{iT} - M^T R_i^T(L)\,M^T = 0 \tag{16}$$

If we note:

$$\tilde{A}_{12} = \begin{bmatrix} A_{12}^{1T} \\ \vdots \\ A_{12}^{nT} \end{bmatrix}, \; \forall \; k \in {1, 2} \quad \tilde{A}_{kk} = \begin{bmatrix} \tilde{A}_{kk}^{1T} \\ \vdots \\ \tilde{A}_{kk}^{nT} \end{bmatrix}, \quad R(L) = \begin{bmatrix} R_1^T(L) \\ \vdots \\ R_n^T(L) \end{bmatrix} \tag{17}$$

$M^T$ is the solution of a Riccati equation of the same type than $L$ :

$$\tilde{A}_{12} + M^T \otimes [\tilde{A}_{11} - R(L)\,M^T] = \tilde{A}_{22}\,M^T \tag{18}$$

The solution of (18) can be obtained by the following iteration scheme:

$$\left. \begin{array}{l} k \in \mathbb{N} \\ \\ M_{k+1}^T = \tilde{A}_{22}^+ [\tilde{A}_{12} + M_k^T \otimes [\tilde{A}_{11} - R(L)\,M_k^T]] \\ \\ \text{with } M_o^T = \tilde{A}_{22}^+ \tilde{A}_{12} \end{array} \right\} \tag{19}$$

Equation (18) is analog to equation (10) so if we define the following matrix:

$$\tilde{A}_o = \tilde{A}_{11} - R(L)\,M_o^T$$

it comes the following lemma:

<u>Lemma 2:</u>    If $\tilde{A}_{11}$ is full of rank and if:

$$\left| \left| \tilde{A}_{22}^{+} \right| \right| \ \le\ \frac{1}{3}\ (\left| \left| A_o \right| \right| + \left| \left| R(L) \right| \right|\ \left| \left| M_o^T \right| \right|)^{-1} \tag{20}$$

then a unique real root of (18) exists satisfying in the sense of the least square :

$$M^T\ =\ M_o^T + P^T$$

with        $$0 \le \left| \left| P \right| \right| \le \frac{2 \left| \left| \tilde{A}_o \right| \right|\ \left| \left| M_o^T \right| \right|}{\left| \left| \tilde{A}_o \right| \right| + \left| \left| R(L) \right| \right|\ \left| \left| M_o^T \right| \right|}$$

## III. SEPARATION OF DYNAMICS

In this part, we consider the bilinear system (1) given in the singular perturbed form:

$$\left[ \begin{array}{c} \dot{x} \\ \epsilon\ \dot{z} \end{array} \right]\ =\ A(u) \left[ \begin{array}{c} x \\ z \end{array} \right] + Bu$$

$$y\ =\ C_1\ x + C_2\ z \tag{21}$$

where        $x \in \mathbb{R}^{q_s}$, $z \in \mathbb{R}^{q_f}$, $q_s + q_f = q$, $0 < \epsilon \ll 1$,

$$A(u)\ =\ \sum_{i=1}^{n}\ u_i(t) \left[ \begin{array}{c:c} A_{11}^i & A_{12}^i \\ \hdashline A_{21}^i & A_{22}^i \end{array} \right]\ ,\ \ B = \left[ \begin{array}{c} B_1 \\ B_2 \end{array} \right]$$

$B_1 \in \mathbb{R}^{q_s \times n}$,    $B_2 \in \mathbb{R}^{q_f \times n}$

$$\forall\ i \in [1,\ldots,n] \left\{ \begin{array}{ll} A_{11}^i \in \mathbb{R}^{q_s \times q_s} \ , & A_{12}^i \in \mathbb{R}^{q_s \times q_f} \\ \\ A_{21}^i \in \mathbb{R}^{q_f \times q_s} \ , & A_{22}^i \in \mathbb{R}^{q_f \times q_f} \end{array} \right.$$

### III.1 <u>Slow System</u>

By setting $\epsilon$ equal to zero, we obtain the slow subsystem as:

$$\dot{x}_s\ =\ A_{11}(u_s)\ x_s + A_{12}(u_s)\ z_s + B_1\ u_s$$

$$0\ =\ A_{21}(u_s)\ x_s + A_{22}(u_s)\ z_s + B_2\ u_s \tag{22}$$

$$y_s\ =\ C_1\ x_s + C_2\ z_s$$

148

The second equation of (22) can be rewritten as:

$$A_{22}(u_s) \, z_s = -[A_{21}(u_s) \, x_s + B_2 \, u_s] \tag{23}$$

To obtain $z_s$, we must impose the following condition:

$$\forall \, t \in [0, +\infty[ \qquad \det A_{22}(u) = 0 \tag{24}$$

If we note by $p(\lambda, u)$ the instantaneous characteristic polynomial of $A_{22}(u)$, we have:

$$p(\lambda, u) = \alpha_{q_f} \lambda^{q_f} + \sum_{j=0}^{q_f - 1} \alpha_j(u) \lambda^j \tag{25}$$

where for $j \in [0, \ldots, q_f]$

$$\alpha_j(u) = \sum_{\substack{j_1, \ldots, j_n = 0 \\ j_1 + \ldots + j_2 = q_f - j}}^{n} \alpha_{j_1, \ldots, j_n} u_1^{j_1}, \ldots, u_2^{j_n}$$

(we will call it a $(q_f - j)$ homogeneous form in the variables $u_1, \ldots, u_n$); $\alpha_{q_f}(u) = 1$).

By application at every time of the Cayley-Hamilton theorem, we have:

$$[A_{22}(u)]^{q_f} + \sum_{j=0}^{q_f - 1} \alpha_j(u) \, [A_{22}(u)]^j = 0 \tag{26}$$

If the condition (24) is satisfied, we get:

$$\frac{[A_{22}(u)]^{q_f}}{\alpha_o(u)} + \sum_{j=0}^{q_f - 1} \frac{\alpha_j(u)}{\alpha_o(u)} \, [A_{22}(u)]^j + I = 0 \tag{27}$$

which can be rewritten as:

$$A_{22}(u) \left[ -\left[ \frac{[A_{22}(u)]^{q_f - 1}}{\alpha_o(u)} + \sum_{j=1}^{q_f - 1} \frac{\alpha_j(u)}{\alpha_o(u)} \, [A_{22}(u)]^{j-1} \right] \right] = I \tag{28}$$

We can enounce the following:

Lemma 3: If there exists a class of inputs U such as:

$\det A_{22}(u) \neq 0 \qquad \forall \, t \geq 0$

then:

$$\left| A_{22}^{-1}(u) \ = \ - \left[ \frac{[A_{22}(u)]^{q_f-1}}{\alpha_o(u)} \ + \ \sum_{j=0}^{q_f-2} \frac{\alpha_{j+1}(u)}{\alpha_o(u)} \ [A_{22}(u)]^j \right] \right.$$

where the $\alpha_j(u)$ are given by (25).

If we look at the structure of:

$$[A_{22}(u)]^j \qquad \forall \ j \in \mathbb{N}$$

we can see that each component of this system-matrix is a j-homogeneous form in the variables $u_i$ so it comes the corollary:

Corollary: If the condition of Lemma 3 is verified, then each
component of $A_{22}^{-1}(u)$ is in the form:

$$\frac{P_{kl}(u)}{Q_{kl}(u)} : \quad (k,l) \in [1,\ldots,q_f]^2$$

where $P_{kl}(u)$ is $(q_f-1)$-homogeneous in the variables $u_i$

$Q_{kl}(u)$ is $q_f$-homogeneous in the variables $u_i$

If we consider (22), it comes out the equations of the slow system:

$$\dot{x}_s \ = \ A_s(u_s) \ x_s + B_s(u_s) \ u_s \ ; \quad x_s(0) \ = \ x(0) \tag{29a}$$

$$y_s \ = \ C_s(u_s) \ x_s + D_s(u_s) \ u_s \tag{29b}$$

$$z_s \ = \ E_s(u_s) \ x_s + F_s(u_s) \ u_s \tag{29c}$$

where

$$E_s(u_s) \ = \ - A_{22}^{-1}(u_s) A_{21}(u_s)$$

$$F_s(u_s) \ = \ - A_{22}^{-1}(u_s) \ B_2$$

$$A_s(u_s) \ = \ A_{11}(u_s) + A_{12}(u_s) \ E_s(u_s)$$

$$B_s(u_s) \ = \ B_1 + A_{12}(u_s) \ F_s(u_s)$$

$$C_s(u_s) \ = \ C_1 + C_2 \ E_s(u_s)$$

$$D_s(u_s) \ = \ C_2 \ F_s(u_s)$$

Following the corollary, the slow system does not appear here under a bilinear form, but we will present in a next part how to simplify it.

III.2 Fast System

Following singular perturbations method, we have:

$$z_f \ = \ z - z_s \ ; \quad y_f \ = \ y - y_s \ ; \quad u_f \ = \ u - u_s \tag{30}$$

and if, in the beginning of the motion, we consider $x = x_s \simeq x(0)$ and $z_s$ constant, it leads to the boundary-layer equation as:

$$\varepsilon \, \dot{z}_f \;=\; A_{22}(u_f) \, z_f + B_2 \, u_f \; ; \quad y_f \;=\; C_2 \, z_f \tag{31}$$

The initial condition must be determined at the initial time, so we have:

$$z_f(0) \;=\; z(0) - E_s(u_s(0) \; x(0) - F_s(0)) \, u_s(0) \tag{32}$$

## III.3  Simplification of the Slow System

The aim of this part is to determine the input variation domain to obtain a valid bilinear slow system.

That point of view differs from those exposed in [GUI, 80]. The non-linearities on the input-vector are introduced by the equation (29c), so if we consider the corollary, the matrix $E_s(u_s)$ and the vector $F_s(u_s).u_s$ can be written in the form:

$$E_s(u_s) \;=\; \frac{1}{\alpha_o(u_s)} \left[ \sum_{\substack{j_1,\dots,j_n=0 \\ j_1+\dots+j_n=q_f}}^{n} E_{j_1,\dots,j_n} \, u_{1_s}^{j_1} \dots u_{n_s}^{j_n} \right]$$

where $E_{j_1,\dots,j_n}$ are constant matrices of $\mathbb{R}^{q_f \times q_s}$

$$F_s(u_s)u_s \;=\; \frac{1}{\alpha_o(u_s)} \left[ \sum_{\substack{j_1,\dots,j_n=0 \\ j+\dots+j_n=q_f}}^{n} F_{j_1,\dots,j_n} \, u_{1_s}^{j_1} \dots u_{n_s}^{j_n} \right] \tag{33}$$

where $F_{j_1,\dots,j_n}$ are constant vectors of $\mathbb{R}^{q_f}$

$$\alpha_o(u_s) \;=\; \sum_{\substack{j_1,\dots,j_n=0 \\ j_1+\dots+j_n=q_f}}^{n} \alpha_{o\,j_1,\dots,j_n} \, u_{1_s}^{j_1} \dots u_{n_s}^{j_n}$$

and these coefficients multiplied by the matrix $A_{12}(u_s)$ introduce non-linearities to get the "exact" reduced model (29). In order to get a bilinear slow model we are going to simplify the terms of the equation (29c) in the following way: every coefficient in $E_s(u_s)$ and $F_s(u_s)u_s$ will be approached by a constant term. For this aim we introduce the notion of input dominance.

## III.3.1  Input dominance

The input $u_i$ of $[u_1,\dots,u_n]$ is said to be dominant on $[t_1, t_2]$, if:

$$\forall \ t \in [t_1, \ t_2] \quad \forall \ k \in [1,\ldots,n] \quad k \neq i \quad \left| \frac{u_k}{u_i} \right| \ll 1$$

Let us consider the following term:

$$\frac{\phi(u)}{\alpha_o(u)} \ = \ \frac{\displaystyle\sum_{\substack{j_1,\ldots,j_n=0 \\ j_1+\ldots+j_n=q_f}}^{n} \phi_{j_1,\ldots,j_n} \, u_1^{j_1},\ldots,u_n^{j_n}}{\displaystyle\sum_{\substack{j_1,\ldots,j_n=0 \\ j_1+\ldots+j_n=q_f}}^{n} \alpha_{o\,j_1,\ldots,j_n} \, u_1^{j_1},\ldots,u_n^{j_n}} \tag{34}$$

Then if $u_i$ is dominant on $[t_1, \ t_2]$ then during this time-interval we have the constant approximant:

$$\frac{\phi(u)}{\alpha_o(u)} \ \simeq \ \frac{\phi_{0,\ldots,0,j_i=q_f,0,\ldots,0}}{\alpha_{o\,0,\ldots,0,j_i=q_f,0,\ldots,0}} \tag{35}$$

So, by this method we can obtain n constant approximants of the equation (29c).

III.3.2 Least-square solution

The other way is to say:

$$\exists \ K \in \mathbb{R} \ ; \quad \forall \ u_i,\ldots,u_n \qquad \phi(u) \ = \ K \ \alpha_o(u) \tag{36}$$

So K is the solution of the equation:

$$\begin{bmatrix} \phi_{q_f,0,\ldots,0} \\ \vdots \\ \phi_{0,\ldots,0,q_f} \end{bmatrix} \ = \ \begin{bmatrix} \alpha_{o\,q_f,0,\ldots,0} \\ \vdots \\ \alpha_{o\,0,\ldots,0,q_f} \end{bmatrix} K \tag{37}$$

that can be noted:

$$\Phi \ = \ \alpha.K$$

So, by using the pseudo-inverse, we have:

$$\boxed{K \ = \ [\alpha^T\alpha]^{-1} \ \alpha^T\phi} \tag{38}$$

This method gives another approximant.

### III.3.3  Mixed method

Combining the two methods, we can build, with r simultaneously dominant inputs, $C_n^r$ models.  So, considering that in the previous method the n inputs are dominant, we can build:

$$\sum_{p=1}^{n} C_n^p = 2^n - 1$$

constant models of equation (29c).

### III.4  Choice of the Model

To choose the equation for the slow system, two points of view can be enounced.  The first is to keep all the models, which lead to a multi-model representation;  this is not the aim of this paper.  The other way is to calculate some statistical moments from every coefficient of these models;  the averages give the coefficients of the slow bilinear system and the variances justify, or not, the choice of a unique bilinear decoupled model.

As we will see in the example, to do a quasi-diagonalisation as a first step of the reduction of order of a bilinear system by the singular perturbation method decreases these variances.

We must keep in mind that the defined input-domains by the proposed approximant method has to be included in the validity domain of dynamical separation.

## IV  IMPLEMENTATION ON AN EXAMPLE

Let us consider now the following system described by:

$$\begin{bmatrix} \dot{x} \\ \dot{z} \end{bmatrix} = \left\{ u_1 \begin{bmatrix} 1 & 2 \\ 6 & 8 \end{bmatrix} + u_2 \begin{bmatrix} 2 & 1 \\ 6 & 10 \end{bmatrix} \right\} \begin{bmatrix} x \\ z \end{bmatrix} + \begin{bmatrix} 1 \\ 2 \end{bmatrix} u_2 \tag{39}$$

where $u_1$ and $u_2$ are scalar inputs.

Considering the matrices in (39), we can apply the singular perturbations method.

### IV.1  Direct Singular Perturbations Method

With $\varepsilon_1 = 0.5$, the second line of (39) gives:

$$\varepsilon_1 \dot{z} = (3u_1 + 3u_2)x + (4u_1 + 5u_2)z + u_2 \tag{40}$$

By doing $\varepsilon_1 = 0$ in (40), we obtain the slow system:

$$\dot{x}_s = (u_{1_s} + 2u_{2_s})x_s + (2u_{1_s} + u_{2_s})x_s + u_{2_s}$$

$$x_s(0) = x(0)$$

$$z_s = -3\left[\frac{u_{1_s} + u_{2_s}}{4u_{1_s} + 5u_{2_s}}\right]x_s - \frac{u_{2_s}}{4u_{1_s} + 5u_{2_s}}$$

(41)

and the fast system:

$$\varepsilon_1 \dot{z}_f = (4u_{1_f} + 5u_{2_f})z_f + u_{2_f} \; ; \quad z_f(0) = z(0) - z_s(0) \qquad (42)$$

To obtain a bilinear slow system, we consider the input dominances:

$$* \; |u_{2_s}| \; \gg \; |u_{1_s}|,$$

then (41) becomes:

$$\dot{x}_s = (-0.2u_{1_s} + 1.4u_{2_s})x_s - 0.4u_{1_s} + 0.8u_{2_s}$$

$$z_s = -0.6x_s - 0.2$$

(41a)

$$* \; |u_{1_s}| \; \gg \; |u_{2_s}|,$$

then (41) becomes:

$$\dot{x}_s = (-0.5u_{1_s} + 1.25u_{2_s})x_s + u_{2_s}$$

$$z_s = -0.75x_s$$

(41b)

$$* \; |u_{1_s}| \; \simeq \; |u_{2_s}|$$

So, we want:

$$\frac{u_{1_s} + u_{2_s}}{4u_{1_s} + 5u_{2_s}} = \alpha \quad \text{and} \quad \frac{u_{2_s}}{4u_{1_s} + 5u_{2_s}} = \beta \,,$$

which corresponds to the system:

154

$$\begin{vmatrix} 1 \\ 1 \end{vmatrix} = \alpha \begin{vmatrix} 4 \\ 5 \end{vmatrix} \quad \text{and} \quad \begin{vmatrix} 0 \\ 1 \end{vmatrix} = \beta \begin{vmatrix} 4 \\ 5 \end{vmatrix} \qquad (43)$$

The solution of this problem, by the use of the pseudo-inverse, is:

$$\alpha = 0.22 \quad \text{and} \quad \beta = 0.12$$

so, (41) becomes:

$$\left. \begin{aligned} \dot{x}_2 &= (-0.32u_{1_s} + 1.34u_{2_s})x_s - 0.24u_{1_s} + 0.88u_{2_s} \\ z_s &= -0.66x_s - 0.12 \end{aligned} \right\} \qquad (41c)$$

To choose a unique bilinear slow system, we must consider now the conditions of separation of dynamics. The fast system-matrix is given by (42):

$$8u_1 + 10u_2$$

and the slow system-matrix is approximated by the form:

$$a\, u_1 + b\, u_2$$

The constant terms  a and b  can be taken as the averages of the coefficients of the slow systems-matrices found in (41a), (41b) and (41c):

$$a = -0.34 \quad ; \quad b = 1.33$$

The calculation of the variances (respectively here $5.10^{-2}$ and $10^{-2}$) justifies such a choice. The validity of the singular perturbations method is conditioned by the ratio between the eigenvalues of the slow system and the fast system, whatever the input-time variations, which can be expressed as:

$$\forall\, t \in \mathbb{R}^+, \ |8u_1 + 10u_2| \gg |-0.34u_1 + 1.33u_2|$$

The simulation results of the original system (39) (OS) and the reduced order model (42) (41c) (DSP) will show that a steady error appears if this condition is not always satisfied.

## IV.2  Singular Perturbation Method After Pseudo-Diagonalisation

### IV.2.1  Pseudo-diagonalisation of (39)

* First step:  triangularisation by:

$$\begin{bmatrix} x_1 \\ x_2 \end{bmatrix} = \begin{bmatrix} 1 & 0 \\ L & 1 \end{bmatrix} \begin{bmatrix} x \\ z \end{bmatrix}$$

We must solve in the least square sense:

$$\begin{bmatrix} 6 \\ 6 \end{bmatrix} + L\left[\begin{bmatrix} 1 \\ 2 \end{bmatrix} - \begin{bmatrix} 2 \\ 1 \end{bmatrix} L\right] = \begin{bmatrix} 8 \\ 10 \end{bmatrix} L \qquad (45)$$

The solution of this problem is $L = 0.7$ and we have:

$$R_1(L) = 0.12 ; \quad R_2(L) = -0.09$$

So, after the variable change the system is given by:

$$\begin{bmatrix} \dot{x}_1 \\ \dot{x}_2 \end{bmatrix} = \left\{ u_1 \begin{bmatrix} -0.4 & 2 \\ 0.12 & 9.4 \end{bmatrix} + u_2 \begin{bmatrix} 1.3 & 1 \\ -0.09 & 10.7 \end{bmatrix} \right\} \begin{bmatrix} x_1 \\ x_2 \end{bmatrix}$$

$$+ \begin{bmatrix} 1 \\ 2.7 \end{bmatrix} u_2$$

* Second step : diagonalisation by:

$$\begin{bmatrix} y \\ v \end{bmatrix} = \begin{bmatrix} 1 & -m \\ 0 & 1 \end{bmatrix} \begin{bmatrix} x_1 \\ x_2 \end{bmatrix}$$

It comes the equation:

$$\begin{bmatrix} 2 \\ 1 \end{bmatrix} + m\left[\begin{bmatrix} -0.4 \\ 1.3 \end{bmatrix} - \begin{bmatrix} 0.12 \\ -0.09 \end{bmatrix} m\right] = \begin{bmatrix} 9.4 \\ 10.7 \end{bmatrix} m \qquad (47)$$

$m = 0.153$ is solution of this equation and we have $R_1(m) = 0.5$ and $R_2(m) = -0.44$, so the variable change:

$$\begin{bmatrix} y \\ v \end{bmatrix} = \begin{bmatrix} 0.89 & -0.153 \\ 0.7 & 1 \end{bmatrix} \begin{bmatrix} x \\ z \end{bmatrix} \qquad (48)$$

transforms (39) into the quasi-diagonal state-space equation:

$$\begin{bmatrix} \dot{y} \\ \dot{v} \end{bmatrix} = \left\{ u_1 \begin{bmatrix} -0.42 & 0.5 \\ 0.12 & 9.42 \end{bmatrix} + u_2 \begin{bmatrix} 1.29 & -0.44 \\ -0.09 & 10.69 \end{bmatrix} \right\} \begin{bmatrix} y \\ v \end{bmatrix}$$

$$\begin{bmatrix} 0.58 \\ 2.7 \end{bmatrix} u_2 \qquad (49)$$

## IV.2.2 Singular perturbations method on (49)

After the quasi-diagonalisation-step, it is more obvious to see if the system is two-time-scale, than in (39). So, we can separate with more accuracy the slow variable and the fast variable. From (49), it comes immediately the condition of separation of dynamics:

$$\forall \, t \in \mathbb{R}^+ ; \quad |9.42u_1 + 10.69u_2| \gg |-0.42u_1 + 1.29u_2| \qquad (50)$$

and this can be known before the input-dominance study.

With $\varepsilon_2 = 0.1$ from (49) we can build the slow and fast models by singular perturbations method:

$$
\left\{
\begin{array}{l}
\varepsilon_2 \dot{v}_f = (0.942u_{1_f} + 1.069u_{2_f})v_f + 0.27u_{2_f} \\[3mm]
v_s = -\dfrac{0.12u_{1_s} - 0.09u_{2_s}}{9.42u_{1_s} + 10.69u_{2_s}} \, y_s - \dfrac{2.7u_{2_s}}{9.42u_{1_s} + 10.69u_{2_s}} \\[4mm]
\dot{y}_s = (-0.42u_{1_s} + 1.29u_{2_s})y_s + (0.5u_{1_s} - 0.44u_{2_s})v_s + 0.58u_{2_s}
\end{array}
\right.
\qquad (51)
$$

If we do the input-dominance study, as previously, the system-matrix of the slow system can be approached by $-0.42u_1 + 1.3u_2$ with a variance of $10^{-4}$ on each coefficient, which increases the validity of the bilinear reduced model.

The slow part can be approached by:

$$
\left.
\begin{array}{l}
\dot{y}_s = (-0.42u_{1_s} + 1.3u_{2_s})y_s - 0.07u_{1_s} + 0.63u_{2_s} \\[4mm]
v_s = -0.005y_s - 0.1
\end{array}
\right\}
\qquad (52)
$$

We will call it "quasi-diagonalised-singularly-perturbed" system (QSP) in the simulation results.

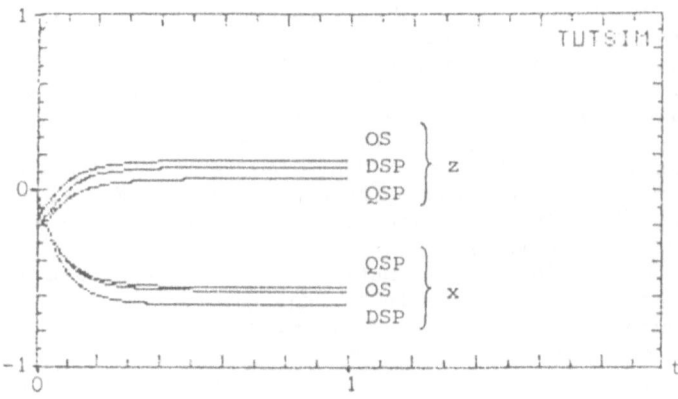

Simulation results of the example

$$u_1 = -1 \; ; \quad u_2 = -10 + 10\,t\,e^{-t}$$
$$x(0) = z(0) = 0$$

CONCLUSION

　　We pointed out, in this paper, the problem of the approximation of a bilinear system by reduced order bilinear models.

　　Determination of fast and slow state-variables is very difficult in the general case and specially here where the eigenvalues depend directly on the inputs. We have proposed a transformation which sets in a quasi-block-diagonal form the original ill-conditioned model. This basis change is solution of a "non-classical" Riccati equation, and minimizes the norms of the off-block-diagonal matrices. It points out the fast and slow components and leads them to a singularly perturbed model by introducing small parameters.

　　When the presence of small parameters is effective, we built up, by the singular perturbation method, a reduced order model where fast and slow parts are decoupled which leads to a parallel structure for analysis and control. But the reduced-order model is not linear for the inputs, in the general case.

　　The introduction of input-dominance allows to lead to bilinear approximant models; and we have shown that to implement, in the first place, the quasi-diagonalisation method makes the application of singular perturbations easier and faster.

REFERENCES

BRA.81　　Braudenbush, Stehle, Practical closed form solution for systems with variable eigenvalues via bilinear system theory. I.J. Circ. Theo. App., vol. 9, No. 1, pp. 103-114.

CEB.84　　Cebuhar, Costanza, 1984, Approximation procedure for the optimal control of bilinear and nonlinear systems. J. of Opt. The. and App., vol. 43, No. 4, pp. 615-627.

CHA.72　　Chang, 1972, Singular perturbations of a general boundary value problem. SIAM J. Math. Appl., vol. 3, pp.520-526.

CHO.76　　Chow, Kokotovic, 1976, Eigenvalue placement in two-time-scale systems. IFAC Symposium on Large Scale Systems, Udine, pp. 321-326.

DER.79　　Derese, Stevens, Noldus, 1979, Design of state observers for bilinear systems. J. d'Aut., vol. 20, No. 4, pp.193-202.

DER.82　　Derese, Noldus, 1982, Optimization of bilinear control systems. IJSS, vol. 13, No. 3, pp. 237-246.

GUI.80　　Guillen, Armada, A singular perturbation method for order reduction of large-scale bilinear dynamical systems. IFAC Symposium on Large Scale Systems, Toulouse, 24-26 June 1980, pp. 229-236.

KAR.78　　Karanam, Frick, Mohler, 1978, Bilinear system identification by Walsh functions. IEEE A.C., vol. 23, No. 4, pp.709-717.

KOK.75　　Kokotovic, 1975, A Riccati equation for block-diagonalization of ill-conditioned systems. IEEE A.C., vol. 20, 812-814.

KOK.76　　Kokotovic, O'Malley, Sannuti, 1976, Singular perturbations and order reduction in control theory. An overview. Automatica, vol. 12, pp. 123-132.

KOR.67　　Korganoff, Pavel-Parvu, 1967, Elements de theorie des matrices carrees et rectangles, Tome 2, Dunod.

MOH.73　　Mohler, 1973, Bilinear control processes. Academic Press.

MOH.80　　Mohler, Kolodziej, 1980, An overview of bilinear system theory and applications, IEEE S.M.C., vol. 10, No. 10, pp. 683-688.

O'MA.82   O'Malley, Anderson, 1982, Decoupling and order reduction for
          linear time-varying systems. Optimal Control Applications
          and Methods, vol. 3, pp. 133-153.
RIN.68    Rink, Mohler, 1968, Completely controllable bilinear systems.
          SIAM J. Control, vol. 6, pp. 477-486.
TZA.84    Tzafestas, Anagnostou, 1984, Stabilization of singularly
          perturbed strictly bilinear systems.   IEEE A.C., vol. 29,
          No. 10, pp. 943-946.

DYNAMIC PROGRAMMING: A PARALLEL IMPLEMENTATION

K. Malinowski[*]  and  J. Sadecki[**]

[*] Institute of Automatic Control, Technical University of
Warsaw, Warsaw, Poland
[**]Higher School of Engineering, Opole, Poland

ABSTRACT

Dynamic Programming algorithms can be implemented on parallel
computing facilities.  The efficiency of such implementation depends
upon the way in which parallel computing tasks are defined, number of
processing units and the capacity of a communication system.  Operation
of a multi-processor system is simulated in order to estimate the
efficiency of several parallel implementations of DP.

1.  INTRODUCTION

    Dynamic Programming (DP) can be used for solving a broad class of
optimisation problems for which Bellman's optimality principle holds.
The major shortcomings of this technique are due to memory requirements
and to the "curse of dimensionality" resulting in very large computing
effort required to solve dynamic optimisation problems.  Several specific
dynamic programming techniques have been developed in order to overcome
these shortcomings, but in general those techniques also require
significant computing effort and/or have limited applicability due, for
example, to convergence conditions (see, e.g., Larson and Korsak 1970).

    The new possibilities of an efficient implementation of dynamic
programming are offered by the use of parallel computing facilities.  In
this paper we consider parallel implementation of the basic, and the most
robust, DP algorithm.  The natural parallelism of this algorithm makes it
possible to split the computation into similar, numerous, parallel tasks
which can be executed on independent processing units.  In view of recent
developments in VLSI technology, memory limitations are no longer important.  One can also consider the multi-processor systems (of MIMD type)
with numerous parallel processing units.  Such systems are being presently
intensively developed (e.g. Dekker 1985).  In this situation the major
bottlenecks in parallel implementations of DP can occur because of a
limited capacity of communication system between the processing units.
This issue is considered in the next sections of this paper.   The
operation of a multiprocessor system is simulated taking into account the
interrupts and data transfers between the processing units.  The
evaluation of efficiencies of several parallel implementations of  DP
algorithm when used for solving a simple optimisation problem is presented.

This allows for comparison of these implementations and for selecting the number of parallel processing units to be used.

## 2. DYNAMIC PROGRAMMING

Let us consider the following discrete-time optimisation problem:

$$P: \quad \min Q = \sum_{k=0}^{K} L(\underline{x}(k),\underline{u}(k),k) + \Phi(\underline{x}(K+1)) , \qquad (1)$$

$$\underline{x}(k+1) = F(\underline{x}(k),\underline{u}(k),k) , \qquad k = 0,1,\ldots,K \qquad (2)$$

$$\underline{x}(k) \in X \quad R^n, \quad \underline{u}(k) \in U \quad R^m , \qquad \underline{X}(0) = \underline{c} , \qquad (3)$$

where $\underline{x}(k) = [x_1(k),\ldots,x_n(k)]^T$, $\underline{u}(k) = [u_1(k),\ldots,u_m(k)]^T$ .

The application of dynamic programming (DP) method is based upon the use of Bellman's optimality principle and results in solving the recursive equation:

$$\hat{S}(\underline{x},k) = \min_{u \in U}\{L(\underline{x},\underline{u},k) + \hat{S}(F(\underline{x},\underline{u},k), k+1)\} \qquad (4)$$

with $\quad \hat{S}(\underline{x},K+1) = \Phi(\underline{x})$ .

Depending upon a particular application the sets X and U may consist of finite numbers of points N and M (discrete systems) or of infinite numbers of points (continuous systems). In the latter case the values of $\hat{S}(\underline{x},k)$ and $\hat{u}(\underline{x},k)$ (where $\hat{u}(\underline{x},k)$ solves equation (4)) are computed for N mesh points in X. In this case the values of $\hat{S}(F(\underline{x},\underline{u},k),k+1)$ in equation (4) are obtained through an appropriate interpolation of the values of $\hat{S}(\underline{x},k+1)$ computed at a previous stage at the N mesh points. Therefore in both above cases the basic DP algorithm involves at each time stage k, $k = 1,\ldots,K$, the computation of N-tuples $\hat{\underline{S}}(k) = [\hat{S}(\underline{x}^1,k),\ldots,\hat{S}(\underline{x}^N,k)]$ and $\hat{\underline{V}}(k) = [\hat{u}(\underline{x}^1,k),\ldots,\hat{u}(\underline{x}^N,k)]$. Once these vectors are known (for $k = 1,\ldots,K$) the solution of problem P can be readily obtained. The major computing effort is required for computing $\hat{\underline{S}}(k)$, $k = 1,\ldots,K$. For large N this task can be overwhelming.

Several modifications of the basic PD technique have been developed in order to decrease the computing time, like: state increment method [Larson,1968], successive approximation technique [Larson & Korsak, 1970], decomposition methods [Collins 1970, Wong 1970], method of polynomial approximation [Bellman 1982], etc. These methods, however, are more complicated in programming and/or may not always be applicable (e.g. successive approximation technique).

The possibility of reducing the computing time by a considerable factor is offered by a parallel implementation of DP on a multi-processor system, (e.g. Casti et al 1973, Larson 1973).

In the next sections we consider several possible parallel implementations of the basic DP technique.

162

## 3. PARALLEL IMPLEMENTATION OF THE BASIC DP ALGORITHM

As mentioned in the previous section the major computing task within the DP algorithm is associated with finding the values of $\hat{S}(\underline{x},k)$ at each stage $k$ and at each discrete point $\underline{x}^j$ ($j = 1,...,N$) of $X$. Let us consider the case in which at each stage the optimal control $\underline{\hat{u}}(\underline{x}^j,k)$ is selected from $M$ possible values (either given $M$ discrete original values, or $M$ mesh points in $U$).

The solution of equation (4) can be represented by the following three nested loops:

(1) for $k = K,K-1,...,1$ compute $\underline{\hat{S}}(k)$, $\underline{\hat{V}}(k)$:

(2) when given $k$, for each $\underline{x}^j$, $j = 1,...,N$, compute $\hat{S}(\underline{x}^j,k)$

and $\underline{\hat{u}}(\underline{x}^j,k)$:

(3) when given $k$ and $\underline{x}^j$, for each $\underline{u}^i$, $i = 1,...,M$

compute $S(\underline{x}^j,\underline{u}^i,k)$

$$= L(\underline{x}^j,\underline{u}^i,k) + \hat{S}(F(\underline{x}^j,\underline{u}^i,k),k+1), \qquad (5)$$

compare* the values of $S(\underline{x}^j,\underline{u}^i,k)$

in order to find $\hat{S}(\underline{x}^j,k) = \min_{i = 1,...,M} S(\underline{x}^j,\underline{u}^i,k)$,

and $\underline{\hat{u}}(\underline{x}^j,k)$ .

Now it can be seen that the iterations at level (1) have to be done in a proper sequence ($k = K,K-1,...,1$) while all iterations at level (2) can be performed in parallel. At level (3) most of the computing (evaluations of equation (5)) can also be done in parallel. The above observation allows to propose the following parallel realizations of the considered DP algorithm when using a multi-processor system with $P$ parallel processing units (PU).

I. Parallel state algorithm (PSA):

(i)   each processing unit $p$ ($PU_p$), $p = 1,...,P$, computes

$\hat{S}(\underline{x}^j,k)$ and $\underline{\hat{u}}(\underline{x}^j,k)$ for $j \in J_p$

($J_1 \cup J_2 \cup ... \cup J_p = \{1,...,N\}$) .

(ii)  $PU_p$, $p = 1,...,P$, sends the values of $\hat{S}(\underline{x}^j,k)$, $j \in J_p$,

to the other PU-s; the values of $\underline{u}(\underline{x}^j,k)$ are stored in a

common (e.g. external) memory.

(iii) steps (i) and (ii) above are repeated for $k = K,K-1,...,1$.

---

\* When solving optimisation problems with continuous control variables $\underline{u}$ the discretisation of $U$ is not necessary and appropriate hill-climbing routines can be used to find $\hat{S}(\underline{x}^j,k)$.

II.   Parallel control algorithm (PCA):

   (i)   $PU_p$, $p = 1,\ldots,P$, computes $S(\underline{x}^j,\underline{u}^i,k)$ from eqn. (5) for

         every $j = 1,\ldots,N$ and for $i \in I_p$

         $(I_1 \cup I_2 \cup \ldots \cup I_p = \{1,\ldots,M\})$.

   (ii)  the PU-s exchange the data required to compare the values

         $S(\underline{x}^j,\underline{u}^i,k)$ so as to find $\hat{S}(\underline{x}^j,k)$, and $\hat{\underline{u}}(\underline{x}^j,k)$,

         $j = 1,\ldots,N$.   Each PU stores then the whole vector $\hat{\underline{S}}(k)$

         (i.e. the values of $\hat{S}(\underline{x}^j,k)$) in its memory.

   (iii) steps (i) and (ii) are repeated for $k = K,K-1,\ldots,1$.

III.  Parallel state and control algorithm (PSCA):

   (i)   $PU_p$, $p = 1,\ldots,P$, computes $S(\underline{x}^j,\underline{u}^i,k)$ from eqn. (5) for

         $(j,i) \in JI_p$

         $(JI_1 \cup \ldots \cup JI_p = \{1,\ldots,N\} \times \{1,\ldots,M\})$

   (ii)  the PU-s exchange the data required to compare the values

         $S(\underline{x}^j,\underline{u}^i,k)$ in order to find $\hat{S}(\underline{x}^j,k)$ and $\hat{\underline{u}}(\underline{x}^j,k)$,

         $j = 1,\ldots,N$.   Each PU stores the whole vector $\hat{\underline{S}}(k)$ in its

         memory.

   (iii) steps (i) and (ii) are repeated for $k = K,K-1,\ldots,1$ .


   The above algorithms can be considered as being synchronous – each
PU at given stage $k$ keeps in its own memory all values of $\hat{S}(\underline{x}^j,k+1)$
and the computations related to stage $k-1$ begin after all PU-s completed
the computing at stage $k$, exchanged the results and stored the new
vector $\hat{\underline{S}}(k) = [\hat{S}(\underline{x}^1,k),\ldots,\hat{S}(\underline{x}^N,k)]^T$.

   It is also possible to consider asynchronous versions of the above
algorithms. In this paper we consider only an asynchronous version of
the parallel state algorithm I:

IA:   Asynchronous parallel state algorithm (APSA):

   Each PU, say $PU_p$, computes for $k = K,K-1,\ldots,1$ the values of
$\hat{S}(\underline{x}^j,k)$ for $j \in J_p$. $PU_p$ keeps in its memory only the values of
$\hat{S}(\underline{x}^j,1)$ for $j \in J_p$, $1 = K-1,\ldots,k+1$. If, during the computation, $PU_p$
requires the value $\hat{S}(\underline{x}^j,k+1)$ for some $j \notin J_p$, say $j \in J_s$, then the
computing process is suspended and the $PU_p$ requests this piece of data
from the appropriate PU ($PU_s$). After obtaining the $PU_s$ suspends the
computing process, sends required data to $PU_p$ and then both PU-s resume
the computing.

   Practical implementations of the above algorithms and their effectiveness depend upon the computing power of the PU-s and upon the
communication system. It is beyond the scope of this paper to describe

possible architectures of multi-processor systems and the types of communication networks (see e.g. Kuck 1977, Dekker 1985). Here we consider a particular multi-processor system with a common bus as depicted in Fig. 1.

4. SIMULATION OF A MULTI-PROCESSOR SYSTEM OPERATION

In order to investigate the proposed variants of parallel dynamic programming techniques when implemented on a multi-processor system shown in Fig. 1, it is necessary either to build such a system or to simulate its operation on a serial machine. The latter approach allows to simulate the behaviour of multi-processor systems composed of the PU-s being presently used in mainframe computers and in minicomputers. The simulation program written with the purpose of investigating the proposed parallel DP techniques has been developed under the assumption that the system (of MIMD type), as shown in Fig. 1, consisted of a required number of PU-s (CPU with local bus and memory) as used in SM-3 minicomputer (equivalent to a PDP-35 machine). It was further assumed that the control processor (CP) was of the same type and that the PU-s were connected to a common bus of the same type as the bus in SM-3 minicomputer. It is also assumed that such extension of the SM-3 computer architecture will not cause the changes of the relevant parameters as the time required to realize interrupts and the time required to transmit the 16-bit data word over the communication bus. These times were estimated as, respectively, $T_{ss} = 1.60 \times 10^{-5}$s and $T_{pp} = 6.66 \times 10^{-5}$s.

Simulation of the operation of the multi-processor system involves book-keeping of times required to perform all arithmetic operations in the PU-s as required by the implemented algorithm and by summing up the times necessary to exchange data between the PU-s .

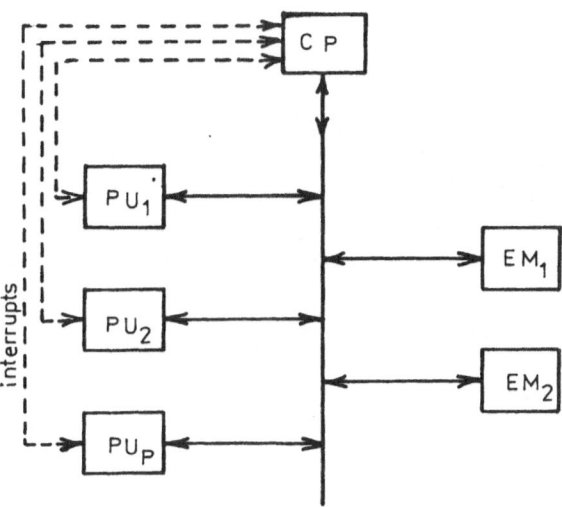

Fig. 1   Multi-processor system with a common bus.

In synchronous algorithms I, II, III the conflicts between the processors requiring access to the common bus are eliminated by prescribing a sequence according to which subsequent PU-s transmit their data. In an asynchronous case of algorithm IA it is assumed that the PU which requires access to the bus has to wait until the bus is free. If two or more PU-s require the access simultaneously then the priority is given to this PU which is nearest to the CP. In more detail the simulation of the considered multi-processor system is described in [Malinowski and Sadecki 1985].

The allocation of tasks to the PU-s depends upon the algorithm, numbers N and M, and upon the number P of parallel PU-s. For example, there are N basic parallel tasks of the same complexity (when each $J_p$ consists of one j) in the parallel state algorithm I and, since they can be allocated to at most P processing units, then the number of tasks performed at each PU is either int (N/P) or int (N/P)+1. Similarly, for algorithms II and III the number of tasks allocated to each of the PU-s is determined by, respectively, the factors M/P and N·M/P. In all cases the difference between the number of tasks allocated to different PU-s is not greater than 1. However, in cases when the overall number of basic tasks is slightly greater than P, several processors may be kept idle over considerable periods of time. It should be noted that with respect to this issue the parallel state and control algorithm III offers the greatest flexibility (N·M possible basic tasks) but this algorithm requires more communications between the PU-s when compared with the parallel state algorithm II.

## 5. EXAMPLE IMPLEMENTATION OF PARALLEL DP ALGORITHMS AND CONCLUSIONS

The developed simulation program was used to evaluate the efficiency of the considered parallel dynamic programming algorithms when used for solving the following optimisation problem:

$$\min Q = \sum_{k=0}^{K} 0.2(x^2(k) + u^2(k)) + 2.5(x(k+1)-2)^2$$

$$x(k+1) = x(k) + 0.2u(k), \qquad x(0) = c$$

$$x(k) \in X = \{x: \ 0 \le x \le 8\}$$

$$u(k) \in U = \{-2, -1.8, \ldots, 2\} \qquad (M = 21)$$

$$k = 0, \ldots, K, \quad K = 50$$

Two discretisations of the set of feasible state values X were considered:

Case 1: $\quad x^j = 0.8 \cdot (j-1), \qquad j = 1, \ldots, 11 \qquad (N = 11)$

Case 2: $\quad x^j = 0.2(j-1), \qquad j = 1, \ldots, 41 \qquad (N = 41)$

The above numbers M = 21 and N = 11 (or N = 41) determine the number of PU-s which can be used in parallel implementation of the considered algorithms:

```
Case 1:   Algorithm  I  and  IA-11
          Algorithm II          -21
          Algorithm III         -11x21 = 231

Case 2:   Algorithm  I  and  IA - 41
          Algorithm II          - 21
          Algorithm  III        - 41x21 = 861
```

Efficiency of parallel implementation is defined as a speed-up factor:

$$e = \frac{T_S}{T_R} ,$$

where $T_S$ - time required to perform computations on a single PU

$T_R$ - time required to perform computations on a multi-processor system.

Under the assumption that $P < N$ (or $P < M$) and that the workload on all PU-s is equal, the value of $e$ can be estimated as:

$$\hat{e}_{PSA} = \frac{NMTP}{NMT+P^3 T_{ss}+P^2 N} \tag{6}$$

for the parallel state algorithm I, and as

$$\hat{e}_{PCA} = \frac{MTP}{MT+(2P^2-P)T_{ss}+(3P^2-2P)T_{pp}} \tag{7}$$

for the parallel control algorithm II (Malinowski and Sadecki 1985). In the above formulas $T$ represents the time required to compute eqn. (5). For algorithm III similar estimate $\hat{e}_{PSC}$ of $e_{PSC}$ depends upon a particular allocation of tasks and for the algorithm IA it is difficult to compute $\hat{e}$ since the algorithm is asynchronous.

The simulation of the execution of parallel algorithms was done for both cases 1 and 2 and for different numbers of parallel processors available. The curves depicted in Fig. 2 and Fig. 3 present the speed up factor $e$ versus the number of processors for case 1 and case 2 respectively. Also approximate estimates $\hat{e}_{PSA}$ and $\hat{e}_{PCA}$ are as shown in order to compare them with the actual speed-up values. The curves representing $\hat{e}_{PSA}$ and $\hat{e}_{PCA}$ show that the parallel state algorithm I should be superior to the parallel control algorithm II. This is due to larger overheads required by algorithm II (more data transfers). The actual values of $e$ for different algorithms are below the optimistic estimates $\hat{e}_{PSA}$ and $\hat{e}_{PCA}$. The major reason is an uneven distribution of workload between the PU-s, some of them being kept idle over large portion of the overall computing time. For those values of $P$, for which the task allocation results in an almost equal workload on all PU-s, the speed-up $e$ is almost the same as the estimated value $\hat{e}$.

In such cases the increase of the number of engaged processors will result in the decrease of speed-up due to unbalanced workload distribution and increased data transfers. For parallel state and control algorithm III (which was simulated only for case 1) the maximum speed-up

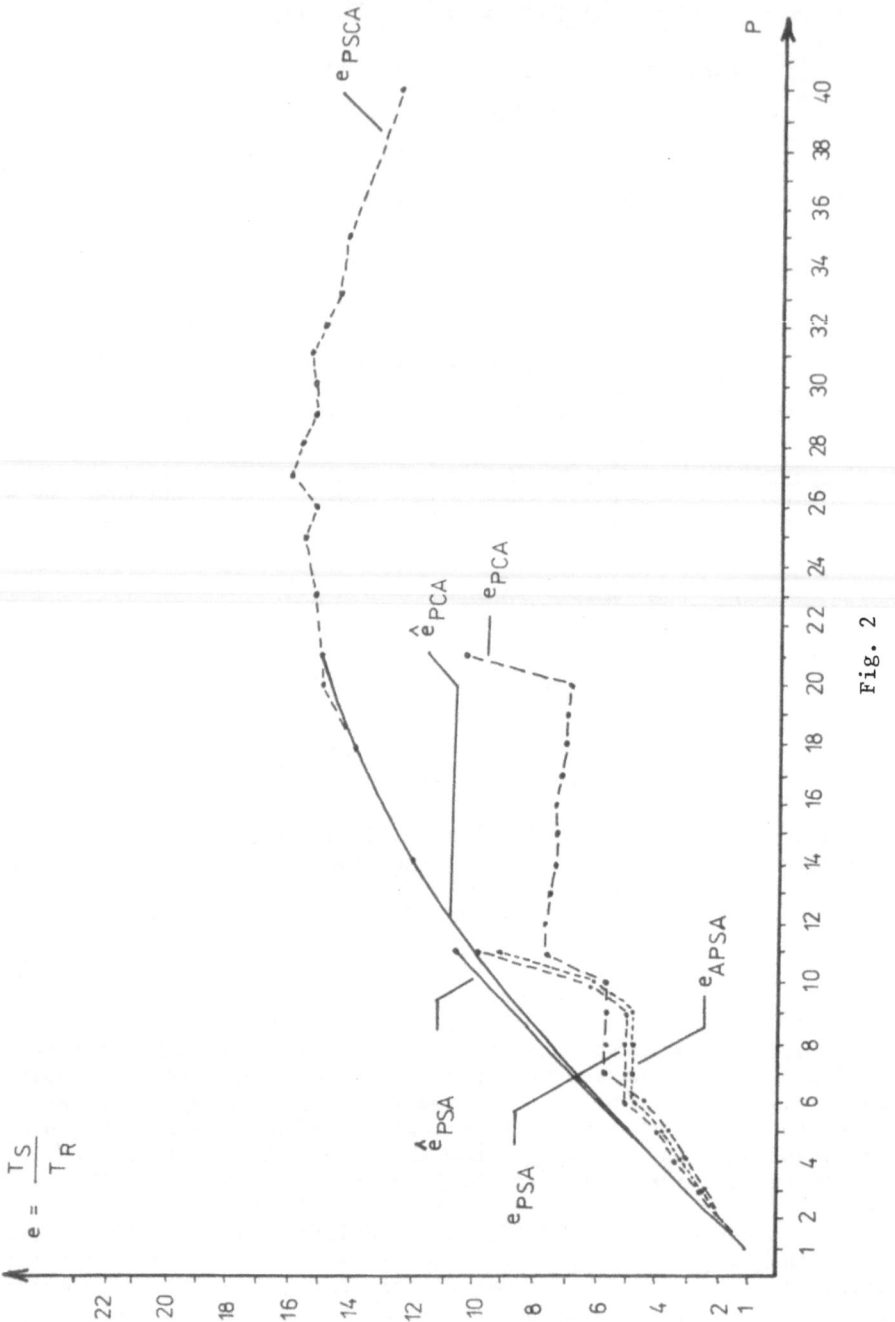

Fig. 2

168

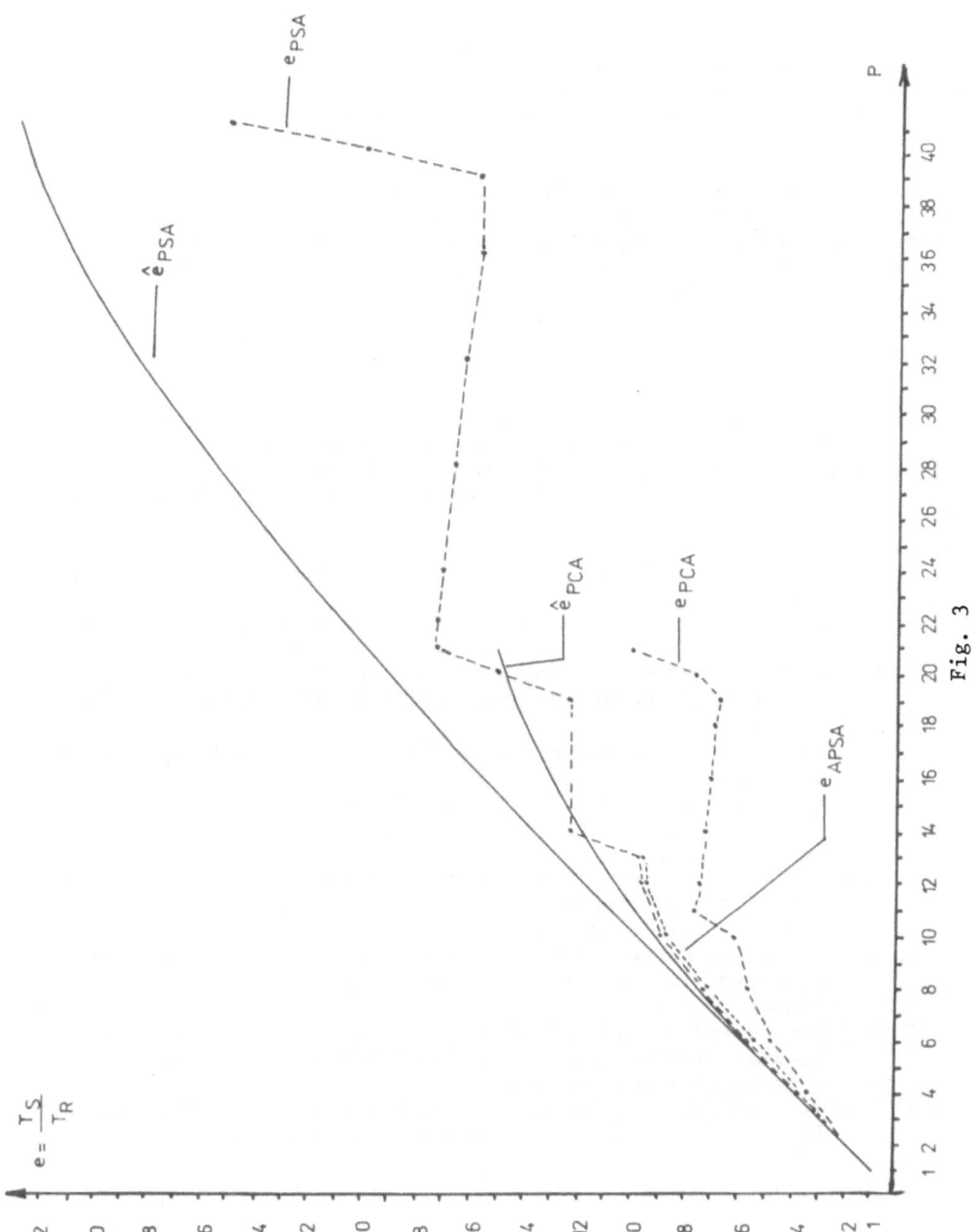

Fig. 3

of 16 was actually achieved for 27 PU-s. Faster communication system (smaller values of parameters $T_{ss}$ and $T_{pp}$) would result in a bigger number of PU-s which could be effectively used while on the other hand increasing the values of $T_{ss}$ and $T_{pp}$ would make it unreasonable (even for the parallel state algorithm) to use more than $P_0$ processors, where $P_0 < N$.

It is interesting to note that the synchronous and asynchronous parallel state algorithms provide for the very similar speed-up values and are better than the parallel control algorithm.

In order to summarize the above it may be observed that the obtained results show that parallel implementations of the DP technique can vastly improve computing time, yet one should expect that for a given optimisation problem, parallel algorithm and a parallel multi-processor system an optimal (maximal) number of the PU-s which can be effectively used can be smaller than the number of available PU-s.

REFERENCES

1.  R. Bellman (1963), Polynomial Appriximation - A New Technique in Dynamic Programming. Math. of Comput., vol. 17.
2.  J. Casti, M. Richardson and R. Larson. (1973), Dynamic Programming and Parallel Computers. Journ. Optimiz. Theory and Appl., vol. 12, Nr. 4.
3.  D.C. Collins (1970), Reduction of Dimensionality in Dynamic Programming by Method of Diagonal Decomposition. Journ. Math. Anal. and Appl., vol. 30.
4.  L. Dekker (1985), Methodology Based Parallel Digital Processors. Proceedings of the International Symposium on Systems Analysis and Simulation, Berlin, GDR, 1985, published in series on Mathematical Research, vol. 27, Akademic-Verlag, Berlin.
5.  O.J. Kuck (1977), A Survey of Parallel Machine Organization and Programming. Comput. Survey, vol. 9, No. 1.
6.  R. Larson (1968), State Increment Dynamic Programming. American Elsevier Publishing Comp., Inc.

7.  R. Larson (1973), Parallel Processing Algorithms for the Optimal Control of Nonlinear Dynamic Systems. IEEE Trans on Comput., vol. C-22, No. 8.
8.  R. Larson and A. Korsak (1970), A Dynamic Programming Successive Approximation Technique with Convergence Proofs. Automatica, vol. 6.
9.  K. Malinowski and J. Sadecki (1985), Parallel Implementation of Dynamic Programming Method, (in Polish). Archiwum Automatyki i Telemechaniki.
10. P.J. Wong (1970), A New Decomposition Procedure for Dynamic Programming. J. Operation Research, vol. 18.

ESL - ADVANCED SIMULATION LANGUAGE FOR PARALLEL PROCESSORS

J. L. Hay

Dept. of Electronic and Electrical Eng.

University of Salford, Salford, Lancs., U.K.

ABSTRACT

This paper reports the results of a software engineering project
entitled "Simulation Algorithms for Parallel Processes" which was
carried out under European Space Agency (ESA) Contracts 4790/81 and
5663/83.  The objective of the project was to produce computer programs
to implement an advanced continuous-system simulation language (CSSL).
The language, which has become known as ESL (ESA Simulation Language), is
characterised by its advanced programming concepts.  These include
separate program units to describe the system and the experiment to be
performed on it;  modular model concepts in the form of submodels to
define independent parts of the system;  a segment facility which allows
sections of the system to be simulated on a parallel processor emulation;
techniques for conveniently describing and handling system discontinuities;  and modern programming structure features with comprehensive
procedural code facilities.  The implementation provides both an
interpreter version for fast turn-round of simulation programs under
development, and a translator version for efficient production runs of
developed programs.

1.  INTRODUCTION

This is a report of a computer software development project entitled
"Consolidation and Extension of the ESL Package" which was carried out in
the Computer Simulation Centre, Department of Electronic and Electrical
Engineering, University of Salford, England, on behalf of the European
Space Research and Technology Centre (ESTEC) at Noordwijk, The
Netherlands.

The objectives of the project were to consolidate and extend ESL
(ESA Simulation Language) an advanced continuous-system simulation
language (CSSL).  This language facilitates the computer simulation, or
modelling, of dynamic systems.  Computer simulation being the technique
of using computers to represent, often in great detail, the performance
of real systems.  The systems involved are often very complex (e.g.
space-craft, control systems, mechanical and electrical plant) and
simulation may be used as an aid to design, to diagnose the cause of
system maloperation, to evaluate performance under fault conditions, for

operator training or simply to gain a deeper understanding of system performance.

The ESL simulation language is characterised by its advanced programming concepts.  These include separate program units to describe the system and the experiment to be performed on it;  modular model concepts in the form of submodels to define independent parts of the system ;  techniques for conveniently describing and handling system discontinuities;  parallel segmentation concepts to enable models to be partitioned and executed concurrently on a multiple-processor emulation; and modern programming structure features with comprehensive procedural code facilities.  The implementation provides both an interpreter version for fast turn-round of simulation programs under development, and a translator version for efficient production runs of developed programs.

The following section presents the background to the development of ESL.

## 1.1  Project Background

In 1981 a group working in the Computer Simulation Centre at the University of Salford, England, published reports (1), (2), (3) and (4) containing outline proposals for a new CSSL specification to replace the 1967 SCi CSSL specification (5).  To distinguish between the new and the old, the new language was called CSSL81 and the original SCi language, CSSL67.  One aim of these proposals was to stimulate discussion within the simulation user community.  The proposals were widely circulated, and in particular were submitted to two International Committees which were studying CSSL standards.  These were Technical Committee TC3 on simulation software of the International Association for Mathematics and Computers in Simulation (IMACS), and the CSSL Committee of the Society for Computer Simulation (SCS).  Valuable feedback emerged from the interest and discussion which the proposals provoked, and the contents of the outline proposal (1) were adopted as the basis for the new language which was developed by the Salford group under ESA contract 4790/81 (6).

The requirements for that project were to design and develop computer programs to implement a continuous-systems simulation language to facilitate the simulation of dynamic systems on modern computers. The ideas presented in the outline software specification (1), and known as CSSL81, did not themselves contribute a complete language specification.  They formed only an outline containing a number of ideas, but leaving numerous questions of detail unspecified.  These ideas provided the basis for defining the requirements to be met in the development of ESL.

It was recognised that the original CSSL67 specification, on which most current simulation languages are based, has many good, well con- structed features.  The CSSL81 proposals attempted to maintain a balance between two contradictory aims.  On the one hand the need to make radical changes in some aspects of CSSL specification, for example the submodel features and the programming of discontinuous events;  on the other hand, the need, as far as it is reasonable, to maintain continuity. These proposals should be seen as evolutionary rather than revolutionary.

The programs which implement the language were designed to provide engineers with a straightforward and concise means of specifying simulation experiments for computer execution.  The user is relieved of the need for an in-depth knowledge of the computer system and the many house-keeping tasks which are normally associated with computer simulation exercises. Mechanisms for good man/machine interaction allow users considerable control over the execution of the simulation problem. An implementation requirement was to produce both an interpreter version for fast turn-round of simulation programs under development, and a translator version for efficient production runs of developed programs. With the interpreter approach the user's program (the simulation specification) is converted into an intermediate code (H-code) which is then interpreted at run-time to perform the specified simulation. The translator, on the other hand, may be used to convert the H-code to FORTRAN 77 to produce a more efficient executable program.

As a result of this contract, a prototype version of the language, which was then named ESL, emerged in 1983.

This paper briefly describes certain aspects of the work of a further ESA contract (5663/83) which began early in 1984 to consolidate and extend the prototype ESL language. The documents (7),(8),(9),(10), (11),(12) and (13) present a complete source of reference for the work undertaken in this project.

1.2  Project Objectives

The consolidation and extension of ESL was undertaken in two distinct phases. In the first phase, ten simulation applications selected from different fields were simulated using ESL. The purpose of this exercise was three-fold. First, to determine whether the facilities provided by ESL were sufficiently comprehensive to undertake a wide range of simulation problems. Second, to correct and extend ESL in the light of problems encountered in performing the simulation applications, and finally to identify a suite of submodels that are common to many simulation problems in order to provide a comprehensive library of submodels. Reference (7) contains a full description of all the applications undertaken, and includes extensive program examples.

The second, extension, phase of the project was mainly directed to introducing the concept of parallel segments into ESL. A segment being a program structure that represents part of a real system that is to be simulated concurrently on a separate processor of a multiprocessor computer system. ESL is currently executed on a conventional sequential processor, and for this project the multiprocessor environment was emulated on a conventional machine. Not only do segments allow the simulation performance of a true multiprocessor machine to be predicted, but they offer many advantages to the simulationist in a conventional processor environment.

The next section presents a brief description of the ESL programming language, a full specification of the language may be found in (8).

2.  OUTLINE DESCRIPTION OF ESL

The main features of ESL are best illustrated by an example. Fig. 1 gives a listing of a program used for testing ESL which uses many of the basic features of the language.

```
0001    --Benchmark for ESL
0002    --
0003    STUDY
0004    --
0005    SUBMODEL INTGL(REAL: out:=REAL: in, IC);
```

```
0006    INITIAL
0007      out:= IC;
0008    DYNAMIC
0009      out':= in;
0010    END INTGL;
0011    --
0012    MODEL VANDERPOL(Real: X,Y:= Real: K);
0013      CONSTANT REAL: XDO/0.0/;
0014      REAL: YD/0.0/,F;
0015      REAL: Z,ERROR;
0016    --
0017    INITIAL
0018      Y:= 0.1;
0019      X:= 0.1; X':= XDO;
0020      PRINT "Model Entry, K =    ",K;
0021    DYNAMIC
0022    --NOTE X,Y and Z are equivalent
0023      F   := K*(1 -Y**2);   --Driving function
0024      YD':= F*YD-Y;
0025      Y'  := YD;
0026      X'':= K*(1-X*X)*X'-X;
0027      Z   := INTGL(YD,0.1);
0028    COMMUNICATION
0029      TABULATE T,X,X',X'';
0030      TERMINATE T >= 10;
0031    TERMINAL
0032      ERROR:= Z-Y ;
0033      PRINT "Error is    ",ERROR;
0034    END VANDERPOL;
0035    --
0036    --EXPERIMENT
0037    --
0038    REAL: X,Y,K;
0039      ALGO:= RK4; CINT:= 1; NSTEP:= 10; TFIN:= 20;
0040      K:= 1;
0041      PRINT "MODEL CALLED WITH K =  ",K;
0042      VANDERPOL(X,Y := K);
0043      PRINT "FINAL VALUES OF X AND Y ARE: ",X,Y;
0044      PRINT "STUDY COMPLETED";
0045    END_STUDY
```

Fig. 1  ESL Example Program

Note that the listing in Fig. 1 contains line numbers for ease of
identification but these are not part of the program.

2.1  Program Structure

The ESL program is called a 'study', which begins with STUDY (line 3)
and ends with END_STUDY (line 45) statements.  A study comprises one or
more models, one experiment and, optionally, one or more submodels,
procedural subprograms or segments.

The general approach adopted in the layout of the program is that
all entities are defined before being used.  The ability to build models
from submodels is a key feature of ESL and the example illustrates this
feature.  A submodel with the name INTGL is defined in lines 5 to 10.

Note the general form of the submodel with the declaration of the
submodel output and input arguments in line 5 (separated by the ':='

174

symbol ), the inclusion of initial and dynamic regions and the END <name>
statement in line 10.  The submodel which defines a simple integrator,
consists essentially of the simple differential equation (out':= in)
contained in line 9.  Note the manner of writing differential equations
and the use of ':= ' for assignments.  The differential equation forms
the dynamic region of this submodel and the state variable (out) is
initialised in the initial region by the statement in line 7 to the value
IC.

The declaration and definition of the submodel is followed by the
model definition introduced by the 'model' statement in line 12.  As
with the 'submodel' statement, this includes the name of the model and a
list of output and input arguments.  A model definition is similar in
most respects to that of a submodel.  It starts with a model declaration
statement which specifies the name and outputs and inputs of the model.
Declarations follow, and finally the model body, which must contain at
least a dynamic region, and may also include initial and terminal regions.
The dynamic region may optionally contain a communication and step sub-
region.  Calls to previously defined submodels may be included in the
dynamic region.

Further submodels and models could be defined at this point, but our
example contains only one model and a submodel.  The remainder of the
study is the 'experiment', which defines the operations to be performed
on the model.

## 2.2  Model Statements

The model statements in the dynamic region, which describe the
physical system being modelled, may include special statements to describe
discontinuities as well as continuous relationships.  A discontinuity is
some event which causes the (continuous) equations describing the system
to suddenly change, that is, they are subjected to a step-change. Normally,
discontinuities cause inaccurate and/or a very slow simulation.   ESL,
however, incorporates an efficient and accurate mechanism to process
discontinuities.

The 'if' clause is an ESL structure for selecting one of a number of
expressions to be assigned to a variable.  The selection, or switching
operation, is based on one or more logical conditions.  The logical conditions describe the points, or discontinuities, at which the switching
operations occur.  For example,

```
y:= if Xin >= Limit then Limit else Xin;
z:= if Xin >= Limit then Limit
    else_if Xin < -Limit the -Limit
    else Xin;
```

In processing the above statements the integration process accurately
locates the point at which the discontinuity, or change, takes place.
For example, when Xin becomes greater than Limit.  This is achieved by
repeating the integration step, which spans the discontinuity, using an
interpolating method to control the step-length to ensure that the
discontinuity will be at the end of the step.  Only after this point has
been located to a specified accuracy will the effect of the discontinuity
be allowed to take place.  Discontinuities, therefore, are forced to lie
between integration steps.

A second ESL statement is provided to assist in the specification of
discontinuities.  This is the 'when' statement which enables action to be
taken at the instant some 'trigger' event occurs.  As with the 'if' clause,

the integration process is controlled to ensure that the event coincides
with an integration step boundary.

```
when x > U1 then
   STATE:= TRUE;
   Y:= 1;
when x < L1 then
   STATE:= FALSE;
   Y:= 0;
end_when;
```

Note the difference between the 'if' and 'when' constructs. The statement
controlled by an 'if' clause is always executed, and a variable is given
a value which depends on the specified conditions. The statements in a
'when' block are executed only at instants at which the trigger event
occurs.

## 2.3 Other Statements

A comprehensive suite of procedural language statements are provided
in ESL which include structured 'if' and 'while' blocks, as well as
simulation orientated input/output statements. This includes:

Print Statements
Tabulate Statement
Prepare statement
Plot statement
Read statement
Interact statement

Seven integration algorithms are provided which include special
stiff-system procedures.

## 3. SEGMENTS

ESL provides facilities to emulate separate processors or complete
computers, which operate concurrently and are synchronised to communicate
after fixed intervals of time.

The ESL program segment structure provides the means of partitioning
a system specification for simulation into a multiprocessor environment.
A complete dynamic system may be specified as being partitioned into a
model, which is regarded as the master segment, and one or more segments.
The model and each segment describe a subset of the complete system and
exchange data after fixed time intervals (the simulated time of one
communication interval).

The underlying concept is that the simulation of the model partition
is proceeding on one processor at the same time as the simulation of
each segment partition on other processors. When the model and all
segment simulations have completed one communication interval, data is
exchanged between the processors and only after this data exchange, or
synchronisation point, is the next communication interval started. The
model, or master segment, initiates the simulation of each segment, and
it is through the model that all data is exchanged between the various
segments.

The concurrent execution of segment simulation is emulated by ESL,
however the result is the same as that which would be obtained in a true
multiprocessor situation.

The program in Fig. 2 presents a simple example which introduces the basic programming concept and the underlying operation.

As with all ESL modules the segment must be specified, or declared, before its invocation in the model. The structure of the segment code is the same as that for a model except that it cannot have a terminal region. The initial region of the segment is used to set the simulation control variables for that segment. That is, the integration algorithm (ALGO), number of sub-steps to complete a communication interval (NSTEP) and the error specifications (INTERR and DISERR). It should not set communication interval (CINT) or the simulation start time (TSTART) as this is the function of the model. The simulation control variables for each segment are unique, and the model and other segments may select alternative values for their control variables. The default values for simulation control variables are the standard defaults and not necessarily those of the model. ESL maintains a separate set of simulation control variables for each segment. In contrast, the simulation control variables for the model may either be set in the experiment or the model's initial region. The model or experiment also has the task of setting the communication interval (CINT) and start time (TSTART) which are common to all segments. The final time (TFIN) should be set by the model, but different TFIN set by a segment can cause the simulation run to stop earlier. The simulation run may also be stopped prematurely by terminate statements in any segment.

The segment invocation must always be in the model communication sub-region and the calling statement has the same form as a model invocation in the experiment. Several segments may be called from the communication region but care must be taken to ensure that they are correctly ordered. If segment A has an input which is produced by segment B then the invocation of segment B must appear before that of segment A. The segments must always be called in every pass of the communication region ('if' statements should not prevent this requirement).

When the same segment is invoked more than once, the second and subsequent invocation are regarded as calls to separate segments, executed on different processors, but with the same segment specification. ESL's dynamic storage allocation used for segments and submodels ensures that no conflicts arise between the separate invocation of the same segment. In these cases, the input/output arguments of the segment invocation should, of course, be different.

On each pass of the model's communication region, all segment invocations must occur in the same order. If this condition is violated (not checked by ESL), unpredictable results will occur!

Examination of the example in Fig. 2 reveals that the model solves a subsystem which produces the result (y) of an exponential response, first increasing from zero and then, at time 6.0, decaying back to zero. Note the use of 'include' statements to access the library submodels (2) REALPL (real-pole), INTEG (integrator) and STEPP (step input at T = 6.0).

The value of y is passed to the segment SEG as an input argument. The output from the segment invocation is x, which is also used in the model code of the dynamic region. Therefore x must be initialised prior to entering the dynamic region for the first time. This initialisation is only effective during model initialisation and is correctly updated, by the segment invocation before the simulation commences.

```
STUDY
INCLUDE "REALPL";
INCLUDE "INTEG";
INCLUDE "STEPP";
--
SEGMENT SEG(REAL: segout:= REAL: segin,Taus);
INITIAL
  NSTEP:= 20;
  ALGO:= RK4;
DYNAMIC
  segout:= REALPL(0.0,Taus,segin);
STEP
--  PREPARE "SEGS1",T,segout,segin;
END SEG;
--
MODEL MOD(REAL: y:= REAL: Tau);
  REAL: x,xf,in;
  REAL: Tauf/0.6/;
  LOGICAL: log;
INITIAL
  x:= 0.0;
  NSTEP:= 15;
DYNAMIC
  log:= STEPP(6.0);
  in:= if log then 0.0 else 1.0;
  y:= INTEG(0,0,(in-y)/Tau);
  xf:= REALPL(0.0,Tauf,x);
STEP
  PLOT T,y,0,TFIN,0,1;
  PREPARE "SEGM1",T,y,x,xf;
COMMUNICATION
-- Segment invocation
  SEG(x:= y,Tauf);
END MOD;
-- EXPERIMENT
  REAL: y,Tau/2.0/;
  ALGO:= RK5;
  TFIN:= 16.0;
-- Model invocation
  MOD(y:= Tau);
END_STUDY
```

Fig. 2  Segmentation example

The segment takes y as its input SEGIN and subjects this to filtering by a real-pole, or lag, to produce the segment output (SEGOUT).

The model receives the segment output in its variable x which it filters by a real-pole to produce xf.

Figure 3 shows the results from the model's point of view. The output from the segment (x) is updated at each communication interval and therefore has the 'staircase' characteristic shown in the graph.

Figure 4 shows the results from the segment's point-of-view. The input to the segment SEGIN is updated at each communication interval and it also has the 'staircase' characteristic.

This example clearly shows the fundamental characteristics of

178

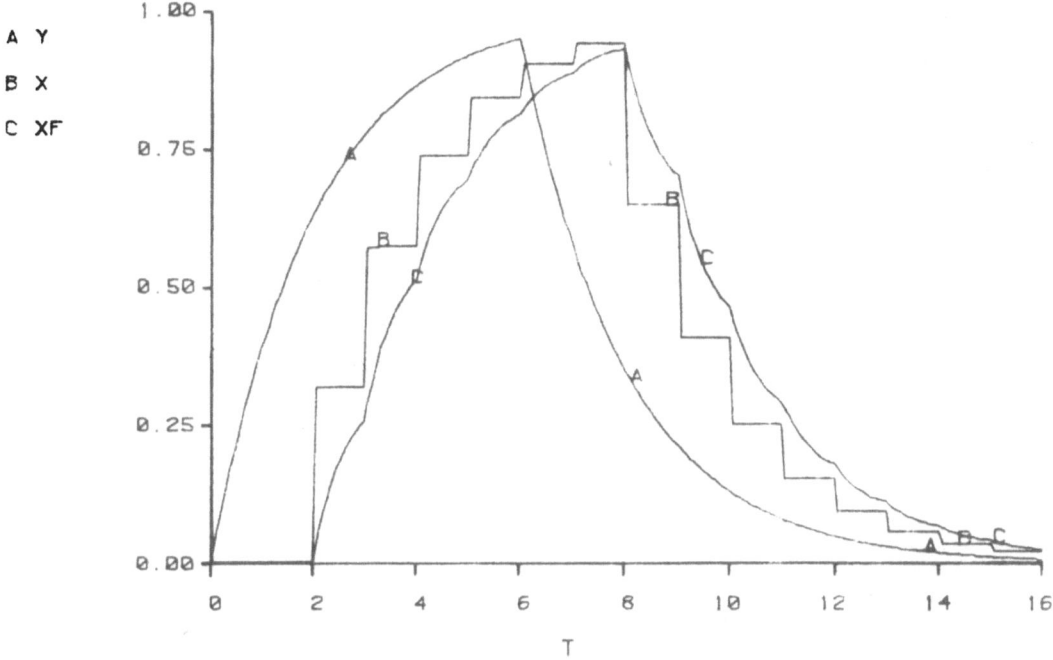

Fig. 3  Output from the model

parallel processing.  That is, the discretisation of data communicated between processors, and the lag, or delay, inherent in this process. In effect, the example passes the value of y to a segment which simply returns it.  A comparison of y and xf shows the effect of communication of data to a segment and then the communication of the segment response back to the model.  That is, there is both a phase and amplitude error which is dependent on the communication interval.

If proper account is taken of this fundamental operation, segmentation can be used to produce very efficient results even on a single processor segment emulation.  The communication bandwidth, determined by the communication interval and the frequency (or rate of change) of data to be communicated, determines the error introduced.  For example, if the period of the highest frequency information communicated between segments is Tp then a communication interval of CINT = Tp/36 will introduce a phase-lag of approximately 10 degrees as the data is transferred between segments.  If the data is processed and returned to the originating segment a further phase delay of 10 degrees will occur.

The following is simply presented to give the user an insight into how ESL processes segments on a single processor.  This information is not needed in order to program segmented parts.  When a segment is invoked during a simulation run, processing starts at the segment's communication interval.  After processing the communication sub-region

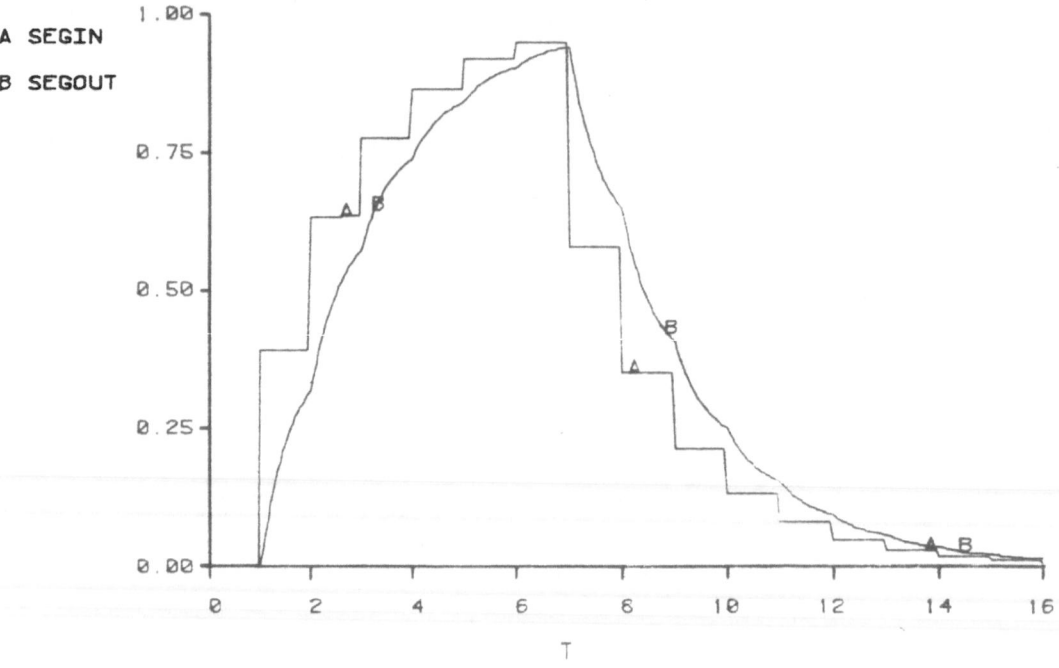

A SEGIN

B SEGOUT

Fig. 4   Output from the segment

code, the results at the current value of time (T) are known and these
are the results (e.g. SEGOUT) to be returned to the model.  The segment
saves these results for subsequent transmission to the model and then
proceeds to simulate its sub-system over the next communication interval
(i.e. to time T+CINT).  At this point, the segment simulation is ahead
of the model.  The new results at time T+CINT are saved, and the results
previously saved at time T are returned to the model.  The model then
'catches-up' with the segments by simulating to the segment, the results
for the current time which were calculated in the previous invocation,
are re-instated and the segment then simulates the next communication
interval.

The parallel segment feature allows fifth-generation computer
concepts to be explored.  This enables a physical system to be divided
into several large sub-systems, or segments.  Each segment can be considered as being concurrently simulated on separate processors which
interchange data after predefined 'frame' intervals.  Of course, ESL
currently emulates the multiprocessor environment to give the same
results as would be obtained in a true multiprocessor situation.

The concept of a segment is not restricted to conventional
simulation, but allows the possibility of a segment representing a
controlling microprocessor, or in fact (in a real-time environment) the
real hardware.  The important point is that ESL has been designed to
incorporate these advanced features, and the appropriate segment interface is already built-in.

## 4. CONCLUSIONS

The purpose of the project described in this report has been to consolidate and extend ESL (ESA Simulation Language) a continuous-systems simulation language (CSSL). At the start of the project a prototype version of ESL existed, and during the course of the project ESL has been extended and improved to a fully operational, commercial-standard, simulation software package.

ESL is characterised by its advanced programming concepts. These include:

- separate program units to describe the system and the experiments to be performed on it;
- modular modelling concepts in the form of submodels and segments to define separate parts of the system;
- techniques to conveniently describe and process discontinuities accurately and efficiently;
- multiprocessor segment emulation facilities allow fifth-generation computer concepts to be explored, and provide an interface for possible 'hardware-in-the-loop' simulations;
- modern programming structural features with comprehensive procedural code facilities.

The introduction of parallel segmentation facilities allows separate partitions, or segments, of a physical system to be simulated on an individual processor of a multiprocessor computer. ESL, which currently runs only on a conventional computer, emulates the multiprocessor system. Even on a conventional processor, segmentation offers speed and organisational advantages for certain large scale problems. Software segmentation is also the first step to 'real-time' operation which can lead to the incorporation of 'hardware-in-the-loop' simulations.

ESL is a new advanced CSSL incorporating many new features, most of which have been strongly advocated in recent debates about the future directions for CSSL's. This project has enabled these modern concepts to be brought to reality by the provision of a complete, fully operational, simulation language.


## 5. ACKNOWLEDGMENTS

The author acknowledges the assistance and encouragement of colleagues J.G. Pearce and L. Turnbull at Salford, and M.G. Guerin, U. Mortensen and S. Mejnertsen of ESTEC, Noordwijk, Netherlands, in the development of ESL.

## 6. REFERENCES

(1)   Hay, J.L., Crosbie, R.E. and Narotam, M.D., Outline software specification, Technical Memo ES81/1, ESA Contract 4155/79, Dept. of Electronic and Electrical Engineering, University of Salford, March 1981.

(2)   Hay, J.L., Crosbie, R.E. and Narotam, M.D., Outline proposals for a New Standard for Continuous System Simulation Languages (CSSL81), Simulation Laboratory Report, Dept. of Electronic and Electrical Eng., University of Salford, March 1981. Republished in TC3-IMACS Newsletter 9, May 1981.

(3)   Crosbie, R.E. and Hay, J.L., Towards new standards for continuous system simulation languages, Proc. 1982 SCSC, pp. 186-190, Denver, Colorado, July 1982.

(4)     Hay, J.L.,  A New CSSL Standard – an implementation view,  UKSC
            Conference on Computer Simulation, Harrogate, 1981.

(5)     The SCi Continuous System Simulation Language,  Simulation, vol. 9,
            No. 6, Dec. 1969.

(6)     Hay, J.L., Crosbie, R.E. and Pearce, J.G.,   Simulation algorithms
            for parallel processes,   Final Report ESA Contract 5663/83,
            Report CSC-1023/00, Dept. Electronic and Electrical Eng.,
            University of Salford, Nov. 1983.

(7)     Hay, J.L., Pearce, J.G., Parke, K.E., Turnbull, L. and
            Crosbie, R.E.,   ESL Application Manual,  Report No.
            CSC-1024, ESA Contract 5663/83, SUIC, March 1985.

(8)     Hay, J.L., Pearce, J.G., Turnbull, L. and Crosbie, R.E.,
            ESL Software User Manual,   Report No. CSC-1019, ESA
            Contract 5663/83, April 1985.

(9)     Hay, J.L., Pearce, J.G., Javey, S. and Crosbie, R.E.,
            ESL Software requirements document,  Report No. CSC-1005,
            ESA Contract 5663/83, April 1985.

(10)    Hay, J.L. and Crosbie, R.E.,   ESL Architectural Design Document,
            Report No. CSC-1006, ESA Contract 5663/83, April 1985.

(11)    Hay, J.L. and Crosbie, R.E.,   ESL Software Transfer Document,
            Report No. CSC-1021, ESA Contract 5663/83, April 1985.

(12)    Hay, J.L., Pearce, J.G. and Turnbull, L.,   ESL Detailed Design
            Document,  Report No. CSC-1022, ESA Contract 5663/85,
            May 1985.

(13)    Hay, J.L., Pearce, J.G. and Turnbull, L.,   Consolidation and
            Extension of ESL,   Final Report, Report No. CSC-1026,
            ESA Contract 5663/85, June 1985.

# SURVEY OF PARALLEL PROCESSING IN SIMULATION

J.G. Pearce, P. Holliday and J.O. Gray

Department of Electronic and Electrical Engineering
University of Salford
Salford, Lancs

ABSTRACT

This paper explores the possibilities of advanced fifth-generation parallel computing applied to continuous-system simulation by examination of current philosophies, and their supporting hardware and software. Already these advanced techniques are being applied in simulation, and this paper examines both commercially available approaches, typified by attached array processors such as Floating Point Systems AP-120B and specialised simulation computers such as Applied Dynamics AD 10, together with more experimental projects such as ESA's MPRS. The specific projects are preceded by a review of parallel processing which particularly refers to the requirements of on-line real-time simulation as well as presenting a general discussion of the more conventional approaches. This is followed by a detailed examination of the various approaches which have been taken to exploit parallel processing. Finally, consideration is given to the software aspects of parallel processing, in particular the implications on Continuous System Simulation Languages.

## INTRODUCTION

Continuous-system simulation makes ever increasing demands on digital computing speed. This arises from the need to simulate very large dynamic systems and the requirement of real-time simulation. Faster computers are required to generate solutions to large problems within a reasonable time scale. In real-time, equipment-in-the-loop simulations, high computing speeds are necessary to complete complex and extensive calculations within a short time-frame. In the past, analogue and hybrid computers were used where high speed solutions of large sets of ordinary nonlinear differential equations were required. Indeed, hybrid computers are still to be found in certain areas of industry, notably the aerospace industry. The high speed of such machines stems from the fact that individual computing elements-integrators, summers, etc.- operate in parallel. Thus all the equations are solved in a parallel manner, the maximum speed of solution being dictated by the bandwidth of the analogue components and the size of problem that can be accommodated, limited only by the number of available amplifiers. It is satisfying to those engineers who carried out their first simulations using analogue computers to see the ideas of parallel computation, inherent in such machines, now being incorporated into digital computers.

This paper identifies the computational requirements of continuous-system simulation in a parallel processing environment and surveys a number of currently available parallel computers and assesses their suitability for simulation. Consideration is also given to the software aspects of simulation using parallel processors and the availability of suitable programming languages. First, however, a brief review of the classification of parallel processing techniques and types of parallel processor is presented. This is followed by a look at the particular computational requirement of continuous-system simulation.

## CLASSIFICATION OF PARALLEL PROCESSORS

Parallel processing techniques are generally classified in terms of instruction-stream and data-stream. Schemes in which a common instruction is executed simultaneously on multiple data is termed Single Instruction stream, Multiple Data stream (SIMD). The ICL-DAP characterises such systems. Alternatively, where several different instructions are executed in parallel on common data, the term Multiple Instruction stream, Single Data stream (MISD) is used. Array processors with multiple floating point multipliers and adders used in parallel to evaluate expressions fall into this category. The most powerful parallel processing scheme is, however, the Multiple Instruction stream Multiple Data stream arrangement (MIMD) where, in effect, an array of complete computers, each solving a part of the whole problem, communicate data between themselves. There are many experimental MIMD computers and GOULD's SCI-CLONE/32 is an example of a commercially available system. Other parallel processing techniques, such as pipelining, are common in most conventional computers.

In this paper, parallel processors are examined under the categories general purpose super-computer, host-peripheral array processor and special purpose simulation processors with a look ahead to procedures based on emerging VLSI technology.

## SIMULATION REQUIREMENTS

Essentially the computer simulation of a dynamical system requires the numerical solution of a set of coupled nonlinear ordinary differential equations. This requires the use of a number of computational techniques including numerical integration algorithms, methods of dealing with discontinuities, matrix techniques, function generation and optimisation methods.

There are a large number of numerical integration algorithms and an extensive literature exists on this topic. However, the Runge-Kutta methods appear to be most popular amongst simulation workers since they offer a robust single-step method. Fixed-step and error controlled variable step methods are available. A further attraction of fixed-step methods is the ease with which they may be adapted to real-time applications.

Where the equations to be solved are stiff, an alternative implicit method may be more appropriate. Such methods often require evaluation of the Jacobian matrix, hence the importance of sparse-matrix techniques.

Many physical systems incorporate features which are conveniently modelled by discontinuous relationships (e.g. switches, limits, backlash, etc.) and an efficient method of handling such relationships is essential to efficient simulation. Various approaches to this problem are possible, the most efficient involves interpolation techniques to pin-point the discontinuity and adjust the integration step length accordingly [1,2].

Often a simulation requires the generation of a nonlinear function of several variables. Typically such a function will be derived from empirical data. Function generation is so important in simulation that special purpose digital processors have been developed for the sole purpose. The AD10 simulation processor (described later) started life as such a device.

All the techniques described above are used in the solution of one run of the simulation, i.e. a solution of the equations over a fixed time period. It is often necessary to repeat the simulation a number of times for example, to optimise certain parameters. Thus consideration must be given to optimisation methods.

Although to a certain extent, Signal Processing and Simulation applications have similar hardware and software requirements, significant differences exist in the design of special-purpose array processors for either procedure.

Signal processing applications typically involve long vectors of data and relatively short computations. An array processor (AP), there-fore, will have a high I/O rate, small program memory, and fast arithmetic units mainly consisting of multipliers and adders. Because the data is arranged in long vectors, very high throughput rates can be achieved. Notwithstanding some notable exceptions, most manufacturers have designed their APs with the above requirements in mind [3].

Simulation applications, however, demand a different architectural philosophy. One of the most fundamental differences is that data will rarely be arranged as a long vector, but will be subjected to relatively long and laborious computations. Large program memories are therefore needed whilst high I/O rates are not as crucial as for signal processing. Large table memories are required for function generation which, together with the corresponding interpolation calculations, are the most time consuming tasks in a simulation process [4,5].

Integration procedures, such as Runge-Kutta, require high-precision floating point word formats to avoid significant accumulation of round-off errors. The same applies to the simulation of distributed parameter systems, characterised by partial differential equations. It frequently involves the manipulation of very large matrices which demands at least 64-bit arithmetic in order to avoid round-off errors. Availability of 64-bit precision is, therefore, a desirable feature.

Simulation applications commonly require divide, square root and trigonometric operations to be extensively performed on the data and so an AP tailored for this market would greatly benefit from these units being incorporated in its hardware.

Most real-time simulations consist of lumped-parameter systems described by ordinary differential equations. They may either be inter-faced with actual physical systems (hardware-in-the-loop) [6] or with human operators (man-in-the-loop).

Common to both applications is the need for the computer system to process the information within certain time limits. The overall speed

of the system (AP+host) must be high enough to meet the real-time bandwidth requirements of the problem.  The lower sampling frequency limit,
called the Nyquist rate, is at 2BW, with a sampling interval  Ts < 1/2BW.
To minimize aliasing errors, produced by band overlaps, the sampling
frequency is usually set to between two and a half and three times the
required bandwidth.

In a real-time simulation there are three distinct stages that must
be performed in each frame.  In the first stage data is gathered, in the
form of an input vector, produced by the sampling and digitization of
analogue signals, relating to the response of the hardware.  The second
stage consists of processing this data, by simulating the mathematical
model in order to produce a corresponding system response.  The third
stage entails the sending of control signals, at the end of each time
frame, in the form of an output vector, to the physical hardware.

The higher the required frequency response and the more complex the
simulated system, the greater the computational power and processing
speed must be.  A cost effective solution to this problem lies in the
attachment of a peripheral array processor to a medium-sized computer,
resulting in an increase in overall processing power of between one to
two orders of magnitude.  The use of on-line APs, employed for such time
consuming tasks as multivariable nonlinear function generation, can
provide a realistic alternative to the utilization of very large digital
computer systems.


GENERAL PURPOSE SUPER-COMPUTER

An example of this type of computer is the CRAY-1.  This computer
achieves its speed through the methods of pipelining and chaining to
enable calculations to be performed rapidly.

The CRAY-1 has a 12.5 nanosecond clock period and a 50 nanosecond
memory cycle time.  Up to 2 million 64-bit words of memory are available.
It is 16-way interleaved and there are twenty-four I/O channels so that
several memory fetches and puts can occur at the same time.  24-bit and
64-bit integers are supported and instructions are either 16-bit or
32-bit in length.

There are twelve functional units:  one floating-point multiply,
one reciprocal approximation, one population count, three integer
multiply, two shift and two logical.  There are 8 64-bit registers and
one 7-bit register.  It is worthy of note that there is no integer divide
(this  is done by software using the floating-point hardware) and that a
full 64-bit floating-point divide must also be programmed.

Instructions are executed from instruction buffers of 64 16-bit
registers and so the need for very tight inner loops is lessened considerably since forward and backward branching within the buffers is possible.

The speed of the CRAY (up to 200 Mflops) is achieved by the use of
the large sets of registers and by the pipelining of some of the
functional units.  For example, the floating-point multiply unit is
divided into seven phases, each of which takes one clock period.  Thus a
complete multiply takes 87.5 nanoseconds.  However, one clock period
after numbers have been put into the multiply unit, a further pair can be
put in and so on.  Thus every 12.5 nanoseconds, a multiply can be
initiated, the result 'coming out of the multiply pipe' seven clock
periods later.  Used in conjunction with a vector of registers, fast
processing becomes a reality.

A further technique used is called chaining. The floating point add unit is divided into six stages. Thus to acquire A1 x B1 + C1 will require 13 clock periods or 162.5 nanoseconds. However, using the multiply 'pipe' we can obtain an Ai x Bi every clock period and, if the timing is obtained correctly, these can come straight out of the multiply pipe and into the add pipe with the corresponding Ci thus producing Ai x Bi + Ci every clock period after the initial set up period. At the same time the other functional units can also be active. Thus there is parallelism within functional units (pipelining) and also between them (chaining). Memory fetches can also occur concurrently and so can be included in the chain thus sometimes incurring no overheads.

The seven-bit register of vector-length register is used to assist this process. It determines how many elements of the register are used to give operands in a pipelining situation. A 64-bit vector-mask register is available for use with this to provide similar facilities to those provided by the activity control register in the ICL-DAP.

The CRAY, along with its rival the CDC CYBER205, is at the forefront of current computer achievements, both in terms of throughput and in terms of cost. It clearly shows the degree to which parallelism can be taken both between and within functional units.

While such powerful computers as the CRAY are ideal for the off-line solution of very large simulation problems, weather forecasting for example, because they are normally operated in a time-sharing batch-mode, they are unsuitable for real-time simulation applications, which require immediate responses to external signals.


PERIPHERAL PROCESSORS

The second group of parallel processors introduced are peripheral processors. These are generally high speed floating point processors which are attached to a host minicomputer or mainframe. The name 'array processor' is often used and reflects the fact that an architecture is used which lends itself to the efficient 'processing of arrays'. However, the first example considered here differs from the others in this respect, being a genuine 'array of processors' (a SIMD system).


ICL DAP

The ICL DAP consists of an array of identical processors. The so-called single-instruction multiple-data-stream (SIMD) approach. The sizes of arrays currently envisaged are 32 x 32, 64 x 64 and 128 x 128.

The individual processor contains a store of 4K bits and three registers, each of which has the capacity for holding a single bit. These are used as activity control, carry and general purpose registers. Data paths are one bit wide. Each processor is linked to its four neighbours in the obvious rectangular grid, which can be considered to be cyclic for processors at the edge. There are row and column highways connecting processors together. The array of processors is controlled by a master control unit which is much more like a conventional CPU with the eight instruction modify registers also being used to send and receive data from the processor array. For this reason they are 32, 64 and 128 bits long as appropriate and can be connected to a row or column highway. An instruction buffer is used for storing instructions for inner loops and so, in programming the DAP, a tight inner loop is essential for the highest speeds.

When processing arrays each number is held as successive bits within the corresponding processor. When processing vectors, a number can be held as one bit in each element of a row. Vector arithmetic can thus be performed significantly faster than matrix arithmetic. Effective means of transposing data between the two storage modes are provided.

The ability to have a totally unfixed word length, in theory at least, leads to considerable potential computational savings when an accuracy is required that does not correspond to those conventionally available. In several of the stages of our simulation algorithm, advantage can be taken of this facility. However, interfacing such 'non-standard-precision work' to the mainframe host is potentially tedious.

The DAP's instruction timings have been designed to give an appropriate and therefore unusual set of timings. Data transfers have been relatively speeded up as have such operations as finding the maximum modulus in a vector or matrix (useful in many matrix procedures for pivoting reasons). The bit structure itself serves to speed up some operations as there are efficient algorithms available. A notable example is the square root. Other procedures can be speeded up by the use of the activity bit as a kind of mask to ensure that operations are performed only on the relevant pieces of data. This is a sort of naive 'contents-addressable filestore' facility.

The DAP resides within the store of an ICL mainframe; say a 2970 or 2980. It appears to the mainframe as a piece of store. There is, therefore, no large overhead for the transfer of data and a user has the capacity of the mainframe available to store data which can be rapidly moved into the DAP.

Timings for operations are crucially dependent on the parameters of the problem. The DAP is for instance close to its best when taking simultaneous square roots of all elements of a 64 x 64 array. Making the array 65 x 65 can affect the timings by a factor of about 4. Each processor takes about 250 microseconds for a 32-bit floating-point multiply. Taking a similar square root takes only 180 microseconds. Thus on a 64 x 64 DAP speeds of 16-20 megaflops (i.e. millions of floating-point operations/sec) are obtainable, more if lower precision is acceptable. As with all the devices considered, however, theoretical speeds and actual results are not necessarily the same. We return to this point when we consider the software.

The cost of a 64 x 64 DAP is said to be about $0.6M. A suitable mainframe to support it however will double to treble this cost. Currently we know of no plans to make DAPs available other than with large ICL mainframes although there is apparently no technical reason why this is the case.

FPS AP-120B

Of all the array processors devised so far Floating Point System's AP-120B (AP-190L) has been most consistently used, particularly in simulation applications [7,8]. Fig. 1 shows the main architectural features, which include a three-stage floating point multiplier, two-stage floating point adder, three-stage memory pipeline and multiple data paths. A convergent rounding is used in the last stage of the pipelines to minimise truncation errors. Separate program source memory and main data memory are provided together with a table memory.

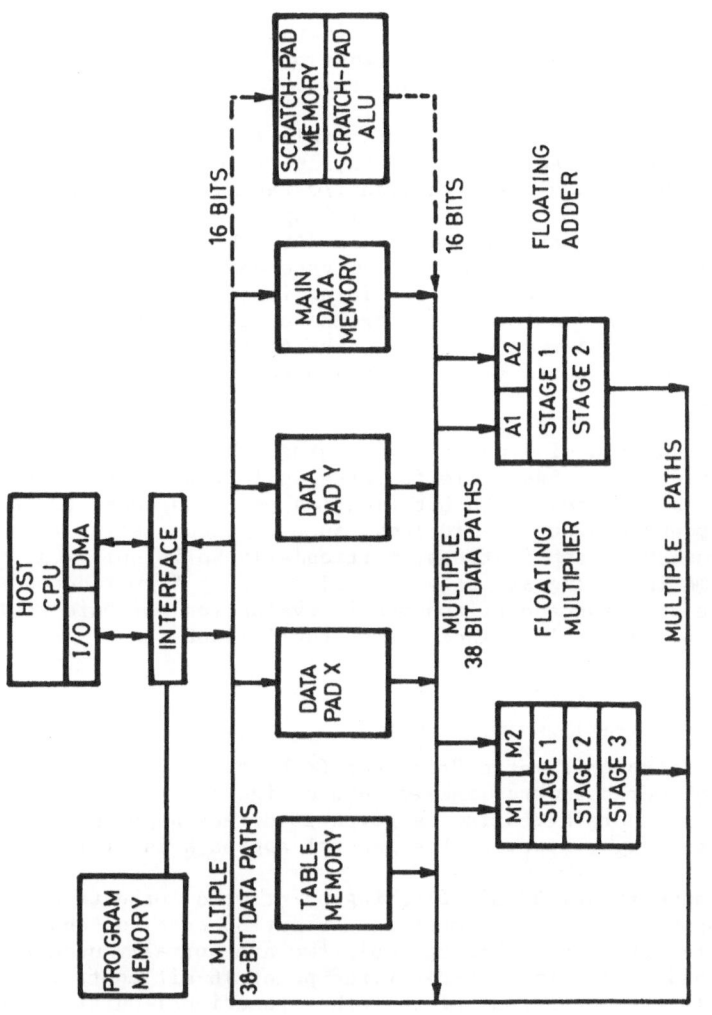

Fig. 1  AP120B Architecture.

189

The AP-120B is a synchronous, parallel pipelined machine with a 167 nanosecond cycle time capable of executing up to 10 commands per 64-bit instruction word, or 60 million instructions per second. The fast table memory unit stores the data for table look-up procedures, such as function generation, with a new value able to be requested every cycle. The processor communicates with the host through a Direct Memory Access (DMA) channel and can accept data at a rate of 6 million (38-bit) words per second.

Originally conceived, in 1976, as a 38-bit array processor for signal processing applications [9], the AP-120B has recently been improved in several ways [10], enhancing its suitability in the simulation field. Particularly relevant are the increases in the program source memory which can now accommodate about 1000 lines of FORTRAN code (8K words), and the Table memory, which now comprises 4.5K words of ROM and 8K of RAM.

In 1980 the General Purpose Programmable I/O Processor (GPIPO) was released, enabling the real-time interfacing of the AP with several supporting peripheral devices with high I/O rates. This move considerably reduced its dependence on the host for I/O operations and improved the overall real-time bandwidth of the system (AP + host). Among the main features of the GPIOP are a 12 Mbyte I/O rate capability, 38-bit FIFO data buffers, four vectored interrupts providing immediate response to external peripheral devices, and a formatting processor to reformat fixed and floating point data entering or leaving the AP. The AP-120B has a dynamic range of $10^{\pm 153}$ which needs to be rounded down to about $10^{\pm 38}$ for most 32-bit systems.

With the addition of some extra features, the AP-120B could be further improved [11]. Desirable features would include an increased word length format as the present 38-bit floating point hardware (28-bit mantissa) introduces accuracy problems, such as accumulating round-off errors, in simulation applications, particularly so in numerical integration. Problems involving stiff systems also suffer from reduced accuracy. Floating point division and trigonometric evaluation hardware would also contribute to the general enhancement of the AP-120B.


CSPI MAP 300

The CSPI Macro Arithmetic Processor (MAP) family of 'array processors' has retained closer hardware links with the signal processing world than its competitors. More attention is given to direct high speed input and output. In many ways, however, the general approach is similar.

A more conventional 32-bit floating-point word for data memory is provided. 64K words of this memory can be put on each bus and up to three buses are permitted, although only two are normally used for data. A 125-nanosecond cycle time, 16-bit fixed-point 16-bit control processor (CSPU) controls the operations of the other functional units in the machine. It is all on memory buses and there is also a priority interrupt system. The CSPU controls an arithmetic processor which has four functional units, two adders and two multipliers. These units are not in themselves pipelined but two can be simultaneously active (see Fig. 2).

As the buses have independent operation and can share memory, it is possible for several processors to cycle-steal, thus increasing the speed of computation.

Memories are multiported by up to 16 I/O scrolls taking a single port. The I/O scrolls (see below) are genuine scrolls and thus, given an

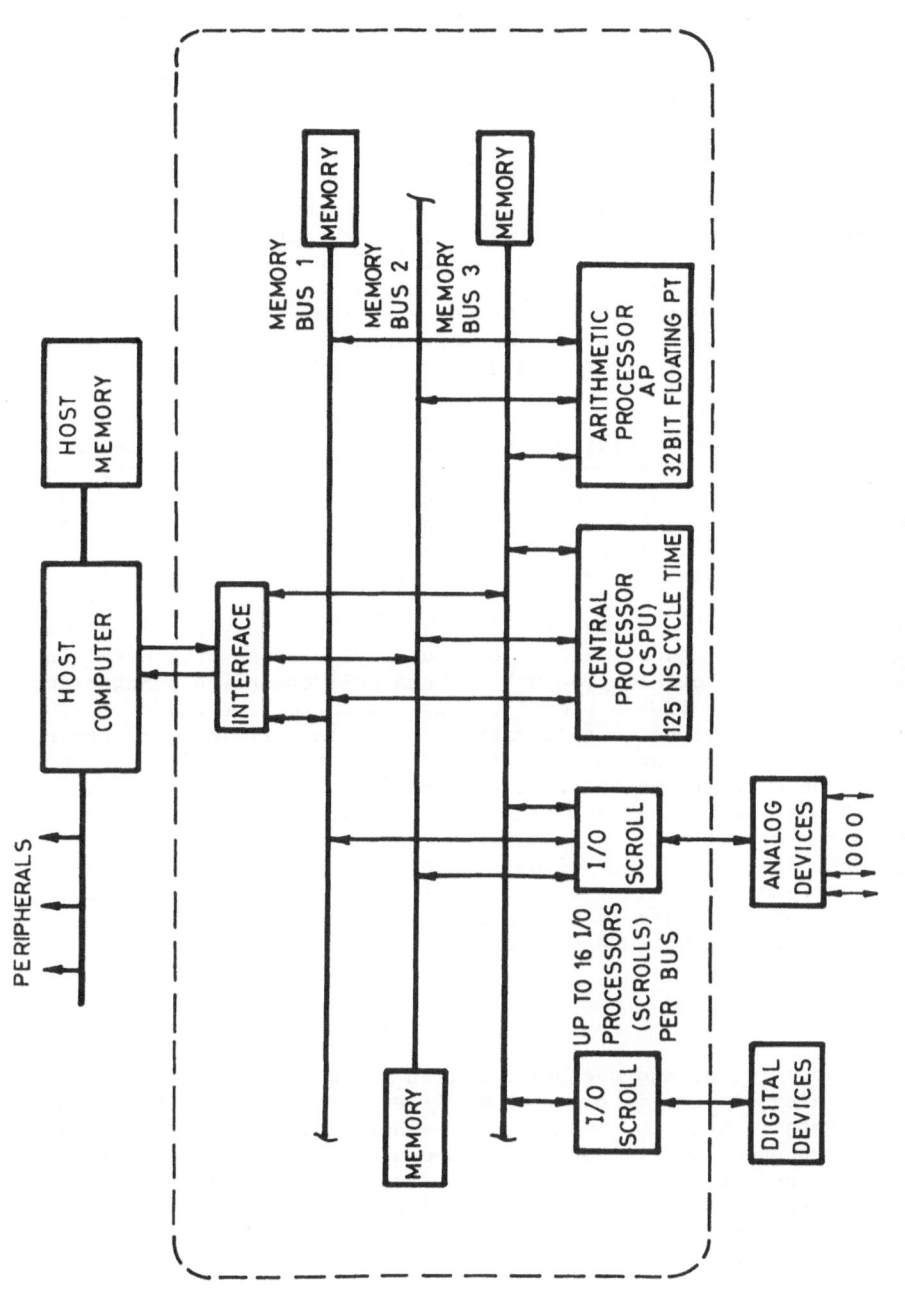

Fig. 2   MAP 300 System Diagram

191

appropriate interface, there is a genuine concept of streams of data of indefinite length being processed to produce other streams. This is, of course, especially relevant in signal processing work although it does have some uses in simulation work also.

Given a 16 or 32-bit word host, the interface board to the host has an easier task than with the AP in terms of format conversion. It is a more sophisticated board in other ways than its AP-120B counterpart, allowing the host to interrupt the CSPU and vice versa at any time to provide suitable synchronisation for transfers. Data transfers are usually controlled by the MAP thus avoiding the large overhead on the host often experienced in AP-120B situations.

The I/O scrolls transfer data in a preprogrammed fashion determined by the CSPU. They are capable of demultiplexing data streams in a fairly sophisticated fashion. Word transfers at 8 MHz can be achieved but, when demultiplexing is in progress, 1 MHz is more normal. A Unibus-compatible variation exists as with the AP-120B and a full duplex RS232C compatible interface at 5 KHz also exists. The most sophisticated board is capable of dealing with the bit-reversal address on an FPT.

A special deal has been concluded between CSPI and SEL whereby the MAP data and program memories can be addressed directly from the SEL. The architecture of the SEL where specific separate banks of data are addressed (in non-virtual mode) facilitates this. SEL will provide packages comprising a SEL machine with one or more MAPs which may or may not be interconnected. This represents a considerable advance as data I/O is a major bottleneck, especially for real-time simulations.

The 64-bit version of the MAP is achieved by merely coupling two of the 32-bit buses together. It has been delivered to several sites and has significant potential. The MAP 300 can deliver only 4.5 Megaflops compared with the AP-120B's 6 million but has better control over I/O. Thus for such operations as full matrix inversions, timings are commensurate. For other operations the MAP 300 is on the slow side perhaps achieving one half of the processing speed of the AP120B. It is priced somewhat lower: a typical configuration with 4K of program memory and 64K of data memory, being about $25,000 cheaper than its rival. It can have attached to it 8 Mbytes of bulk data on an auxiliary port. This is a further attempt to achieve very high I/O rates for real-time simulation work. The range of hosts is not as extensive as for the AP120B but includes PDP11, VAX, Data General and SEL.

ANALOGIC AP400

This more recent and cheaper addition to the spectrum of available hardware achieves speed in a somewhat different fashion to the MAP and AP-120B. Data is held in blocks; the architecture is shown in Fig. 3. A control processor (CSPI) contains its own microprocessor unit to perform data transfers to support pipeline requirements. It has an interrupt vector structure. A host interface provides DMA or programmed transfer to and from the array processor as with the larger processors. The arithmetic pipeline processor takes eight 24-bit inputs through three pipeline stages. The first stage is data/instruction characterization, the second multiplication and the third arithmetic/logic operations or accumulations. Up to 4 outputs are produced. This rather complex structure, when programmed fully, provides a high degree of parallelism but full exploitation is often difficult. The operations are all essentially fixed point for reasons outlined below. The three stages are pipelined. The control bus is 8-bits wide and is pipelined with a

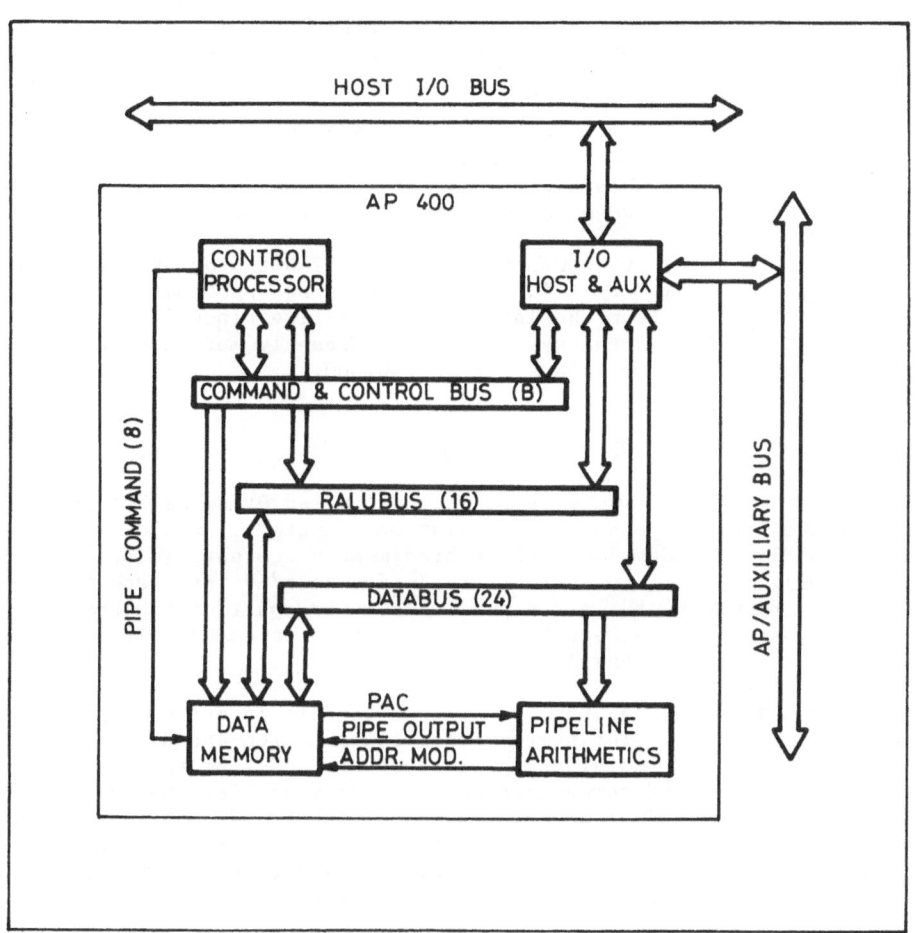

Fig. 3   AP400 System Architecture.

160-nanosecond clock period. The data bus is 24-bits wide and cycle-stealing occurs on it when the host or control wish to use the bus. The final bus is used to transfer addresses etc.

A data block is assumed to consist of commensurable data (approximately the same order of magnitude). Thus a common exponent and the length of the block are held at its head, followed by 24-bit mantissae to complete the block. Once again a strong signal processing trend is shown and there is a large overhead in storing and computing with a large number of individual entities. Up to 2.1 Mflops can be obtained for multiplications and 6.3 Mflops for additions/subtractions, but these speeds are only possible with common exponents and the resultant loss of precision can be catastrophic. For a 'correct application' processing speeds can rival the AP-120B but these applications are not common. No double precision facilities currently exist.

The data is easily transferred from a 16/32 bit host, since mantissae can be transferred wholesale when compatible. A further auxiliary bus accessed through the I/O card provides not only for UNIBUS, or similar inputs, but can also be used to keep up with real-time.

A typical configuration would have 2K of 22-bit program memory (160 nsec), 64K of 24-bit memory (160 nsecs). There are problems involved in overlaying the former from the latter. A real-time clock is provided (unlike the AP-120B) and the system is significantly more compact than its larger rivals. It would cost about $40,000.

SPECIAL PURPOSE SIMULATION PROCESSORS

Both Gould's SEL computer systems and Applied Dynamics International are among the leading vendors of simulation computers. Gould, in conjunction with CSPI, have developed the shared memory concept, (models VPS-3300 and VPS-6400), in a bid to overcome real-time host/AP interface bottlenecks [12]. By making the array processor memory part of the host memory, the need to pass data between the two processors is completely eliminated, eliminating a major processing bottleneck.

The same strategy has been adapted in Gould's latest system, the SCI-CLONE/32 computer. The system consists of a high-speed bus-centered architecture in which individual processors are nodes on the bus. Peripherals, memory and communication interfaces all form part of the bus and not of the independent processors. The high-speed (26MHz) synchronous bus provides the communication paths between the two pipelined processing units and both the shared memory and peripheral devices. These devices may comprise both I/O related hardware and peripheral array processors.

ADI's latest 'System 10 Plus' [13,14] is a purposely-designed digital computer optimized for the time-critical simulation of large, complex dynamic systems (Fig. 4). Its hardware consists of the well-established AD10 processor (Fig. 5) together with a new Floating Point Expansion Unit, FX. The AD10 was originally conceived as a fixed point fast multivariable function generator for use with hybrid computer systems. Subsequent improvements, such as the Numerical Integration Processor (NIP) has enabled the AD10 to solve large sets of coupled nonlinear ordinary differential equations, and considerably enhance its suitability for simulation applications.

The FX processor, with its 20 to 30 Mflops capability, 100 nanosecond instruction cycle time and 40 MHz bus bandwidth, significantly increases the computational power and speed of the whole system.

AD 10                                    FX

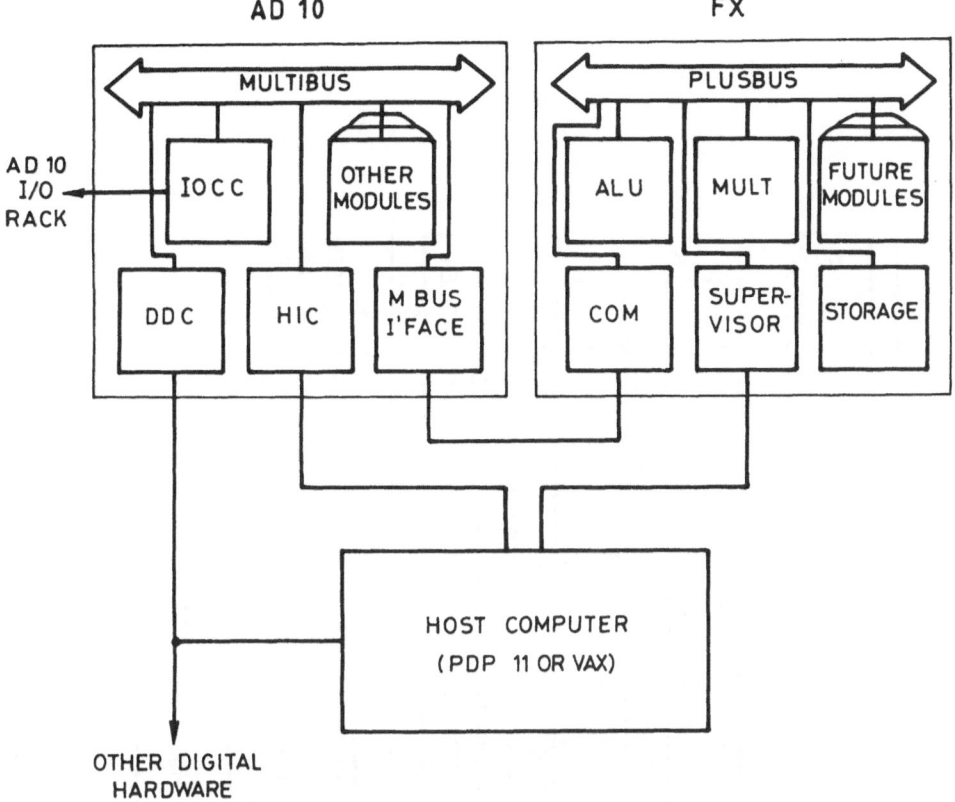

SYSTEM 10 PLUS : HARDWARE SUBSYSTEM

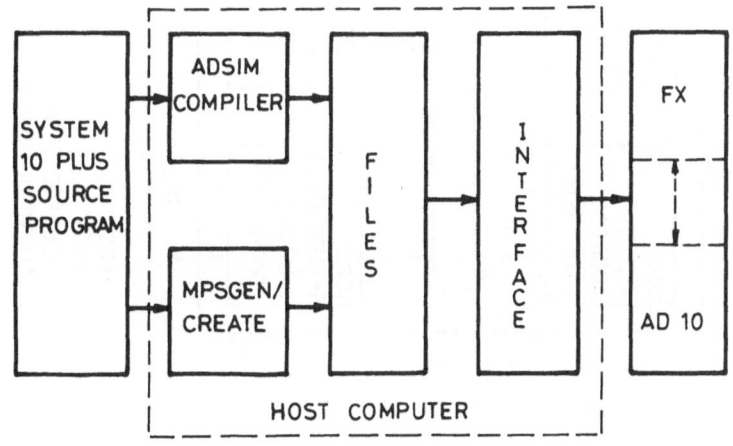

SYSTEM 10 PLUS : SOFTWARE SUBSYSTEM

Fig. 4    ADI System 10 Plus

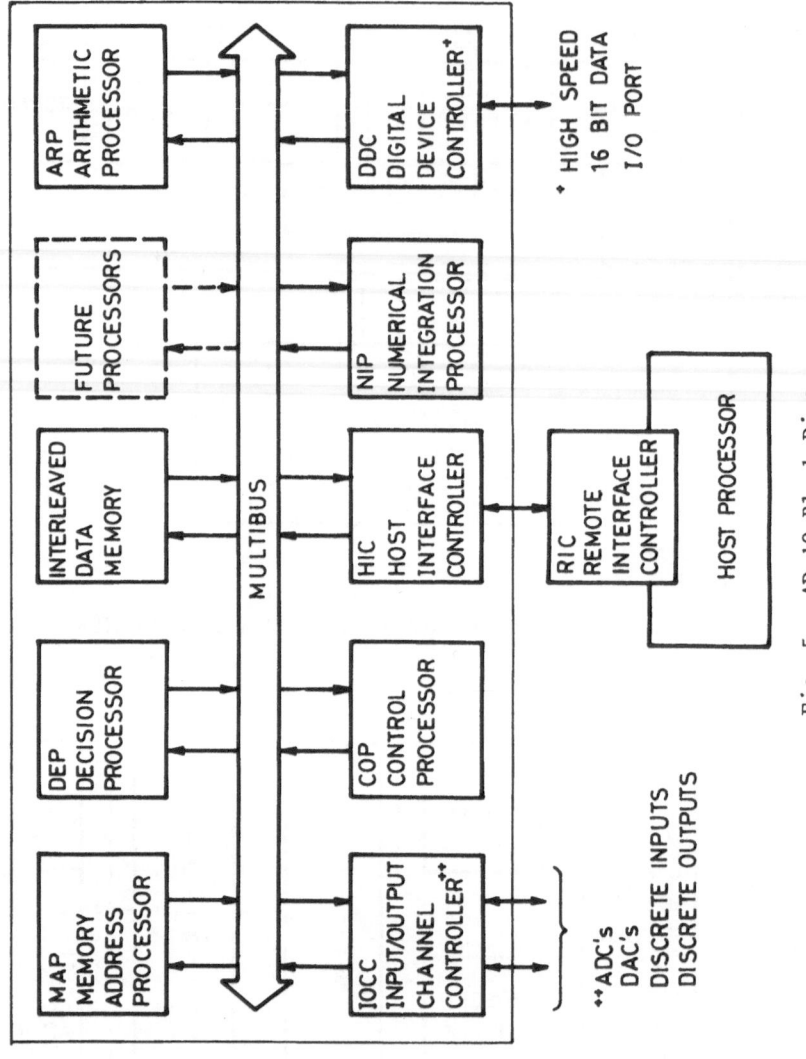

Fig. 5   AD 10 Block Diagram.

196

Simulations which were previously performed on single AD 10 systems will now typically run up to four times faster. The System 10 Plus is able to handle simulations that are an order of magnitude larger and more complex than before, resulting in increased accuracy, realism and dynamic fidelity. By keeping pipeline operations short the FX can handle both scalar and vector operations efficiently. Hardware features are incorporated to facilitate the efficient handling of discontinuities, and to increase accuracy, both 53 and 65-bit floating point word formats are used. This is especially relevant when handling simulatin models involving stiff systems having time constants which may differ by several orders of magnitude.

Both units are capable of either combined or stand-alone operations but they are, nevertheless, attached simulation computers and as such require the services of a host computer.

Large simulation models, such as the one developed by the Department of Nuclear Energy of the Brookhaven Laboratory (USA) are natural candidates for the System 10 Plus. This particular simulation, one of the largest ever performed on AD10's, studies the feasibility of performing faster (10 times) than real-time simulations to investigate various courses of action prior to the actual implementation of a procedure in a real nuclear power plant emergency situation [13]. The model comprises almost 1700 algebraic variables and 116 state variables, and typically runs 11 times slower than real time on a large mainframe computer.

An example of an MIMD architecture is the Multiprocessor Reconfigurable Simulator (MPRS) [15] proposed by the European Space Agency (ESTEC). The MPRS supports up to 32 processors connected in parallel via a common bus system comprising up to four identical buses each supporting up to eight processors. In addition an I/O processor provides communication with operators and external systems. A 5-port common memory is accessible from the system processors and the I/O processors via the system buses and the I/O bus respectively. Each processor also has local memory which is partly mapped on to the common memory. 16-bit data transfers are supported throughout the system. The MPRS concept calls for the possibility of different types of processors to be mixed in the system.

A comprehensive operating system will provide a support environment (based on concurrent Pascal) capable of supporting parallel processes and a simulation environment (based on sequential Pascal) with high-level programming facilities for applications programs. A kernel is provided in each processor which implements the basic primitives needed by the support environments.

This system offers a sound base on which to build simulation applications software including a simulation language.

Programs are allocated to processors by a System Generation Program (Sysgen) which includes in each store image a set of 'Kernel tables' which are used by the Kernel to manage communication between different processors.

User programs may be written in Sequential Pascal and run concurrently with the programs for different processors being scheduled by the SSE. The SSE provides facilities whereby concurrent programs can communicate with each other and gain access to system peripherals and common memory.

A program DEFSIM is used to define a simulation. This requires the user to specify (a) the name of each task in the simulation, (b) the sequential program defining each task, (c) the parameters for each task and the name of the simulation.

The SSE supports communications between processors and with common memory. For interprocessor communication the source processor executes a SENDMESS associated with a particular destination. The executing task is suspended until WAITMESS associated with the appropriate source is executed by the destination task. If the WAITMESS is executed first the waiting task is suspended until the corresponding SENDMESS is executed. Queues are dealt with in FIFO order.

MPRS Common Memory is divided into pages of 256 words each. Initially these pages cannot be accessed by any task. A number of SSE primitives are provided which allow a task to gain access to a page or to release it. Access to a page may be shared between tasks. A page must be locked by a task before it can write to it. A locked page can only be accessed by the task which locked it. Conflicts are resolved in FIFO order. Timing primitives are also provided with a maximum resolution of 100mSec.

The facilities described above provide an excellent software environment for programming simulation tasks on a multiprocessor system. There is as yet no information available concerning the performance of this software. Furthermore, no applications software has yet been implemented (e.g. integration routines, function generators, etc.).

SOFTWARE REQUIREMENTS

At the lowest level a parallel processor may be programmed in its assembly language. Only by this means is it possible to realise the full parallel capability and maximum efficiency of the hardware, and time critical routines, such as integration algorithms, are best written in this way. However, such an approach is time-consuming and error-prone and quite unsuitable for the preparation of general simulation programs. Nevertheless, specific real-time simulations, which demand the highest possible computational speed to keep within the allocated frame-time, have been written entirely at assembler level.

Most of the larger processors are provided with a high-level language compiler. The Cray, DAP and the AP-120B possess FORTRAN compilers, although certain restrictions sometimes apply. Usually a knowledge of the processor architecture is required to achieve the most efficient code. For example, on the Cray, inner program loops may be written in such a manner as to allow vectorisation. A FORTRAN program for the AP-120B consists of two parts, one executing in the host and written in host FORTRAN, and the other executing in the AP and written in AP FORTRAN. Maths libraries are available and these contain a selection of numerical integration and matrix routines, thus partially satisfying the simulation requirements.

An alternative software system for the AP-120B is TOAST, developed by System Software Factors. This product was developed initially to support flight simulation on PDP11/AP-120B systems, is FORTRAN based but produces more efficient code that AP FORTRAN.

A high-level general purpose language is not the ideal environment in which to write simulation programs. There have been in existence for some time now a number of special purpose simulation languages for conventional computers. In particular, Continuous System Simulation Languages (CSSL's) such as CSMP and ACSL greatly simplify the engineer's task when simulating a dynamic system. These languages incorporate a selection of integration methods and take care of the program structure and housekeeping operations, leaving essentially the specification of the system equations to the user. Built-in facilities may be provided for discontinuity detection and function generation together with a simple means of specifying graphical and tabulated output.

The way forward on the software front for parallel processors is through the incorporation into CSSL's of features to use efficiently the parallel architectures, without involving the user in their detailed design. The ESL simulation language, described in an earlier paper [16], includes the concept of parallel segments, each running with its own integration routine and step length and communicating with each other at specified times. Although currently running on conventional computers in which parallel operations are simulated, the parallel segments of this language will map directly onto MIMD architectures. Indeed, ESL originated out of a study of simulation algorithms for parallel processors. Many problems remain, such as how to partition efficiently a large simulation between available processors, but the ultimate aim must be to provide the user with a simulation environment in which full advantage can be made of parallel computers.


FUTURE TRENDS

Peripheral array processors have developed from being fast but expensive arithmetic units to rather sophisticated independent machines able to perform relatively complex tasks. Already a stage has been reached where the host is merely required for initial program development, down-loading of programs and periodic analysis of data provided by the AP. The direct interfacing of I/O devices [17] has significantly reduced both the frequency of host/AP communication and the dependence on the host for the Input and Output of data. It has also increased maximum throughput performance, particularly in real-time applications requiring rates in excess of 100 KHz where the host's limited I/O capabilities, together with host/AP interface bottlenecks, have often hindered higher data transfer rates to take place [12].

Array processor manufacturers have, thus far, mainly catered for the signal processing market, with simulators able to adapt the inherent raw processing power of APs to their requirements. While both applications pose similar demands there are, nonetheless, significant differences in memory size, mathematical hardware and I/O requirements to warrant the future development of APs tailored for simulation applications.

The advent of VLSI technology, and its more recent advances, will surely influence the future development of array processor architectures [18]. One possibility is for array processors to incorporate in their architecture, many hundreds of APs, in parallel, in order to increase their computational power by several orders of magnitude. A particular VLSI development which will influence the approach to the design of parallel processors is the Transputer. For example, the newly available IMS T414 from Inmos, integrates a 10 MIPS 32-bit microprocessor, four communication links, 2 Kbytes of RAM, a 32-bit memory interface and memory controller onto a single CMOS chip. It is claimed that 300 of these devices suitably linked will provide equivalent performance to a Cray XMP-1.

The continuing reduction in size and cost of APs, coupled with an increase in performance, has made possible the emergence of single chip signal processors. These are mostly special-purpose chips designed to perform pre-defined functions, such as speech analysis. They are more cost-effective but less flexible than fully-fledged APs. A similar development has been the design, by such firms as AMD and WITEK, of special-purpose AP components. They include high-speed parallel multipliers, multiport pipeline processors and high speed sine/cosine function generators. These components can be used to produce a wide range of array processors, each tailored for a specific application in order to achieve optimum performances.

With the progressive reduction in size, array processors will eventually be incorporated into the host system as shared memory co-processors, sharing its resources and significantly reducing most problems associated with host/AP interfacing.

Other recent developments include the integration of APs in multiprocessor systems comprising processors of differing characteristics, by utilizing special controllers, such as Aptec's Dimensional Processing System [19]. On the other hand, several array processors can be interconnected, in parallel [20], in order to achieve very high throughput rates for heavy compute-bound real-time applications.

It is expected that peripheral array processors will also be used to enhance the resources of local area networks and workstation environments. A good example of which is the Ethernet network system being installed at UCLA [21]. It will make more computing power available to individual users, in a shared resources environment. This is a much more viable proposition of efficient sharing of an array processor than the impractical time-sharing option, as commonly applied to conventional digital computers.

As far as software is concerned, improved development tools and higher level languages, possibly machine-dependent, will provide easier access, shorter program development times, faster execution through optimizing compilers, and much more productive exploitation of system resources.


CONCLUSION

Parallel processing techniques are able to provide the computing power required for very large continuous-system simulation and real-time simulation applications. These techniques allow efficient execution of the numerical methods associated with continuous-system simulation, such as integration algorithms and multivariable function generation, and enable parallel solution of the system equations to take place.

Currently available computers employing parallel techniques range from the very powerful, general purpose super-computer, such as the CRAY-1, through host-attached processor arrangements typified by the FPS AP-120B, to special purpose simulation computers such as ADI's AD 10, Gould's SCI-CLONE/32 and ESA's experimental MPRS. The most fruitful approaches for simulation lie in multiprocessor systems (MIMD arrangements), which allow a large simulation to be partitioned between a number of autonomous processors, operating in parallel and communicating with each other as and when necessary. VLSI developments, in particular the emergence of powerful single chip computers such as the transputer, provide the means of constructing very powerful, low cost parallel simulation computers.

It is essential that software keeps pace with hardware developments and much effort is being expended at the lower level to provide operational software and techniques for automatic partitioning of system equations. However, a suitable user-interface must be established by adapting the present continuous-system simulation languages to take full advantage of parallel hardware.

The superiority of the analogue computer for continuous-system simulation lay in its inherent parallel operation. This same principle incorporated into digital computers, will provide the simulation computers of the future.

REFERENCES

1.  Carver, M.B., 1978, Efficient integration over discontinuities in o.d.e. simulations. Maths and Comp. in Simulation.
2.  Crosbie, R.E. and Hay, J.L., 1974, Digital techniques for the simulation of discontinuities. Proc. Summer Computer Simulation Conf., Houston, July 1974, AFIPS Press.
3.  Alexander, P., 1981, Array Processor design concepts. Computer Design, December 1981, pp. 163-172.
4.  Karplus, W.J., 1977, Peripheral processors for high-speed simulation. Simulation, vol. 29, No. 5, pp. 143-153.
5.  Alexander, P., 1979, The array processor as an intelligent simulation co-processor. Proc. Summer Computer Simulation Conf., pp. 2-13.
6.  Chen, Y.P. and McAlpine, G., 1984, A real-time hardware-in-the-loop missile simulation using the DPS-2400 as an executive controller. Peripheral array Processors, Simulation Series 14:2. The Society for Computer Simulation, pp. 115-124.
7.  Crosbie, R.E., Hay, J.L., Javey, S., Narotam, M.D. and Slater, J.B., 1981, Simulation studies with modern computer structures. ESA Contract No. 4155/79/NL/PP(SC), Final Report, Salford University Industrial Centre.
8.  Kemmler, K. and Martson, W., 1978, The array processor AP-120B/190L for Simulation Applications, Proc. Military Electronics Defence Expo, pp. 43-53.
9.  Charlesworth, A.E., 1981, An approach to scientific array processing: the architectural design of the AP-120B/FPS-164 Family. IEEE Computer, 14, pp. 18-27.
10. Kushner, E.J., 1982, Recent developments in the hardware and software provided by floating point systems. Peripheral Array Processors, Simulation Series 11:1, The Society for Computer Simulation, pp. 39-47.
11. Kushner, E.J., 1984, Parallel processing: the approach selected by floating point systems for providing a new generation of cost effective array processors and scientific computers. Peripheral Array Processors, Simulation Series 14:2, The Society for Computer Simulation, pp. 27-37.
12. Borgioli, R.C., 1982, Real-time performance considerations in array processing. Peripheral Array Processors, Simulation Series 11:1. The Society for Computer Simulation, pp. 49-60.
13. Fadden, E.J., 1982, The System 10 Plus: A major advance in scientific computing. Peripheral Array Processors, Simulation Series 11:1, The Society for Computer Simulation, pp. 61-75.
14. Fadden, E.J., 1984, The System 10 Plus: Broader horizons. Peripheral Array Processors, Simulation Series 14:2. The Society for Computer Simulation, pp. 53-70.
15. Crosbie, R.E. and Slater, J.B., 1981, Simulation Studies with modern computer structures, Report 2, Part II, Salford University Industrial Centre Ltd.
16. Hay, J.L., 1985, ESL - Advanced simulation language for parallel processors. Proc. of 1st European Workshop on Parallel Processing Techniques for Simulation.

17. Wiley, P., 1979, Interfacing peripherals directly to an array processor. Computer Design, vol. 18, No. 8, pp. 158-164.

18. Cohen, D., 1982, The impact of VLSI on peripheral array processors. Peripheral Array Processors, Simulation Series 11:1. The Society for Computer Simulation, pp. 33-38.

19. McAlpine, G.L., 1982, Dimensional processing system: A controller for multi-processor architectures. Peripheral Array Processors, Simulation Series 11:1. The Society for Computer Simulation, pp. 139-157.

20. Burns, J.F., 1981, Greater throughput with multiple array processors. Computer Design, pp. 207-211.

21. Karplus, W.J., 1984, The changing role of peripheral array processors. Simulation Series 14:2. The Society for Computer Simulation, pp. 1-13.

INTEGRATED SYSTEM OPTIMISATION AND PARAMETER ESTIMATION TECHNIQUE

USING A DISTRIBUTED HIERARCHICAL COMPUTER SYSTEM

S. Chen, P.D. Roberts and D.S. Wadhwani

Control Engineering Centre

The City University,  London

ABSTRACT

   The practical implementation of an integrated system optimization and parameter estimation technique for hierarchical control of steady state systems has been investigated using a distributed hierarchical computer system.  Problems associated with the on-line implementation of the technique have been discussed and methods for dealing with these problems are suggested and have been tested in the real-time situation. It is demonstrated that a double  iterative version of the technique has important advantages under real time operation.

1.  INTRODUCTION

   The advent of inexpensive and more powerful microcomputers has facilitated the implementation of on-line optimization schemes to industrial processes.  Most optimization algorithms require a process model and their performance clearly depends on how accurate the model is to reality.  Unfortunately, for many industrial processes an accurate model may be very difficult, time consuming, or impractical to obtain.

   It has been shown (Brdyś, Chen and Roberts, 1984; Roberts, 1979) how an algorithm can be derived which is capable of obtaining the process optimum steady-state operating condition using an approximate model.  The model inaccuracies are overcome using an integrated system optimisation and parameter estimation (ISOPE) technique.  By combining this technique with the price correction mechanism (Findeisen and co-workers, 1980) several hierarchical decomposition-coordination algorithms have been formulated (Chen, Brdyś and Roberts, 1985; Michalska, Ellis and Roberts, 1984).

   The hierarchical ISOPE techniques share many common advantages (such as simplifying the overall optimisation task, enabling parallel calculation and being suitable for implementation using the distributed computer system) with other existing hierarchical optimisation techniques.  On the other hand, the hierarchical ISOPE methods have an important advantage over several other hierarchical methods (such as the price method with feedback and the direct method with feedback) in that, in an efficient manner and based only on an approximate model, they attain the optimal steady state operating condition rather than a  suboptimal one.   This benefit is obtained at the cost of demanding more information from the real process in the form of measurement derivatives with respect to the controller set points. A further advantage occurs, however, in that due to their adaptive nature, the ISOPE techniques enable simpler models to be used, and consequently, have a great advantage in the situation where process knowledge is highly uncertain.

Previous research has applied the price method with feedback and the direct method with feedback using the distributed hierarchical computer system in the computer control laboratory of the City University (Roberts and co-workers, 1984). In this paper the hierarchical ISOPE technique given by Chen, Brdyś and Roberts (1985) is implemented using the same computer system, where the aim is to investigate the behaviour of the technique in a real-time environment and to study methods for dealing with some problems associated with on-line implementation. The efficiency of the two versions (a single iterative algorithm and a double iterative algorithm) of this technique is also compared.

## 2. HIERARCHICAL INTEGRATED SYSTEM OPTIMIZATION AND PARAMETER ESTIMATION TECHNIQUE

In previous work Chen, Brdyś and Roberts (1985) suggested two forms of an integrated system optimization and parameter estimation technique for hierarchical control of steady state processes: a single iterative algorithm and a double iterative algorithm. In this section these two algorithms are briefly described.

It is assumed that the real system including its follow-up controllers is described in a decomposed way by a set of subsystem input-output mappings

$$F_i^* : \quad C_i \times U_i \rightarrow Y_i, \quad i = 1, \ldots, N$$

as follows

$$y_i = F_i^*(c_i, u_i), \quad i = 1, \ldots, N,$$

where $N$ denotes the number of subsystems, $C_i$, $U_i$ and $Y_i$ are finite dimensional spaces, $c_i$, $u_i$ and $y_i$ are the ith subsystem's control, interaction input and interaction output vectors, respectively. The subsystems are interconnected through the coupling equations

$$u_i = H_i y = \sum_{j=1}^{N} H_{ij} y_j, \quad i = 1, \ldots, N,$$

where $H_i$ and $H_{ij}$ are interconnection matrices. The local constraint set takes the form

$$(c_i, u_i, y_i) \varepsilon \; CUY_i \underset{=}{\Delta} \{(c_i, u_i, y_i) \varepsilon \; C_i \times U_i \times Y_i :$$

$$G_i(c_i, u_i, y_i) \leq 0\}, \quad i = 1, \ldots, N,$$

where $G_i$ is the ith subsystem constraint function vector. The system as a whole can be written as

$$y = F^*(c, u) \tag{1}$$

$$u = Hy \tag{2}$$

$$(c, u, y) \varepsilon \quad CUY \triangleq \{(c,u,y) \ \varepsilon C \times U \times Y : \ G(c,u,y) \leq 0 \} \tag{3}$$

Next, we require that for each $c \ \varepsilon \ C$, the equation $y = F^*(c, Hy)$ has a unique solution

$$y = K^*(c), \tag{4}$$

where $K^* : C \rightarrow Y$ and $K^*(c) = (K_1^*(c),\ldots,K_N^*(c))$.

In general the above real system relations are not known exactly and, consequently, we have only their approximate models

$$F_i : C_i \times U_i \times A_i \rightarrow Y_i,$$

$$y_i = F_i(c_i, u_i, \alpha_i), \qquad i = 1,\ldots,N,$$

where $A_i$ is a finite dimensional space and $\alpha_i$ is the ith subsystem model parameter vector. As before, the global model equations can be written as

$$y = F(c, u, \alpha) \tag{5}$$

The interconnection matrix and the local constraint set are assumed to be known exactly.

Finally, a known local performance index , associated with each subsystem

$$Q_i : C_i \times U_i \times Y_i \rightarrow R, \qquad i = 1,\ldots,N,$$

is required to be minimized. The overall performance function

$$Q : C \times U \times Y \rightarrow R,$$

is assumed to be

$$Q(c,u,y) = \sum_{i=1}^{N} Q_i(c_i,u_i,y_i) \tag{6}$$

The task of determining the optimum controller set points $c$ for a real system can be defined as the following real steady state optimizing control problem

$$\min_{c,u} Q(c,u,y) ,$$

(ROCP) s.t. $y = K^*(c)$ ,

$$u = Hy , \tag{7}$$

$$G(c,u,y) \leq 0 .$$

An equivalent optimal control problem of (ROCP) (see Brdyś and Roberts, 1984) is

$$\min_{c,v,u,\alpha} q(c,u,\alpha),$$

(EOCP) s.t. $F(v, HK^*(v), \alpha) = K^*(v)$,

$$u = HF(c, u, \alpha),$$

$$g(c, u, \alpha) \leq 0, \tag{8}$$

$$v = c,$$

where $q(c, u, \alpha) = Q(c, u, F(c, u, \alpha))$ and $g(c, u, \alpha) = G(c, u, F(c, u, \alpha))$. The Lagrangian associated with (EOCP) can then be written as

$$L(c, u, v, \alpha, p, \lambda, \eta, \xi) = q(c, u, \alpha) + p^T[u - HF(c, u, \alpha)] + \lambda^T(v - c)$$

$$+ \eta^T[F(v, HK^*(v), \alpha) - K^*(v)] + \xi^T g(c, u, \alpha), \tag{9}$$

where $p$, $\lambda$, $\eta$ and $\xi$ are Lagrange multipliers. In particular, $p$ is known as the price vector and $\lambda$ is known as the modifier vector. Assuming that all required derivatives exist and regularity conditions are satisfied then the Kuhn-Tucker necessary optimality conditions of (EOCP) are:

$$\nabla_c L = \frac{\partial^T q(c, u, \alpha)}{\partial c} - \frac{\partial^T F(c, u, \alpha)}{\partial c} H^T p - \lambda + \frac{\partial^T g(c, u, \alpha)}{\partial c} \xi = 0, \tag{10}$$

$$\nabla_u L = \frac{\partial^T q(c, u, \alpha)}{\partial u} + \frac{\partial^T [u - HF(c, u, \alpha)]}{\partial u} p + \frac{\partial^T g(c, u, \alpha)}{\partial u} \xi = 0, \tag{11}$$

$$\nabla_v L = \lambda + \left( \frac{\partial^T F(v, HK^*(v), \alpha)}{\partial v} - \frac{\partial^T K^*(v)}{\partial v} \right) \eta = 0, \tag{12}$$

$$\nabla_\alpha L = \frac{\partial^T q(c, u, \alpha)}{\partial \alpha} - \frac{\partial^T F(c, u, \alpha)}{\partial \alpha} H^T p + \frac{\partial^T F(v, HK^*(v), \alpha)}{\partial \alpha} \eta$$

$$+ \frac{\partial^T g(c, u, \alpha)}{\partial \alpha} \xi = 0, \tag{13}$$

$$\nabla_p L = u - HF(c, u, \alpha) = 0, \tag{14}$$

$$\nabla_\lambda L = v - c = 0, \tag{15}$$

$$\nabla_\eta L = F(v, HK^*(v), \alpha) - K^*(v) = 0, \tag{16}$$

$$\nabla_\xi L = g(c, u, \alpha) \leq 0, \tag{17}$$

$$\xi^T \nabla_\xi L = \xi^T g(c, u, \alpha) = 0, \qquad \xi \geq 0. \tag{18}$$

The modifier formula can be derived from equations (12) and (13)

$$\lambda = \lambda(c,u,v,\alpha,p,\xi) = \left( \frac{\partial^T F(v,HK^*(v),\alpha)}{\partial v} \right.$$

$$- \frac{\partial^T K^*(v)}{\partial v} \left. \right) \left( \frac{\partial^T F(v,HK^*(v),\alpha)}{\partial \alpha} \right)^{-1} \left( \frac{\partial^T q(c,u,\alpha)}{\partial \alpha} + \frac{\partial^T g(c,u,\alpha)}{\partial \alpha} \xi \right.$$

$$\left. - \frac{\partial^T F(c,u,\alpha)}{\partial \alpha} H^T p \right). \tag{19}$$

The ith local parameter estimation problem is defined as determining the value of the ith parameter vector so that

$$F_i(v_i,H_iK^*(v),\alpha_i) = K_i^*(v), \qquad i = 1,\ldots,N . \tag{20}$$

The ith local model based optimisation problem can be defined as

$$\min_{c_i,u_i} L_i(c_i,u_i,\alpha_i,\lambda_i,p) , \qquad i = 1,\ldots,N , \tag{21}$$

$$\text{s.t.} \quad g_i(c_i,u_i,\alpha_i) \leq 0 ,$$

where

$$L_i(c_i,u_i,\alpha_i,\lambda_i,p) = q_i(c_i,u_i,\alpha_i) - \lambda_i^T c_i + p_i^T u_i$$

$$- \sum_{j=1}^{N} p_j^T H_{ji} F_i(c_i,u_i,\alpha_i) .$$

The following two iterative algorithms can be used to find a solution which satisfies conditions (10) to (18) .

## Single iterative algorithm (SIA)

The ith local control problem which consists of the ith local parameter estimation problem and the ith local optimisation problem is defined:

$\text{LCP}_i$

$i=1,\ldots,N$

$\left\{\begin{array}{l}\end{array}\right.$

(i) For given $v_i$ and measurements $K_i^*(v)$ and $H_iK^*(v)$ find the model parameters $\alpha_i$ which satisfy

$F_i(v_i,H_iK^*(v),\alpha_i) - K_i^*(v) = 0$ ,

(ii) For given $\alpha_i$, p and $\lambda_i$, find the controller set points $\hat{c}_i$ and interconnection inputs $\hat{u}_i$ such that

$(\hat{c}_i,\hat{u}_i) = \arg \min_{c_i,u_i} L_i(c_i,u_i,\alpha_i,\lambda_i,p)$,

s.t. $g_i(c_i,u_i,\alpha_i) \leq 0$ .

The task of the coordinator is to ensure that the solution which satisfies conditions (10) to (18) is obtained in an efficient manner. This task can be stated as:

$$
\text{CP} \left\{ \begin{array}{l} \text{Find } p^* \text{ and } \lambda^* \text{ such that} \\[2mm] \text{the conditions (12) to (15) are satisfied.} \end{array} \right.
$$

The overall problem is suitable for employing an iterative procedure where $v$ and $\xi$ are improved based on $\hat{c}$ and $\hat{\xi}$ given by the local units, $p$ is improved based on the interaction imbalance $\Delta \hat{u} = \hat{u} - HF(\hat{c}, \hat{u}, \alpha)$ and $\lambda$ is updated by formula (19).

The following iterative strategies for updating $v$, $\xi$ and $p$ are proposed:

$$
v_i^{k+1} = \psi_{v_i}(\hat{c}_i^k, v_i^k) = v_i^k + K_{v_i}(\hat{c}_i^k - v_i^k), \qquad i = 1,\ldots,N, \tag{22}
$$

$$
\xi_i^{k+1} = \psi_{\xi_i}(\hat{\xi}_i^k, \xi_i^k) = \xi_i^k + K_{\xi_i}(\hat{\xi}_i^k - \xi_i^k), \qquad i = 1,\ldots,N, \tag{23}
$$

$$
p^{k+1} = \psi_p(p^k, \Delta \hat{u}^k) = p^k + K_p(\hat{u}^k - HF(\hat{c}^k, \hat{u}^k, \alpha^k)), \tag{24}
$$

where $K_{v_i}$, $K_{\xi_i}$ and $K_p$ are diagonal gain matrices. The kth iteration is described as follows:

(a)  The local control units update $v_i$, $i = 1,\ldots,N$, then apply $v_i^k$ to the real subsystems and obtain the corresponding steady-state measurements of local outputs $K_i^*(v^k)$ and interconnection inputs $H_i K^*(v^k)$. Determine $\alpha_i^k$ by solving

$$
F_i(v_i^k, H_i K^*(v^k), \alpha_i) = K_i^*(v^k), \quad i = 1,\ldots,N \ .
$$

Perform additional perturbations about $v_i^k$ and take the corresponding real measurements. The values of $\alpha_i^k$, $v_i^k$ and all real measurements are sent to the coordinator for computing finite difference approximations of the derivatives

$$
\frac{\partial^T F(v^k, HK^*(v^k), \alpha^k)}{\partial v} - \frac{\partial^T K^*(v^k)}{\partial v}
$$

(b)  The coordinator computes $\xi^k$ and $p^k$, calculates modifiers according to (19)

$$
\lambda^k = \lambda(\hat{c}^{k-1}, \hat{u}^{k-1}, v^k, \alpha^k, p^k, \xi^k),
$$

and sends $\lambda_i^k$ and $p^k$ to the ith local unit.

(c)  The local units perform their own optimisation task

$$
(\hat{c}_i^k, \hat{u}_i^k) = \arg \min_{c_i, u_i} L_i(c_i, u_i, \alpha_i^k, \lambda_i^k, p^k)
$$

s.t.  $\quad g_i(c_i, u_i, \alpha_i^k) \le 0, \quad i = 1,\ldots,N \ .$

The values of $\hat{c}_i^k, \hat{u}_i^k,$ and associated Lagrange multiplier $\hat{\xi}_i^k$ are then sent to the coordinator.

It is noted that the requirement to obtain real process derivatives produces a serious practical consideration particularly when the measurements are contaminated by noise. For this reason it is often advisable to apply simple software filtering to the computation of the modifiers $\lambda$ (Roberts and Ellis, 1981). Figure 1 shows the information structure of the algorithm. The overall convergence is achieved when successive solutions of $v$ and $\xi$ are unchanged and interaction balance is satisfied. In practice, the overall process may be terminated when every element of $v$, $\hat{u}$ and $\xi$ satisfies

$$\mid v_{ij}^k - \hat{c}_{ij}^k \mid < \beta_{v'} \tag{25}$$

$$\mid \xi_{i1}^k - \hat{\xi}_{i1}^k \mid < \beta_{\xi'} \tag{26}$$

$$\mid \hat{u}_{is}^k - (H_i F(\hat{c}^k, \hat{u}^k, \alpha^k))_s \mid < \beta_{p'} \tag{27}$$

where $\beta_{v'}$, $\beta_\xi$ and $\beta_p$ are some desired tolerances. $(H_i F)_s$ denotes the sth element of $H_i F$.

A special situation occurs when the system constraints do not depend on the outputs. In this case $\dfrac{\partial^\Gamma g(c,u,\alpha)}{\partial \alpha} \xi = 0$, and the iterative strategy (23) and the criterion (26) are not needed.


## Double Iterative Algorithm (DIA)

This algorithm is formed by separating the function of improving the modifier $\lambda$ and local parameter estimation from that of improving the price $p$ and local optimisation. The latter gives rise to an interior procedure during which information is interchanged only between local optimisation units and the coordinator, while the former, which requires real system information feedback, is solved at less frequent intervals in an outer iterative loop. The overall iterative procedure is described as follows:

With $\alpha$ and $\lambda$ fixed, the interior procedure is the entirely model-based problem

$$\text{ILOP}_i: \quad i=1,\ldots,N \quad \begin{cases} \text{Find } \hat{c}_i \text{ and } \hat{u}_i \text{ such that} \\[6pt] (\hat{c}_i, \hat{u}_i) = \underset{c_i, u_i}{\arg\min}\, L_i(c_i, u_i, \alpha_i, \lambda_i, p), \\[6pt] \text{s.t.} \quad g_i(c_i, u_i, \alpha_i) \leq 0, \end{cases}$$

$$\text{ICP}: \quad \begin{cases} \text{Find } \hat{p} \text{ such that} \\[6pt] \hat{u} - HF(\hat{c}, \hat{u}, \alpha) = 0, \end{cases}$$

This problem is solved iteratively using an open-loop interaction balance method. When interaction balance has been achieved, $\alpha$ and $\lambda$ are adjusted by the outer iterative loop and then the interior procedure restarts with new $\alpha$ and $\lambda$. The outer loop iterates $\alpha$ and $\lambda$ similarly to the single iterative algorithm as described in steps (a) and (b) except that

$\hat{c}^{k-1}, \hat{u}^{k-1}$ and $p^k$ are the solution and corresponding price vector of the interior procedure. The overall process is terminated when v and $\xi$ determined in the outer procedure remain sufficiently unchanged between successive iterations. The information structure of the algorithm is shown in Fig. 2. The following remarks are made.

Both algorithms operate in a decentralized manner at the local level and hence can easily be implemented using hierarchical computer networks such as the one to be discussed later, with a central computer (supremal level) serving as the coordinator and local computers (infimal level) serving as the local control units. Parallel local optimisation and parameter estimation calculations are achieved.

It has been demonstrated using analysis and simulation studies by Chen, Brdys and Roberts (1985) that the double iterative algorithm is preferred to the single iterative algorithm in that it can significantly reduce the time for determining the optimum steady state condition. This conclusion will be confirmed in the later experimental results.

Two types of information interchange which occur during the iterative procedure are defined. The first type occurs between the local control units and the coordinator after each model-based optimization iteration, which we will denote as off-line information interchange. The second type occurs between the real subsystems and the coordinator (through the local control units) after each time that all the controller set points have been changed, which we will refer to as on-line information interchange. It is assumed that the total number of information exchanges is roughly equal to the sum of the off-line and on-line information interchanges.

## 3. DISTRIBUTED HIERARCHICAL COMPUTER SYSTEM

One of the main research areas in the Control Engineering Centre of The City University is concerned with the application of hierarchical control techniques to industrial processes. Much of this research is conducted in the laboratory using a distributed two-level computer network (Fig. 3), which operates several pilot-scale processes simultaneously. The infimal level of the network contains four I-MIC (8085 based) microcomputers and a DEC LSI11/02 minicomputer which are used, depending on the application, for data acquisition, direct digital control, local parameter estimation and local optimization. A DEC LSI11/23 minicomputer is used at the supremal level to coordinate the computers at the infimal level.

At present the I-MICs control a pilot scale freon vaporiser, a mixing process and an analogue computer simulating interconnected plant. The LSI11/02 controls a pilot scale travelling load furnace. Further details of the network and its applications are described by Stevenson and co-workers (1984). The part of the network used to implement the technique given in Section 2 includes the LSI11/23, which serves as the coordinator, two I-MICs, which are used as two local control units, and an EAL Pace TR48 general purpose analogue computer, which simulates an interconnected two-subsystem dynamic process (Fig. 4).

The LSI11/23 has a full complement of 256kb of RAM and runs the TSX-PLUS time-shared operating system. Programmes and data are stored on twin 20Mbyte Winchester discs and a 1.2Mbyte floppy disc. Other peripherals available at this level include an Intecolor 8001S colour graphics terminal and a Tektronix graphics terminal with hard copy unit.

The LSI11/23 is programmed mainly in FORTRAN IV with the real-time
support routines written in FORTRAN or are FORTRAN callable.

The I-MIC has 16kb of RAM.  The software for the I-MICs is mostly
written in CONTROL BASIC with a few routines, such as the link communication routines, written in Intel 8085 machine code in situations where the
execution speed of the high level interpretive language is insufficient.
Each I-MIC is interfaced to its respective sub-process using plug-in
memory-mapping interface cards.

Communication between the I-MICs and the LSI11/23 takes place over
20mA current loop serial lines at 1200bd.  Data transfer is initiated by
the local computers and takes place according to a protocol used in the
laboratory (see Stevenson and co-workers, 1984).


4.  ON-LINE IMPLEMENTATION ASPECTS

Using the distributed computer system, parallel computation can be
performed at the local level once coordination variables have been
received from the supremal level.  However, synchronisation and  inter-
process communication problems arise at the local units because the
iteration in each unit is finished within a different time interval.

Because steady state measurements are needed in order to estimate
parameters and, particularly, to compute finite difference approximations
of the derivatives with respect to the controller set points, bad quality
measurements will affect the iterative procedure.  For this reason
synchronisation is recommended to enable controls to be sent to the real
process simultaneously.  Roberts and co-workers (1984) considered two
methods for synchronising the local units - elapsed time and semaphore.

The 'elapsed time' method requires an estimate of the time taken for
each local unit to finish one iteration.  Each unit then waits a sufficient time to enable the slowest unit to complete its task before sending
the controls to the real subprocesses.  Although decentralisation can be
obtained at the local level, it is difficult to determine the waiting
time for every iteration of each unit.  This arises because the computation time required at each iteration is, in general, unpredictable and
often varies due to differing communication delays caused by the time-
shared operating system used at the supremal level.

The second method is the 'semaphore' method.  After each optimization
computation (or other actions, such as taking measurements) each local
unit sends a completion flag to the coordinator and waits.  A task of the
coordinator is to check that all the completion flags from the local units
have been received before transmitting a start flag to each unit, all of
which then send controls simultaneously to the real subprocesses. In real
implementation, data sent to the coordinator may well serve as completion
flags and data sent from the coordinator may be used as a start flag.
This synchronisation scheme has been shown to work efficiently in previous
research (Roberts and co-workers, 1984) and, therefore, is used in the
implementation of the hierarchical ISOPE technique investigated in the
present paper.

After controls have been applied to the real subprocesses, all local
control units should wait a sufficient time until the plant has settled
down, at which point steady state measurements are taken.  It is difficult
to reduce this on-line waiting time interval without risking the quality
of the steady state measurements.  However, if the number of controller
set point changes can be reduced the time for determining the optimal

solution may be less. The double iterative algorithm is aimed at achieving this purpose.

## 5. STUDY EXAMPLE AND EXPERIMENTAL RESULTS

An interconnected two-subsystem plant (Fig. 5) is simulated using a TR48 analogue computer. The steady state of the process is governed by the following equations:

$$y_{11}^* = 1.4c_{11} - 0.6c_{12} + 1.8u_{11} ,$$

$$y_{21}^* = 1.3c_{21} - 1.1c_{22} + 1.1u_{21} ,$$

$$y_{22}^* = 2.3c_{22} - 0.7c_{23} - 1.1u_{21} .$$

The model equations and coupling equation are:

$$y_{11} = c_{11} - c_{12} + 2u_{11} + \alpha_{11} ,$$

$$y_{21} = c_{21} - c_{22} + u_{21} + \alpha_{21}$$

$$y_{22} = 2c_{22} - c_{23} - u_{21} + \alpha_{22} .$$

$$\begin{pmatrix} u_{11} \\ u_{21} \end{pmatrix} = \begin{pmatrix} 0 & 1 & 0 \\ 1 & 0 & 0 \end{pmatrix} \begin{pmatrix} y_{11} \\ y_{21} \\ y_{22} \end{pmatrix} .$$

The performance indices and constraint sets are

$$Q_1(c_1, y_1) = (y_{11}-1)^2 + c_{11}^2 + c_{12}^2 ,$$

$$Q_2(c_2, y_2) = 2(y_{21}-2)^2 + (y_{22}-3)^2 + c_{21}^2 + c_{22}^2 + c_{23}^2 .$$

$$CUY_1 \triangleq \{(c_1,u_1,y_1) \in R^4: |c_{11}| \leq 1, |c_{12}| \leq 1 \text{ and } 0.8 - c_{12} - 0.6u_{11}$$
$$\geq 0\} ,$$

$$CUY_2 \triangleq \{(c_2,u_2,y_2) \in R^6: |c_{2i}| \leq 1 , \quad i = 1,2,3\} .$$

where $c_1 = (c_{11},c_{12})$, $c_2 = (c_{21},c_{22},c_{23})$, $u_1 = u_{11}$, $u_2 = u_{21}$,

$y_1 = y_{11}$, $y_2 = (y_{21},y_{22})$ .

The aim is to use the iterative hierarchical ISOPE technique discussed in Section 2 to determine the optimum (minimum) value of the real system performance

$$Q(c,y^*) = Q_1(c_1,y_1^*) + Q_2(c_2,y_2^*) ,$$

in spite of the fact that the model used to solve the optimisation problem is not a faithful representation of reality.

The real process settling time period is set at 30 seconds. An item of on-line information interchange, beginning when the local control units apply the controls to the real process until the start flag of the next action is received, will take slightly longer than 30 seconds. It is observed that it takes approximately 10 seconds for the system to complete a typical off-line information interchange (each local unit performs its optimization calculation, sends its results to the coordinator and then waits until it is informed by the coordinator that all other units have completed their tasks), which is nearly a third of the time needed to perform an item of on-line information interchange.

Due to lack of computing capability in the I-MICs (integer arithmetic, only one-dimensional array) standard numerical optimization algorithms cannot be loaded into the I-MICs to solve the local optimization problems. Fortunately because of the decomposed nature of the technique used, each local optimization problem is much simpler than the overall optimisation problem and can be solved analytically. Consider the two local optimization problems:

Local optimization problem 1:

$$\min_{c_{11},c_{12},u_{11}} \{(c_{11}-c_{12}+2u_{11}+\alpha_{11}-1)^2 + c_{11}^2 + c_{12}^2 - \lambda_{11}c_{11} - \lambda_{12}c_{12}$$

$$+ p_{11}u_{11} - p_{21}(c_{11}-c_{12}+2u_{11}+\alpha_{11})\} ,$$

s.t. $|c_{11}| \leq 1$, $|c_{12}| \leq 1$ and $0.8 - c_{12} - 0.6u_{11} \geq 0$ .

Local optimization problem 2:

$$\min_{c_{21},c_{22},c_{23},u_{21}} \{2(c_{21}-c_{22}+u_{21}+\alpha_{21}-2)^2 + (2c_{22}-c_{23}-u_{21}+\alpha_{22}-3)^2$$

$$+ c_{21}^2 + c_{22}^2 + c_{23}^2 - \lambda_{21}c_{21} - \lambda_{22}c_{22} - \lambda_{23}c_{23}$$

$$+ p_{21}u_{21} - p_{11}(c_{21}-c_{22}+u_{21}+\alpha_{21})\},$$

s.t. $|c_{2i}| \leq 1$, $i = 1,2,3$ .

Because both problems are convex the first-order necessary optimality conditions are also sufficient. The Kuhn-Tucker necessary conditions, therefore, can be employed to solve these optimization problems. There are eighteen possible solutions for problem 1 and twenty-seven possible solutions for problem 2, depending upon the values of the parameters $\alpha_{11}$, $\alpha_{21}$ and $\alpha_{22}$, the modifiers $\lambda_{1i}$, $i = 1,2$ and $\lambda_{2j}$, $j = 1,2,3$, and the prices $p_{11}$ and $p_{21}$ . The optimization computation in each local decision unit is reduced to deciding which solution is applicable according to the values of the given parameters, modifiers and prices.

The suitable values of the gain matrices $K_v$, $K_p$ are found by experiment as $K_v = 0.4I$ and $K_p = \text{diag}\{0.8, 0.9\}$. The desired tolerances $\beta_v$ and $\beta_p$ are chosen as $\beta_v = 0.006$ and $\beta_p = 0.009$ according to the computing accuracy. Starting from zero initial controls, typical results obtained using the technique discussed in Section 2 are shown in Table 1 where the results are compared with the real optimum values. To compare the efficiency of the single iterative algorithm and the double iterative algorithm, the time required to determine the real optimum solution and the communication requirement of both versions are listed in Table 2.

TABLE 1    Comparison of results

|  | $c_{11}$ | $c_{12}$ | $c_{21}$ | $c_{22}$ | $c_{23}$ | $Q(c,y^*)$ |
|---|---|---|---|---|---|---|
| real optimum | −0.717 | 0.118 | 0.900 | 1.000 | −0.830 | 5.93 |
| SIA | −0.719 | 0.118 | 0.905 | 0.998 | −0.830 | 5.93 |
| DIA | −0.718 | 0.121 | 0.894 | 0.997 | −0.826 | 5.93 |

TABLE 2    Comparison of efficiency

| Algorithm | iterations of modifiers | Optimisation iterations | set point changes | total information interchanges | computation time (minutes) |
|---|---|---|---|---|---|
| SIA | 22 | 23 | 110 | 133 | 71 |
| DIA | 12 | 116 | 60 | 176 | 52 |

## 6.  CONCLUSIONS

It has been demonstrated that by iteratively updating model parameters and, more importantly, modifying local optimisation computations, the hierarchical ISOPE technique is capable of overcoming the model-reality differences and finding the real optimum solution in on-line implementation. The aim of the double iterative algorithm (DIA) is to reduce on-line information interchange and hence, to reduce the time required for determining the optimum steady state control, even at the cost of increasing off-line information exchange. This feature is well demonstrated by the results shown in Table 2.

The DIA is particularly desirable for the processes which have long settling time periods. While the real process settling time period is often beyond our control, using more powerful computers at the local level and increasing the efficiency of the communication link will reduce the time required for off-line information interchange, which also favours the DIA.

Although in this small example, the total number of information interchanges needed by the DIA is slightly greater than that required by the SIA, it has been shown (Chen, Bryds and Roberts, 1985) that this is not generally true. For many examples, the DIA does not increase the total number of information exchanges and, in larger examples, may even reduce it significantly.

As with the majority of decentralised methods, to guarantee good performance, the hierarchical ISOPE technique requires synchronised operation. In situations where measurement noise is intolerable filter techniques can be employed to the measurements and to the computed modifiers. Further research needs to be conducted to study these effects and also to investigate how the system transient affects the performance of this technique when the measurements are taken before the process entirely reaches its steady state.

The I-MICs are currently being replaced by more powerful microcomputers in order to increase the efficiency of the distributed computer network and to accommodate more research activities.

## REFERENCES

Brdys, M., S. Chen and P.D. Roberts, 1984. An extension to the modified two step algorithm for steady state system optimisation and parameter estimation. Research Report CEC/MB/SC/PDR-14, Control Engineering Centre, The City University, London.

Brdys, M. and P.D. Roberts, 1984. Optimal structures for steady-state adaptive optimising control of large scale industrial processes. Research Report, CEC/MB-PDR/1, Control Engineering Centre, The City University, London.

Chen, S., M. Brdys and P.D. Roberts, 1985. An integrated system optimisation and parameter estimation technique for hierarchical control of steady state systems. Research Report, CEC/SC-MB-PDR/19, Control Engineering Centre, The City University, London.

Findeisen, W., F.N. Beiley, M. Brdys, K. Malinowski, P. Tatjewski and A. Wozniak, 1980. Control and Coordination in Hierarchical Systems, Wiley, London.

Michalska, H., J.E. Ellis and P.D. Roberts, 1984. Joint coordination method for the optimizing control of large scale systems. Research Report, CEC/HM-JEE-PDR/2, Control Engineering Centre, The City University, London.

Roberts, P.D., 1979. An algorithm for steady-state system optimisation and parameter estimation. Int. J. of Systems Science, vol. 10, No. 7, pp. 719-734.

Roberts, P.D. and J.E. Ellis, 1981. Refinements to an algorithm for combined system optimisation and parameter estimation. Proc. of the 1981 UKSC Conference on Computer Simulation, Harrogate, England, pp. 245-251.

Roberts, P.D., C.W. Li, I.A. Stevenson and D.S. Wadhwani, 1984. On-line distributed hierarchical control and optimisation of large scale processes using a microcomputer based system. Preprints of the First European Workshop on the 'Real Time Control of Large Scale System', University of Patras, Greece, pp. 432-441.

Stevenson, I.A., D.S. Wadhwani, J.E Ellis, C.W. Li and P.D. Roberts, 1984. A distributed computer network and its application to systems control. 6th European Conference on Electrotechnics, "Eurocon'84", Computers in Communication and Control, Brighton, United Kingdom, pp. 189-193.

Kratos Computer Systems Ltd., I-MIC manual.

S & H Computer Systems Inc., TSX-PLUS manual.

# PARALLEL KALMAN FILTER BANK DESIGN FOR ADAPTIVE IMAGE RESTORATION

S. Tzafestas and M. Skolarikos

Control and Automation Group, Computer Engineering
Division, National Technical University
Zografou, Athens, Greece

ABSTRACT

One of the basic problems in image reconstruction and restoration is to improve the visual quality of the degraded data of the image at hand. In many practical cases, the observed image is a degraded version of the ideal (original) image due to noise and blur. The problem which is solved here is that of finding an optimal estimate of the ideal image on the basis of the observed function that describes the degraded image and a specific optimality criterion. The image is modelled by a linear state space model involving space invariant additive Gaussian white noise. The adaptive image restoration is performed using a parallel bank of filters (partitioning approach) for estimating the state of the image model when the covariance function of the observed image has a separable exponential form depending on an unknown parameter "a". The method has so far been tested with simulated images.

## 1. INTRODUCTION

Recorded images are degraded versions of real images due to blur and noise. Blur is a process that affects the images in a deterministic way, and is due to the relative motion of the object and imaging system (camera). The noise is a stochastic process due to internal errors and inaccuracies in the electronic measuring devices and imaging sensors and/ or to corruption introduced during transmission.

Image restoration (or enhancement) is the process of restoring an image which has been distorted by blur and/or noise, and has attracted the interest of many researchers. Comprehensive reviews and bibliographies in the field of image restoration may be found in the works of Andrews and Hunt [1-3]. As can be seen in these reviews, presently there exists a large number of image restoration approaches such as Wiener filtering, constrained least squares filtering, homomorphic filtering, recursive Kalman filtering, etc. The approach which is followed in the present paper is the Kalman filtering one which was proved to be very popular in image restoration during the last years. The initial investigation of this approach includes the works of Nahi [4], Silverman, Aboutalib and Murphy [5-7], Habibi [8], Woods and Radewan [9], Panda and Cak [10], Suresh and Shenoi [11], and Baroncelli et al [12]. Most of these works lead to space invariant filters which smooth out edges due to a compromise between noise and resolution. This difficulty can be overcome by using adaptive restoration schemes, which of course are computationally more demanding. Some works proposing adaptive image restoration schemes were done by Trussell and Hunt [13], Lim [14], Rajala and Figueiredo [15] and Chan and Lim [16]. In [13-15] the image region is partitioned into disjoint subregions in which different stationary image models are assumed. In [16] a cascade of four 1-D adaptive filters oriented in the four major correlation directions of the image is used, the main objective being to improve the performance of some existing 2-D image restoration techniques. Other adaptive image restoration techniques are based on a simple image model and employ a moving 2-D window to continuously estimate the model parameters and adjust a nonlinear 2-D filter [17-18].

In this paper we provide an alternative adaptive image restoration technique which is based on the partitioning approach to adaptive Kalman filtering [19-21] and leads to a parallel bank of Kalman filters. The adaptivity of the resulting restoring filter concerns the parameter "a" of the exponential and separable covariance function $r_y(k,1)$ of the image field $y(0,j)$ under restoration.

Here we describe the case where the image is distorted only by noise. The case where the recorded image also contains blur will be simply outlined here, and will be presented in detail elsewhere. Numerical and experimental work with real images is in progress.

## 2. STATE SPACE IMAGE MODELLING

If $y(x_1,x_2)$ is the 2-dimensional brightness function that describes the ideal image and $h(x_1,x_2)$ is the point spread function (PSF) of the linear space-invariant blur, the blurred image $f(x_1,x_2)$ is given by

$$f(x_1,x_2) = h(x_1,x_2)*y(x_1,x_2) \qquad (1)$$

where $*$ represents the convolution operation. Now, if $v(x_1,x_2)$ is an additive Gaussian-white noise that affects the image, the recorded (observed) image $z(x_1,x_2)$ is described by

$$z(x_1,x_2) = f(x_1,x_2) + v(x_1,x_2) \qquad (2)$$

The restoration problem is to determine an optimal estimate $\hat{y}(x_1,x_2)$ of the ideal image brightness function $y(x_1,x_2)$ on the basis of the degraded image observation $z(x_1,x_2)$ and some optimality criterion. Of course it is assumed that the PSF $h(x_1,x_2)$ of the blur as well as the statistical parameters of the noise $v(x_1,x_2)$ are completely known.

In the following we assume, without loss of generality, that the PSF of the blur is a Kronecker delta function, i.e.

$$h(x_1,x_2) = h(x_1,x_2;\xi_1,\xi_2) = \delta(x_1-\xi_1)\delta(x_2-\xi_2)$$

in which case the blurred image $f(x_1,x_2)$ in (1) reduces to

$$f(x_1,x_2) = \iint\limits_{-\infty}^{\infty} \delta(x_1-\xi_1)\delta(x_2-\xi_2)y(\xi_1,\xi_2)d\xi_1 d\xi_2 = y(x_1,x_2)$$

and (2) becomes

$$z(x_1,x_2) = y(x_1,x_2) + v(x_1,x_2) \tag{3}$$

The way of treating a general blur PSF, which can also be unknown will be illustrated in Section 4.

Our state-space image model is a discrete one and is based on the following fundamental assumptions.

A1. The observed (recorded) image $z(i,j)$, $(i,j)\epsilon(N\times N)$ where $N\times N$ are the dimensions of the image, has the form

$$z(i,j) = y(i,j) + v(i,j) \tag{4}$$

where $y(i,j)$ is the ideal image under restoration, which is assumed to be a zero-mean 2-D Gaussian stochastic process with covariance

$$r_y(k,1) = E\{y(i+k,j+1)y^T(i,j)\} \tag{5}$$

The noise $v(i,j)$ is assumed to be a zero-mean 2-D Gaussian white process, spatially invariant, uncorrelated with the process $y(i,j)$ and with covariance $\sigma^2\delta_{i+k,i}\delta_{j+1,j}$ ($\delta_{m,n}$ = Kronecker delta).

A2. The ideal image scalar process $y(i,j)$ can be represented by an horizontally invariant state space model of which the state and noise vectors are correlated without any variation from point to point in the vertical direction. This state space model is

$$\underline{x}(i,j+1) = \underline{F}\ \underline{x}(i,j) + \underline{w}(i,j), \underline{x}(i,0) = x_i \tag{6}$$

$$y(i,j) = \underline{H}\ \underline{x}(i,j) \tag{7}$$

where $\dim \underline{x} = \dim \underline{w} = M$, $\dim \underline{F} = M\times M$, $\dim \underline{H} = 1\times M$ (M is the order of the model) and $\underline{F}$, $\underline{H}$ are constant matrices. The zero-mean Gaussian processes $\underline{x}(i,j)$ and $\underline{w}(i,j)$ have the properties

$$
\left.
\begin{aligned}
E\{\underline{w}(i+k,j+1)\underline{w}^T(i,j)\} &= Q(k)\delta_{0,1} \quad \text{with} \quad Q(-k) = Q^T(k) \\
E\{\underline{x}(i+k,j+1)\underline{w}^T(i,j)\} &= 0 \quad \text{for} \quad 1 \geq 0 \\
E\{\underline{x}(i+k,j+1)\underline{x}^T(i,j)\} &= P(k) \quad \text{with} \quad P(-k) = P^T(k)
\end{aligned}
\right\} \tag{8}
$$

From (6) and (8) it follows that $P(k)$ satisfies the following Lyapunov equation

$$P(k) = \underline{F}P(k)\underline{F}^T + Q(k) \tag{9}$$

The covariance $r_y(k,1)$ of $y(i,j)$ is determined from (5)-(7) as follows:

$$
\begin{aligned}
r_y(k,1) &= E\{y(i+k,j+1)y^T(i,j)\} \\
&= \underline{H}\ E\{\underline{x}(i+k,j+1)\underline{x}^T(i,j)\}\underline{H}^T \\
&= \underline{H}\ E\{[\underline{F}^1\underline{x}(i+k,j)+\underline{w}(i+k,j)]\underline{x}^T(i,j)\}\underline{H}^T \\
&= \underline{H}\ \underline{F}\ E\{\underline{x}(i+k,j)\underline{x}^T(i,j)\}\underline{H}^T
\end{aligned}
$$

since $E\{\underline{w}(i+k,j)\underline{x}^T(i,j)\} = 0$ .

Now taking into consideration the properties (8) one obtains

$$r_y(k,1) = \begin{cases} \underline{H}\ \underline{F}^1\underline{P}(k)\underline{H}^T & \text{for } 1 \geq 0, \ k \geq 0 \\ \underline{H}\ \underline{F}^1\underline{P}^T(k)\underline{H}^T & \text{for } 1 \geq 0, \ k \leq 0 \end{cases} \tag{10}$$

A3. The state-space model depends on the form of the covariance function $r_y(k,1)$. Here we assume that $r_y(k,1)$ has the form

$$r_y(k,1) = a^{|k|+|1|}, \quad 0 < a < 1 \tag{11}$$

which is separable. This is a very simple selection but it was proved experimentally that covers most practical cases [12].

From (11) it follows that the state space model of the image has the form (6) and (7) with $\underline{F} = a$ and $\underline{H} = 1$, i.e.

$$x(i,j+1) = a\ x(i,j) + w(i,j) \tag{12}$$

$$y(i,j) = x(i,j) \tag{13}$$

In order to describe fully the image model it is necessary to determine the expressions of $\underline{P}(k)$ and $\underline{Q}(k)$ in terms of the parameter a.

From (10) and (11) we get

$$\underline{H}\ \underline{F}^1P(k)H^T = a^{|k|}a, \quad 1 \geq 0$$

or, since $\underline{F} = a$ and $\underline{H} = 1$, $a^1P(k) = a^{|k|}a^1$.

Thus

$$P(k) = a^{|k|} \tag{14}$$

Finally, from (9) one finds that $a^{|k|} = a^2a^{|k|} + Q(k)$, i.e.

$$Q(k) = (1-a^2)a^{|k|} \triangleq Q(k,a) \tag{15}$$

A4. In order to reduce the computational burden it is assumed that the scanning of the image is column-to-column, i.e. at each step of the sequential Kalman filtering a whole column is estimated (not a single picture element). Thus the following state vector is defined

$$\underline{X}(j) = [\underline{x}(1,j),...,\underline{x}(N,j);\underline{x}(1,j-1),...$$

$$...\underline{x}(N,j-1);\underline{x}(1,j-M+1),...,x(N,j-M+1)]^T,$$

where M is the order of the model, and hence

$$\underline{X}(j+1) = \bar{\underline{F}}\ \underline{X}(j) + \underline{W}(j), \quad \underline{Z}(j) = \bar{\underline{H}}\ \underline{X}(j) + \underline{V}(j) \tag{16}$$

where

$$\bar{\bar{H}} = \mathrm{diag}[\bar{H},\bar{H},\ldots,\bar{H}], \quad \bar{\bar{F}} = \mathrm{diag}[\bar{F},\bar{F},\ldots,\bar{F}]$$

with $\bar{H} = 1$, $\dim \bar{\bar{H}} = N \times NM$, $\bar{F} = a$, $\dim \bar{\bar{F}} = NM \times NM$ .

Also

$$
\begin{aligned}
E\{\underline{X}(j)\underline{X}^T(j)\} &= \bar{\bar{P}}(j) \\
E\{\underline{W}(j)\underline{W}^T(j)\} &= \bar{\bar{Q}}(j) \\
E\{\underline{V}(j)\underline{V}^T(j)\} &= \bar{\bar{R}}(j) = \sigma^2 I
\end{aligned}
\right\} \tag{17}
$$

The matrices $\bar{\bar{P}}$ and $\bar{\bar{Q}}$ are symmetric Toeplitz matrices with elements the functions $P(k)$ and $Q(k)$ as defined in (14) and (15), i.e.

$$
\bar{\bar{P}} = \begin{bmatrix}
\underline{P}(0) & \underline{P}^T(1) \ldots\ldots \\
\underline{P}(1) & P(0) & \underline{P}^T(1)\ldots \\
\underline{P}(2) & \underline{P}(1) & \underline{P}(0) \ldots \\
\ldots\ldots\ldots\ldots \\
\ldots\ldots\ldots\ldots
\end{bmatrix}, \quad
\bar{\bar{Q}} = \begin{bmatrix}
\underline{Q}(0) & \underline{Q}^T(1) \ldots\ldots \\
\underline{Q}(1) & \underline{Q}(0) & \underline{Q}^T(1)\ldots \\
\underline{Q}(2) & \underline{Q}(1) & \underline{Q}(0) \ldots \\
\ldots\ldots\ldots\ldots \\
\ldots\ldots\ldots\ldots
\end{bmatrix}
$$

Due to the assumption that the image model is horizontically invariant, and that the state and noise vectors are correlated in the vertical direction, we can assume that $M = 1$, i.e. one can examine the processing of only one column. This property allows one to minimize the computational and memory requirements.

## 3. FORMULATION AND SOLUTION OF THE ADAPTIVE IMAGE RESTORATION PROBLEM

The classical sequential Kalman filter has found considerable attention in the framework of image processing and restoration. From among the great number of papers existing in the technical literature, here we shall briefly summarize the ones mentioned in the introduction [5-12].

This will help the reader to appreciate the value of the present approach. In [5] Aboutalib and Silverman have derived some basic results on how the sequential Kalman filter could be applied for the restoration of images degraded by linear motion blur and additive noise. The image covariance was assumed separable exponential and known, whereas the scanning was assumed line-by-line. The methodology of [5] was extended by the same authors in [6], whereas in [7] two practical Kalman-filter-based suboptimum restoration schemes (strip restoration and constrained optimal restoration) were given and applied to actual degraded image. In [8], Habibi provides a dynamic model of images possessing separable, exponential and known autocorrelation function, and presents a recursive Kalman-type 2-D filter derived through a Bayesian estimation procedure. In [9], Woods and Radewan have proposed the Kalman band filter for the restoration of images degraded only by noise. Similar procedures based on Kalman band filters were employed by Pand and Cak in [10] and Suresh and Shenoi [11]. In [11] a different state space image model was also derived using autoregressive spectral analysis for the case of non-separable image covariance function. Finally in [12] one can find a useful presentation of the problems related with the recursive estimation of 2-D images, as well as a Kalman-type algorithm which is used for the restoration of x-rays.

All of the works summarized above were based on the assumption that the autocovariance function of the image field under restoration is known. Here the case where this autocovariance function is of the separable exponential form [11] with the parameter "a" unknown will be considered.

Thus our problem is to *recursively estimate* $\underline{X}(j)$ *in the model, under the condition that the covariance parameter "a" is unknown.*

If the parameter "a" was known then one could use directly the Kalman filter which is a copy of the system model (16) with a feedback term from the output estimation error.

To treat the present problem we quantize the parameter interval $0 < a < 1$, i.e. we assume that "a" can take a set of values $\{a_s, s = 1, 2,...,S\}$ such that $0 < a_1 < a_2 <...< a_S < 1$.

For each one value another image model is obtained, and thus another Kalman filter is designed, which provides an optimal estimate of the image only for the associated value of "a".

Then to generate an optimal (parameter adaptive) image estimate we use the well known partitioned Kalman filter [19-21] which is implemented by a parallel bank of filters each obtained for a particular value of "a".

The a-dependent filter equations are:

$$\hat{\underline{X}}(j|j;a_s) = \bar{\underline{F}}(a_s)\hat{\underline{X}}(j-1|j;a_s)$$
$$+ \underline{K}(j;a_s)\{\underline{Z}(j)-\bar{\underline{H}}\,\bar{\underline{F}}(a_s)\hat{\underline{X}}(j-1|j-1;a_s)\},$$
$$\hat{\underline{X}}(0|0;a_s) = 0$$
$$\underline{K}(j;a_s) = \underline{\Sigma}(j|j-1;a_s)\bar{\underline{H}}^T\{\bar{\underline{H}}\,\underline{\Sigma}(j|j-1;a_s)\bar{\underline{H}}^T+\sigma^2 I\}^{-1} \qquad\left.\right\}\quad (18)$$
$$\underline{\Sigma}(j|j-1;a_s) = \bar{\underline{F}}(a_s)\underline{\Sigma}(j-1|j-1;a_s)\bar{\underline{F}}^T(a_s)+\bar{\underline{Q}}(a_s)$$
$$\underline{\Sigma}(j|j;a_s) = \underline{\Sigma}(j|j-1;a_s) - \underline{K}(j;a_s)\bar{\underline{H}}\,\underline{\Sigma}(j|j-1;a_s)\;,$$
$$\underline{\Sigma}(0|0;a_s) = \underline{P}_0$$

where $\underline{\Sigma}(j|j-1;a_s)$ is the covariance matrix of the state prediction error, and $\underline{\Sigma}(j|j;a_s)$ is the covariance of the state filtered estimate.

The total (adaptive) estimate $\hat{\underline{X}}(j|j)$ is obtained by averaging the estimates $\hat{\underline{X}}(j|j;a_s)$ with weights the *a posteriori* probabilities $p(a_s|Z_j)$ of the parameter values $a_s$, $s = 1,2,...,S$ conditional upon the measured set

$$\underline{Z}_j = \{\underline{Z}(1),\underline{Z}(2),...,\underline{Z}(j)\} \qquad (19)$$

That is

$$\hat{\underline{X}}(j|j) = \sum_{s=1}^{S} \hat{\underline{X}}(j|j;a_s)p(a_s|\underline{Z}_j) \qquad (20)$$

The implementation of (20) is performed by using a parallel bank of a-dependent Kalman filters as shown in Fig. 1.

The *a posteriori* probability density $p(a_s|\underline{Z}_j)$ is obtained through

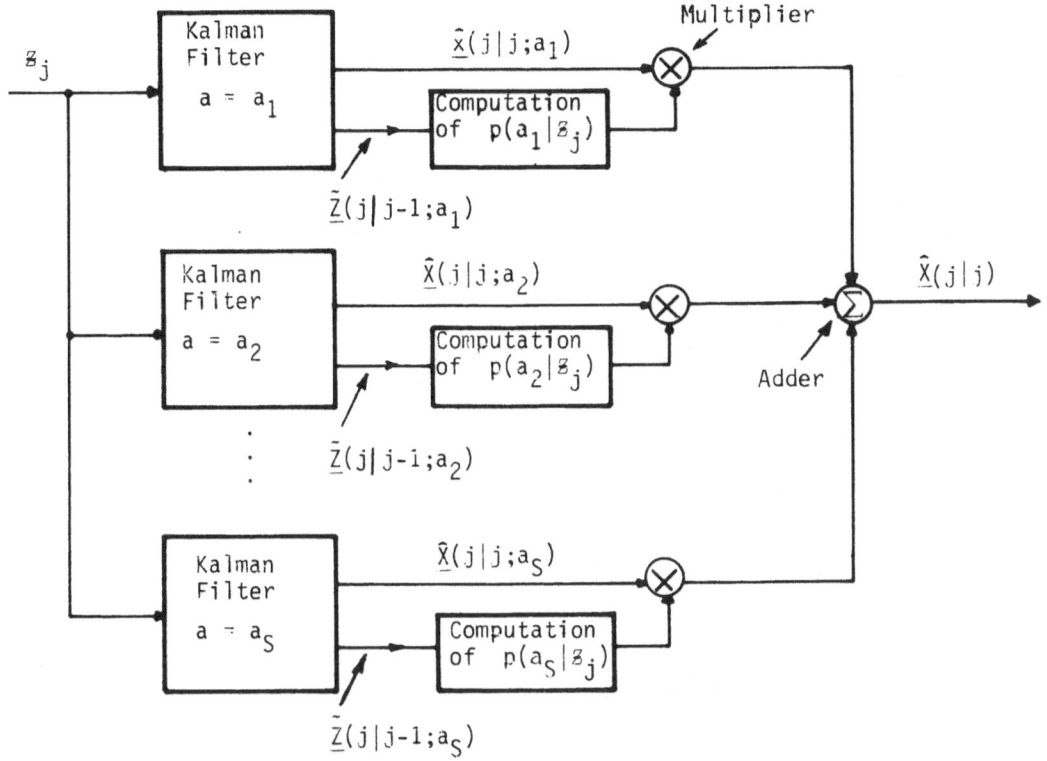

Fig. 1   Parallel bank of Kalman filters for the
generation of $\hat{\underline{X}}(j|j)$.

$$p(a_s|\mathcal{Z}_j) = p(\mathcal{Z}_j, a_s)/p(\mathcal{Z}_j)$$

$$= p(Z_j, \mathcal{Z}_{j-1}; a_s)/p(Z_j, \mathcal{Z}_{j-1})$$

$$= \frac{p(Z_j, a_s|\mathcal{Z}_{j-1})p(\mathcal{Z}_{j-1})}{p(Z_j|\mathcal{Z}_{j-1})p(\mathcal{Z}_{j-1})}$$

$$= \frac{p(Z_j|\mathcal{Z}_{j-1}; a_s)p(a_s|\mathcal{Z}_{j-1})}{\sum_{s=1}^{S} p(Z_j|\mathcal{Z}_{j-1}; a_s)p(a_s|\mathcal{Z}_{j-1})} \tag{21}$$

Now $p(Z_j|\mathcal{Z}_{j-1}; a_s)$ is assumed Gaussian with:

Mean value:   $\hat{\underline{Z}}(j|j-1; a_s)$

Covariance:   $E\{\underline{\tilde{Z}}(j|j-1; a_s)\underline{\tilde{Z}}^T(j|j-1; a_s)\} = \underline{\Omega}(j|j-1; a_s)$ \tag{22}

where

and

$$\tilde{\underline{Z}}(j|j-1;a_s) = \underline{Z}(j) - \hat{\underline{Z}}(j|j-1;a_s) \qquad (23a)$$

$$\hat{\underline{Z}}(j|j-1;a_s) = \bar{\bar{H}}\,\bar{\bar{F}}(a_s)\hat{\underline{X}}(j-1|j-1;a_s) \qquad (23b)$$

Thus

$$\underline{\Omega}(j|j-1;a_s) = \bar{\bar{H}}\,\Sigma(j|j-1;a_s)\bar{\bar{H}}^T + \sigma^2\underline{I} \qquad (23c)$$

This means that $p(Z_j|\mathbf{Z}_{j-1};a_s)$ is given by

$$p(Z_j|\mathbf{Z}_{j-1};a_s) = \frac{1}{\sqrt{2\pi}}\,|\underline{\Omega}^{-1}(j|j-1;a_s)|^{\frac{1}{2}}$$

$$\times \exp\{-\frac{1}{2}\tilde{\underline{Z}}^T(j|j-1;a_s)\underline{\Omega}^{-1}(j|j-1;a_s)\tilde{\underline{Z}}(j|j-1;a_s)\} \qquad (24)$$

and so

$$p(a_s|\mathbf{Z}_j) = C|\underline{\Omega}(j|j-1;a_s)|^{-\frac{1}{2}}\exp\{-\frac{1}{2}\tilde{\underline{Z}}^T(j|j-1;a_s)\underline{\Omega}^{-1}(j|j-1;a_s)$$

$$\times \tilde{\underline{Z}}(j|j-1;a_s)\}p(a_s|\mathbf{Z}_{j-1}) \qquad (25)$$

where C is a constant independent of a, which is selected such that

$$\sum_{s=1}^{S} p(a_s|\mathbf{Z}_j) = 1 \qquad (26)$$

To implement the sequential equation (25) we need an initial condition

$$p(a_s|0) = p(a_s) \qquad (27)$$

This can be found using the maximum entropy principle (MEP), i.e. by maximizing the entropy

$$H = -\sum_{s=1}^{S} p(a_s)\log p(a_s) \qquad (28a)$$

subject to

$$\sum_{s=1}^{S} p(a_s) = 1 \qquad (28b)$$

Using a Lagrange multiplier $\lambda$ one has to maximize

$$J = -\sum_{s=1}^{S} p(a_s)\log p(a_s) - \lambda\{\sum_{s=1}^{S} p(a_s)-1\}$$

with respect to $p(a_s)$, without any constraint. The solution is given by: $\partial J/\partial p(a_s) = \log p(a_s)+1+\lambda = 0$, i.e.

$$p(a_s) = e^{-(\lambda+1)} \qquad (29)$$

Replacing (29) into the condition (28b) one finally finds that
$S \exp[-(1+\lambda)] = 1$, whence

$$p(a_s) = \exp[-(1+\lambda)] = 1/S \tag{30}$$

Equation (30) shows that the initial probability $p(a_s)$ which satisfies the MEP is constant (uniform distribution).


## 4. EXTENSION TO COVER GENERAL MOTION BLURS AND IMAGE PARTITIONING

The above parallel adaptive filtering technique provides the best estimate of the image field at hand for the case where the image degradation is only due to the presence of observation noise as in (4). The case of images degraded by motion blurs can be easily treated.

One simply has to incorporate the appropriate state-space model of the blur, which is of the form (see (6)) [5-7]:

$$\underline{x}'(i,j+1) = \underline{A}\,\underline{x}'(i,j) + \underline{B}\,y(i,j), \quad \underline{x}'(i,0) = \underline{x}_i \tag{31}$$

$$s(i,j) = \underline{C}\,\underline{x}'(i,j) + \underline{d}\,y(i,j) + v(i,j) \tag{32}$$

where $s(i,j)$ is the image degraded by blur and noise, and $v(i,j)$ is a zero-mean white noise with variance $\sigma_v^2$ .

Combining the object model (6)-(7) with the degradation model (31)-(32) yields the overall image model

$$\underline{\xi}(i,j+1) = \bar{\underline{A}}\,\underline{\xi}(i,j) + \bar{\underline{B}}\,\underline{w}(i,j) \tag{33}$$

$$s(i,j) = \bar{\underline{C}}\,\underline{\xi}(i,j) + v(i,j) \tag{34}$$

where

$$\underline{\xi}(i,j) = \begin{bmatrix} \underline{x}'(i,j) \\ \hline \underline{x}(i,j) \end{bmatrix}, \quad \bar{\underline{A}} = \begin{bmatrix} A & B\,H \\ \hline 0 & F \end{bmatrix}, \quad \bar{\underline{B}} = \begin{bmatrix} 0 \\ \hline I \end{bmatrix},$$

$$\bar{\underline{C}} = [\underline{C},\ \underline{d}\,H] \tag{35}$$

The adaptive filtering technique of the paper is directly applicable to the image model (33)-(35) where not only the parameter $\underline{F} = a$ is allowed to be unknown, but also parameters involved in the blur degradation matrices $\underline{A}$, $\underline{B}$, $\underline{C}$ and $\underline{d}$.

Having estimated $\underline{\xi}(i,j)$ the adaptive estimate of the ideal picture elements $y(i,j)$ is obtained using the formula

$$\hat{y}(i,j|j) = [0\ \underline{H}]\hat{\underline{\xi}}(i,j|j), \quad j = 0,1,2,\ldots,N-1 .$$

As in the pure noise case, the degraded image can be scanned column-to-column (or line-by-line).

We close this section by remarking that the present approach can be combined with the adaptive image restoration method, where the image y(i,j) is partitioned in a set of disjoint regions $\Omega_p; p=1,2,\ldots,K$. Each of these regions is described by a two dimensional stationary random field with an unknown autocorrelation function of the type (11). This decomposition results in more computational requirements, but is expected to lead to better overall image estimates.

The lack of good initial condition information for the starting points in the regions can be balanced by employing 2-D interpolation schemes to obtain better estimates of the initial state vectors (see [15]).

## 5. CONCLUDING REMARKS

As it is seen from (20) and Fig. 1 the present image restoration technique has a parallel nature at the system level. Of course each filter of the parallel bank of Fig. 1 can be implemented by existing techniques of parallelizing the Kalman filter.

So far the present technique was used to restore simulated images with success. Up to ten values of $a_s$ were used in quantizing the a-parameter set.

Just for illustration we give here the results obtained with two values $a_1 = 0.8$ and $a_2 = 0.9$ with *a priori* probabilities $p(a_1) = p(a_2) = 0.1$. Application of the method gave $p(a_1|z_{10}) = 0.22$ and $p(a_2|z_{10}) = 0.78$. The final a-dependent estimates of the state vector are:

$$\hat{\underline{X}}(10/10;a_1) = \begin{bmatrix} 0.20825 \\ 0.59082 \\ 0.89734 \\ 1.32077 \\ 1.20364 \\ 0.71148 \\ 0.72755 \\ 0.19790 \\ 0.49393 \\ 0.66777 \end{bmatrix} \quad \text{and} \quad \hat{X}(10/10;a_2) = \begin{bmatrix} -0.21522 \\ -0.24229 \\ 0.19240 \\ -0.65630 \\ -0.44991 \\ -0.30749 \\ -0.03714 \\ -0.16012 \\ -0.06553 \\ -0.01037 \end{bmatrix}$$

The final adaptive estimate $\hat{\underline{X}}(10/10) = \hat{\underline{X}}(10/10;a_1)p(a_1/z_{10}) + \underline{X}(10/10;a_2)p(a_2/z_{10})$ and the true state $X(10/10)$ are:

$$\hat{X}(10/10) = \begin{bmatrix} -0.12204 \\ -0.05900 \\ 0.34748 \\ -0.22134 \\ -0.08611 \\ -0.08332 \\ 0.13109 \\ -0.08147 \\ 0.05755 \\ 0.13882 \end{bmatrix} \quad , \quad X(10/10) = \begin{bmatrix} -0.11560 \\ -0.05173 \\ 0.34750 \\ -0.22134 \\ -0.08613 \\ -0.08335 \\ 0.13108 \\ -0.08150 \\ 0.05751 \\ -0.13880 \end{bmatrix}$$

We see that even with this two-value parameter quantization the accuracy obtained is good. Results are currently obtained with real images.

An alternative approach to adaptive image restoration is the one which uses an ARMA model of image field combined with maximum likelihood parameter estimation and appropriate filtering [22-23]. It is useful to remark here that the formulation of state space models of 2-dimensional (or m-dimensional) images having general (non separable) autocorrelation functions is a very difficult problem due to the lack of a universal factorization technique of 2-D polynomials. Some useful results in this direction can be found in [24-25].

Finally, we mention that the present parallel adaptive filtering technique can also be used in cases where the image field is described by partial differential equation (distributed parameter) models [21,26-28].

REFERENCES

1. H.C. Andrews and B.R. Hunt, 1977, Digital image restoration. Englewood Cliffs, NJ: Prentice Hall.
2. H.C. Andrews, 1974, Digital image restoration: A survey. Computer, vol. 8, pp. 36-45.
3. B.R. Hunt, 1975, Digital image processing. Proc. IEEE, vol. 63, pp. 693-708.
4. N.E. Nahi, 1972, Role of recursive estimation in statistical image enhancement. Proc. IEEE, vol. 60, pp. 872-877.
5. A.O. Aboutalib, L.M. Silverman, 1975, Restoration of motion degraded images. IEEE Trans. Circuits and Systems, vol. CAS-22, pp. 278-286.
6. A.O. Aboutalib, M.S. Murphy and L.M. Silverman, 1977, Digital restoration of images degraded by general motion blurs. IEEE Trans. Auto Control, vol. AC-22, pp. 294-302.
7. M.S. Murphy and L.M. Silverman, 1978, Image model representation and line-by-line recursive restoration. IEEE Trans. Auto Control, vol. AC-23, No. 5, pp. 809-816.
8. A. Habibi, 1972, Two dimensional Bayesian estimate of images. Proc. IEEE, vol. 60, pp. 878-883.
9. J.W. Woods and C.H. Radewan, 1977, Kalman filtering in two dimensions. IEEE Trans. on Inf. Theory, vol. 23, pp.473-482.
10. D.P. Panda, A.C. Cak, 1977, Recursive least squares smoothing of noise in images. IEEE Trans. on Acoust., Sp. Signal, 25, 520-524.
11. B.R. Suresh, B.A. Shenoi, 1981, New results in 2-D Kalman filtering with applications to image restoration. IEEE Trans on Circuits and Syst., vol. 28, No. 4, pp. 307-319.
12. L. Baroncelli, A. Del Bimbo, G. Zappa, 1980, 2-D Kalman filtering with applications to the restoration of scintigraphic images. Digital Signal Processing (V. Cappellini and A.G. Constantinides Eds.), pp. 183-186, Academic Press.
13. H.J. Trussell and B.R. Hunt, 1978, Sectioned methods for image restoration. IEEE Trans. Acoust., Speech, Signal Processing, vol. ASSP-26, p. 157.
14. J.S. Lim, 1980, Image restoration by short space spectral subtraction. IEEE Trans. Acoust., Speech, Signal Processing, vol. ASSP-28, p. 191.
15. S.A. Rajala and R.J.P. de Figueiredo, 1981, Adaptive nonlinear restoration by a modified Kalman filtering approach. IEEE Trans. Acoust., Speech, Signal Processing, vol. ASSP-29, p. 1033.
16. P. Chan and J.S. Lim, 1985, One-dimensional processing for adaptive image restoration. IEEE Trans. Acoust., Speech, Signal Processing, vol. ASSP-33, No. 1, p. 117.
17. J.F. Abramatic and L. Silverman, 1982, Nonlinear Restoration of Noisy images. IEEE Trans. Pattern Anal. Machine Intell., vol. PAMI-4, p. 141.

18. J.S. Lee, 1980, Digital image enhancement and noise filtering by use of local statistics. IEEE Trans. Pattern Anal. Machine Intell., vol. PAMI-2, p. 165.

19. D.T. Magill, 1965, Optimal adaptive estimation of sampled stochastic processes. IEEE Trans. Auto. Control, vol. AC-10, p. 434.

20. D.G. Lainiotis, 1971, Optimal adaptive estimation: structure and parameter adaptation. IEEE Trans. Auto. Control, vol. AC-16, p. 160.

21. S.G. Tzafestas, 1980, Partitioning approach to distributed parameter filtering. Ric. Automatica, vol. 11, No. 1, p.51.

22. T. Katayama, 1982, Restoration of images degraded by motion blur and noise. IEEE Trans. Auto. Control, vol. AC-27, p. 1024.

23. S.G. Tzafestas, 1985, Three dimensional image restoration using ARMA modelling and ML estimation. Digital Signal Processing, (A. Luque, A.R. Figueiras Vidal and V. Cappellini, Eds.), vol. II, pp. 227-232, North Holland.

24. S.G. Tzafestas and N.J. Theodorou, 1984, Multidimensional State space models: A comparative overview. Mathematics and Computers in Simulation, vol. 26, p. 432.

25. N.J. Theodorou and S.G. Tzafestas, 1985, Reducibility and factorizability of multivariable polynomials. C-TAT: Control Theory and Advanced Technology, vol. 1, No. 1, p. 25.

26. A.K. Jain and J.R. Jain, 1978, Partial differential equations and finite difference methods in image processing. Part I - Image representation, J. Optimiz. Theory and Advanced Technology, vol. 23, p. 65, Sept., 1977; Part II - Image restoration, IEEE Trans. Auto Control, vol. AC-23, No. 5, p. 817, Oct., 1978.

27. S.G. Tzafestas, 1981, Partitioned adaptive filtering and control of distributed systems with space-dependent unknown parameters. Mathematics and Computers in Simulation, vol. 23, p. 206.

28. P. Stavroulakis and S.G. Tzafestas, 1982, Multipartitioning in distributed-parameter adaptive estimation. Int. J. Systems Science, vol. 13, p. 301.

# A MULTIMICROPROCESSOR FOR PARALLEL PROCESSING

G. Authie* and D. El Baz

Laboratoire d'Automatique et d'Analyse de Systemes du
C.N.R.S., 7 Avenue du Colonel Roche, 31077 Toulouse Cedex
France
*also I.N.S.A., Departement de Genie Electrique
 31077 Toulouse Cedex, France

## ABSTRACT

We present a low cost efficient multimicroprocessor designed to study
parallel algorithms like fixed point parallel algorithms or parallel
algorithms for distributed systems control. This multiprocessor is a
collection of 8 processors sharing a common memory. Its architecture
presents some original features that guarantee data consistency and good
efficiency. Experimental results are given for synchronous and asynchronous algorithms applied to the Dirichlet problem.

## INTRODUCTION

Today there is more and more an increasing need for very fast computing machines; as a consequence some multiprocessors are now becoming
available. A multiprocessor is a collection of processors that execute
various processes in parallel and share a common memory (see [3][5]).
Although some theoretical work has been done on the convergence of parallel
fixed point algorithms for multiprocessors (Chazan and Miranker [2],
Miellou [4]) there is considerable work available concerning the analysis
of parallel algorithms, i.e. reliability, efficiency. Among this research
little has been done through real simulation on multiprocessors (Baudet
[1]). Most result from simulations on a single processor.

In this paper we present a low cost efficient multimicroprocessor
designed to study parallel algorithms. The field of application of this
machine is very wide: fixed point problems, dynamic programming, large
scale systems simulation, distributed system control, robotics, etc. This
multimicroprocessor consists of 8 identical processors that can perform
various tasks simultaneously and that share a common memory. A supervision unit controls the synchronization of the tasks. This multimicro
computer presents quite original features that ensure data consistency
and a good efficiency.

In the first paragraph we present the multimicroprocessor features.
The second paragraph deals with the architecture of a processor.
Paragraphs 3 and 4 are devoted respectively to the arbitration technique

for accessing the shared memory and to the multiprocessor supervision.
In paragraph 5 we describe the multiprocessor software. In paragraph 6
we comment on some experiments carried out on the multiprocessor. We
analyse various parallel algorithms (e.g. synchronous or asynchronous
algorithms) which are applied to the resolution of the well-known Dirichlet
problem for Laplace's equation.


1. MULTIMICROPROCESSOR FEATURES

   The main features of our machine are:

- An architecture designed to perform simultaneously different tasks,
  mainly scientific computations. The multimicroprocessor is a
  collection of 8 processors, its architecture is illustrated in Fig. 1.

- A time shared bus to access a 30 K words common memory (the memory
  words are of 32 bits). The arbitration technique which is simple and
  original guarantees a mutual exclusion for accessing the memory.

- A design of the multimicroprocessor as a peripheral of a host computer.
  The host computer is used to prepare applications.

- An exchange and supervision module which controls the synchronization
  of the processors and the exchange of data between the host computer
  and the common memory or the local memory of the processors.

- A very low cost. Each processor CPU is a microprocessor Motorola
  MC 6809. The machine is rather designed to study parallel algorithms
  than to perform very fast computations.

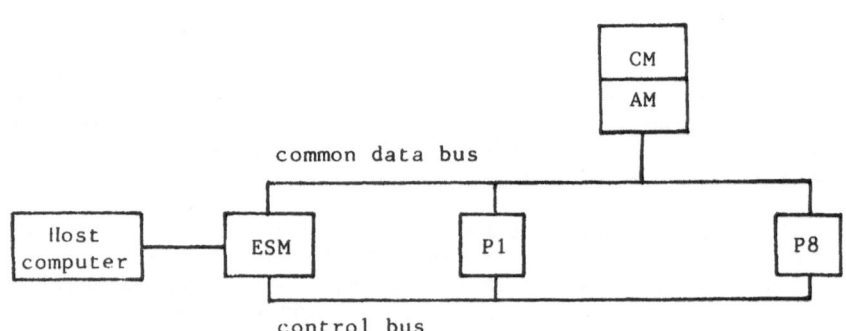

ESM : Exchange and supervision module

CM  : Common Memory (30 K words;  1 word = 32 bits)

AM  : Arbitration module

P   : Processor

Fig. 1   Multimicroprocessor architecture

## 2. ARCHITECTURE OF A PROCESSOR

Each processor is a full microcomputer with a classical architecture (see Fig. 2).   A processor has its own private bus, on this bus are

1.   a CPU (MC 6809)

2.   an arithmetic processing unit AM 9511, which performs computations on 4 bytes floating point operands.  This unit permits to increase the performances of the processor.

3.   a local memory consisting of 26 K bytes of random access memory (this memory is used to store routines and data) and 4 K bytes of REPROM (which contains the processor monitor).

4.   an interface between the local data bus (8 lines dedicated to data) and the common data bus (32 lines dedicated to data).  This is one of the most original features of our machine, its main interest is to ensure data consistency.

5.   A timer which permits to point out the time of some events (like the end of an iteration), this is very useful to analyse parallel algorithms.

6.   an input-output connection to control an asynchronous communication interface adapter MC 6850 of MOTOROLA.  This permits to connect a peripheral with a RS 232 line.  With this feature each processor can supervise or control a peripheral.

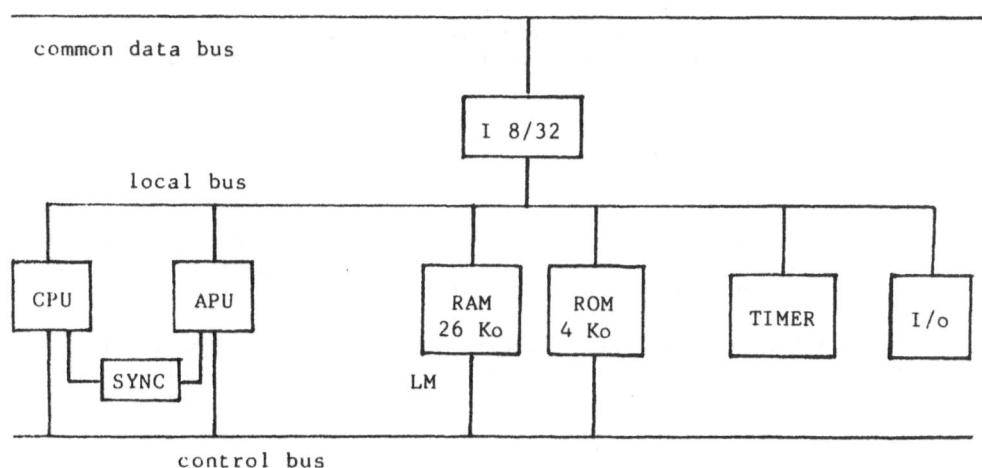

APU     :  Arithmetic processing unit (AM 9511)

SYNC    :  Synchronization of the CPU clock.

LM      :  Local Memory.

I/O     :  Input-output connection for an asynchronous interface adapter MC 6850.

I 8/32  :  Interface between the local data bus and the common data bus.

Fig. 2   Processor architecture

# 3. ARBITER

Each processor performs computations with its arithmetic processing unit on 32 bits floating point numbers. The processors cooperate by exchange of data through the common memory. The shared memory is organized in 32 bits words. Data consistency is guaranteed and contention is reduced since there are interfaces between the local bus (8 lines dedicated to data) and the time shared bus (32 lines dedicated to data). The interfaces do not eliminate contention, two processors or more may want to access the shared memory simultaneously. So there must be some arbitration among the competing processors.

## 3.1 Arbiter Structure

We have chosen a synchronous arbiter in order to simplify its design. In the beginning of each CPU cycle the arbiter picks all the requests $\bar{R}_i$, $i = 1,...,p$. The multimicroprocessor address space is divided up in 2 parts of 32 K words each, one is dedicated to the local memory and the other to the common memory. As a result we simply have $\bar{R}_i = (A_{15})_i$ where $(A_{15})_i$ is the most significant digit of the address in processor i. This implies that an access to the common memory does not essentially differ from an access to the local memory. The set of the requests constitutes an input address in the decision memory of the arbiter.

In the middle of the CPU cycle the arbiter has read the grant orders $\bar{G}_i$ and the waiting orders $(MRDY_i)$ $i = 1,...,p$. These orders are then presented to the service units.

The grant service unit enables then a processor to access the shared memory during the cycle where the request was emitted. The waiting service unit stops during a cycle the processors not serviced. The arbiter is synchronized on this signal.

## 3.2 Arbitration Policy

An arbitration policy is the set of decisions $\bar{G}_i$ and $MRDY_i$ , $i = 1,...,p$, which corresponds to the various requests picked in the beginning of a cycle. The arbiter is in fact a microprogrammed device with a 256 words control memory (the words are of 16 bits).

We have programmed a fixed priority policy corresponding to the processor number.

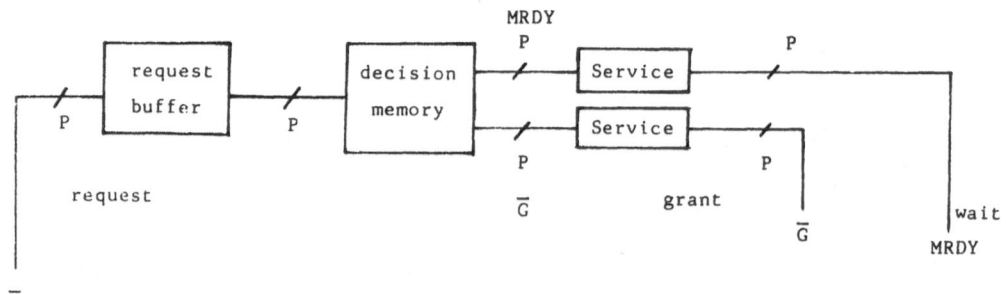

Fig. 3    Centralized arbiter.

## 4. EXCHANGE AND SUPERVISION MODULE (ESM)

The ESM module has two functions:

### 1. Communication between the host computer and the multimicroprocessor

Applications are prepared in the host computer (programs and data). A communication protocol has been designed to transfer packets of information between the host computer and the multiprocessor by a serial line. There are two types of packets:

- data packets to transfer data and programs
- control packets to send commands from the host computer to the multi processor and to send acknowledgment messages when transmitting data packets.

### 2. Control and supervision of the multiprocessor

The ESM module has the greatest priority level when accessing the common memory. It doesn't access directly local (private) memory; the loading or reading of local memory is obtained through common memory. Transfer of information between common memory and a given processor is controlled by the processor when receiving an order from the ESM module.

The ESM module can access the status register of each processor (the different states are: busy, idle, waiting for synchronization, problem solved) and is able to change the status of each processor by sending orders on the control bus (see Fig. 1). Then it is possible to implement all types of computation schemes (synchronous and asynchronous ones).

## 5. MULTIPROCESSOR SOFTWARE

The multiprocessor software consists of a monitor and an operating system.

### 5.1 The Monitor

The monitor is located in each processor. It interprets the controls sent by the exchange and supervision module and executes them.

### 5.2 The Operating System

The operating system is composed of two parts:

- in the host computer, programs permit the preparation of applications and send all the necessary information to the exchange and supervision module.
- in the exchange and supervision module programs receive information from the host computer and control the whole multimicroprocessor.

## 6. MULTIMICROPROCESSOR VALIDATION

In this paragraph we present some experiments carried out to validate our machine. Asynchronous iterative algorithms are implemented on the multimicroprocessor to solve a set of algebraic linear equations.

## 6.1  Asynchronous Iterative Algorithms

Various problems in numerical analysis, optimization or control can be formulated as fixed point problems $x^* = F(x^*)$, where $F$ is a mapping from $R^n$ to $R^n$. Those problems can be solved by iterative methods. Iterative algorithms start with a given vector $x(o)$ and generate a sequence $x(k)$ of vectors of $R^n$ with $k \in N^*$ the set of positive integers. Iterative methods can be defined using the general definition of asynchronous iterations which is due to Baudet [1].

We first give some notations.

### Notations

The components of vector $x$ of $R^n$ will be denoted by $x_i$ $i = 1, \ldots, n$. The components of the operator $F$ will be denoted $f_i(x)$ or $f_i(x_1, \ldots, x_n)$ $i = 1, \ldots, n$ .

Definition:  An asynchronous iteration corresponding to the operator $F$ and starting with a given vector $x(o)$ is a sequence $x(k)$ $(k \in N^*)$ of vectors of $R^n$ defined recursively by:

$$x_i(k) = \begin{cases} x_i(k-1) & \text{if } i \notin J_k \\ \\ f_i(x_1(s_1(k)), \ldots, x_n(s_n(k))) & \text{if } i \in J_k \end{cases} \qquad (1)$$

where $J = \{J_k / k \in N^*\}$ is a sequence of nonempty subsets of $\{1, \ldots, n\}$ and $S = \{(s_1(k), \ldots, s_n(k)) / k \in N^*\}$ is a sequence of elements in $N^{*n}$.

In addition, $J$ and $S$ are subject to the following conditions, for each $i = 1, \ldots, n$.

1.  $s_i(k) \le k-1$,  $k \in N^*$

2.  $s_i(k)$, considered as a function of $k$, tends to infinity as $k$ tends to infinity;

3.  $i$ occurs infinitely many often in the sets $J_k$ $k \in N^*$ .

An asynchronous iteration corresponding to $F$, starting with $x(o)$ and defined by $J$ and $S$ will be denoted by $(F, x(o), J, S)$.

Condition (1) implies that only components of previous iterates are used in the evaluation of a new iterate.
Condition (2) implies that the values used in the evaluation of a new iterate are more and more recent.
Condition (3) states that no component is abandoned in the computations.

The definition of asynchronous iterations permits to represent classical iterative methods like Jacobi and Gauss-Seidel or new algorithms for parallel computation like asynchronous algorithms.  For example, we have:

Jacobi: $\forall k \in N^*$, $J_k = \{1, \ldots, n\}$ and $\forall k \in N^*$, $\forall i \in \{1, \ldots, n\}$
$$s_i(k) = k-1$$

Gauss-Seidel: $\forall\ k \in N^*$ $J_k = 1 + (k-1) \mod n$ and $\forall\ k \in N^*$,

$$\forall\ i \in \{1,\ldots,n\}\ s_i(k) = k-1.$$

These methods (which correspond to sequences $J$ and $S$ well defined) are called synchronous methods since the processors must be synchronized to generate such sequences. On the contrary if the processors are not synchronized the sequences $J$ and $S$ are not defined and the algorithm is called asynchronous.

## 6.2  Synchronization of the Tasks

Each processor  i  which participates to an application executes cyclicaly a sequence of tasks $\{T_i(q)\}^N_1$ where N depends on the nature of the problem. A task consists of an evaluation of the new values of a subset of components denoted J. Whatever  q  the tasks $T_i(q)$ (associated to processors  i = 1,...,p)  start synchronously. Each task $T_i(q)$ is segmented in 2 parts, each one is followed by a synchronization (see Fig. 4).

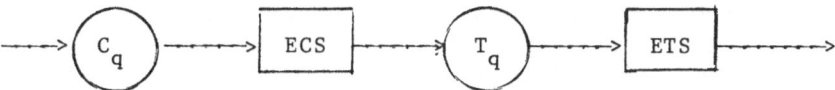

$C_q$ : Computation state

$T_q$ : Transmission state

ECS: End of computation semaphore (stop the execution of the task until all the processors have finished their computations).

ETS: End of task semaphore (stop the processor until all the processors have finished the access to the common memory).

The semaphore is controlled by the exchange and supervision module. When a processor has signalled the end of a computation state or the end of a transmission state he waits for a synchronization.

## Implementation of asynchronous algorithms

The exchange and supervision module ensures no synchronization of the tasks.

## 6.3  An Example:  Resolution of a Linear System of Equations

As an example let us take the implementation of the Dirichlet problem for Laplace's equation on a square domain using the method of finite differences. We chose a 16 x 16 square grid with potential fixed on the edge as shown in Fig. 4 .

Note that only 196 components in the interior of the grid must be evaluated. Let  $x_i(k)$  denote the potential node  i  at iteration  k. For all methods, we start with  $x_i(o) = 0$,  i = 1,...,196. It is well known that in such a problem the evaluation of  $x_i$  is done by adding the potential of the 4 neighbours of  i  on the grid and then dividing the sum by 4.

<u>Termination criterion</u>:   Let   $X(j) = \{x_i / x_i$ is evaluated by processor   $j\}$
and let   $x_i[k]$   denote the kth evaluation of   $x_i$ .   When

$$|x_i\ [k+1] - x_i\ [k]| \ < \ 10^{-4} \ , \ \forall\, x_i \in X(j),$$

then the local termination criterion is satisfied on processor   $j$.

The supervisor stops the application when the local termination
criterion is satisfied for every   $j$   used for the resolution of the
problem.

<u>Speed-up and efficiency</u>:   Let   $\Theta(1)$   be the running time using one
processor and   $\Theta(p)$   be the running time using   $p$   processors.
$S(p) = \Theta(1)\ /\ \Theta(p)$   is called the speed-up and   $E(p) = S(p)/p$   is the
efficiency of the parallel algorithm.


## Experimental results

Three methods have been experimented on the machine.   Each method
has been run in synchronous and asynchronous mode.

<u>Serial-Parallel method</u>:   For   $p = 1$   it is the classical Gauss-Seidel
method.   For   $p > 1$,   $p$   components are evaluated simultaneously and the
evaluation of a new set of   $p$   components starts when all the components
of the previous set have been evaluated.


<u>Jacobi method</u>:   Obviously, in this method, components can be evaluated
simultaneously.   Each processor evaluates a subset of the components.
A new iteration is started when all the components of the previous
iteration have been evaluated.

<u>Red-black decomposition method</u>:   Because of the particular structure of
the grid, the nodes can be coloured in red and black so that 2 red (or
black) nodes are not adjacent.   An algorithm evaluating alternatively the
red and then the black components can be efficiently parallelized (Young
[6]).

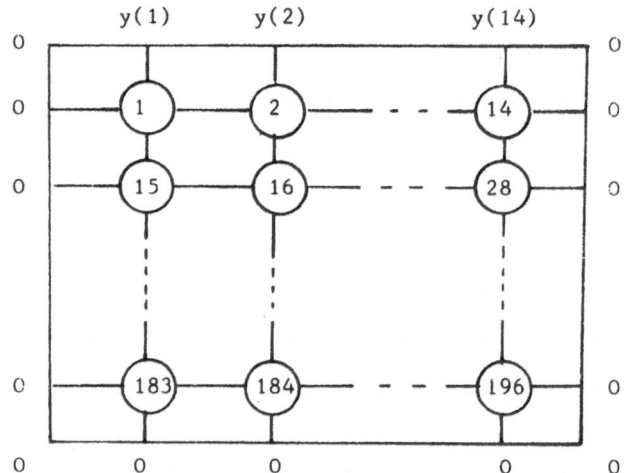

Fig. 4    Grid for the Dirichlet problem.   Potential on
the top edge is given by

$$y(i) = \tfrac{1}{9}(15-i), \quad i = 1,\ldots,14 \ .$$

Table 1    Serial-Parallel Method

| Number of processors p | Number of iterations | Running Time (sec) | Speed-up S(p) | Efficiency E(p) |
|---|---|---|---|---|
| 1 | 152 | 42.758 (37.840) | 1 (1) | 1 (1) |
| 2 | 153 | 21.894 (19.221) | 1.953 (1.969) | 0.976 (0.984) |
| 4 | 154 | 11.333 (9.578) | 3.773 (3.951) | 0.943 (0.988) |
| 8 | 156 | 6.125 (4.913) | 6.981 (7.702) | 0.873 (0.963) |

Table 2    Jacobi Method

| 1 | 277 | 64.103 (64.059) | 1 (1) | 1 (1) |
|---|---|---|---|---|
| 2 | 277 | 32.917 (32.418) | 1.947 (1.976) | 0.974 (0.988) |
| 4 | 277 | 17.027 (16.121) | 3.765 (3.974) | 0.941 (0.993) |
| 8 | 277 | 8.996 (8.181) | 7.126 (7.830) | 0.891 (0.979) |

Table 3    Red-Black decomposition method

| 1 | 155 | 35.631 (35.582) | 1 (1) | 1 (1) |
|---|---|---|---|---|
| 2 | 155 | 18.140 (18.136) | 1.964 (1.962) | 0.982 0.981 |
| 4 | 155 | 9.508 (9.551) | 3.747 (3.725) | 0.937 (0.931) |
| 8 | 155 | 5.085 (4.994) | 7.007 (7.125) | 0.876 (0.891) |

The results are reported in Tables 1, 2 and 3. For each method, the problem was solved using 1, 2, 4 and 8 processors. The results obtained with an asynchronous scheme are reported in parenthesis. Note that:

- For all experiments, the efficiency is very good and never lower than 0.87.

- For p = 1 the difference between the running time for synchronous mode is the total book-keeping synchronization time with no delay to go through the semaphore. From the tables it can be seen that this time is 80 µs per semaphore; in the serial-parallel method 152x196x2 semaphores are crossed per iteration, but in the Jacobi method they are only 277x2.

- Note that efficiency is always greater for asynchronous methods because there is no delay to cross semaphores (idle time). Efficiency is not 1 because some computations are duplicated (a new evaluation of a component can use some components which have not been re-evaluated).

- The rate of convergence for the Jacobi method is approximately twice as much as for the Gauss-Seidel method.

- The efficiency is a decreasing function of the number of processors for the two reasons: (i) conflicts to access the common memory and waiting time for synchronization are increasing functions of the number of processors, and (ii) for some algorithms the number of iterations to achieve the solution is also an increasing function of the number of processors (as in Table 1).


CONCLUSION

The architecture of our multimicroprocessor was designed to reduce book-keeping synchronization time and time to access the shared memory. The experiments carried out on our machine clearly show the correctness of our choice and the efficiency of the multimicroprocessor. It can be seen that the ratio cost/efficiency for this machine is very good, since its cost price is 30,000 French francs with 8 processors and the efficiency is close to 1 (see the results in Section 6).

REFERENCES

1.  Baudet, G. (1978). Asynchronous iterative methods for multi-processors. J.A.C.M., vol. 25, No. 2, 226-244.
2.  Chazan, D. and Miranker, W. (1969). Chaotic relaxation. Linear algebra and its applications. vol. 2, 199-222.
3.  Fathi E.T. and Krieger (1983). Multiple microprocessor systems: what, why and when. Computer, 23-92.
4.  Miellou, J.C. (1975). Algorithmes de relaxation chaotique a retards. RAIRO, vol. 8, R1 - 52-82.
5.  Paker, Y. (1983). Multi-microprocessor systems. Academic Press.
6.  Young, D. (1971). Iterative solution of large linear systems. Academic Press.

PARALLEL CALCULATION MODELLING WITH DATA FLOW PETRI NETS

M. Barbagelata   and   P. Abellard

Laboratoire d'Automatique et d'Informatique

Appliquees de Toulon, Universite de Toulon
83130 La Garde, France

SUMMARY

The model of parallel calculation representation used is derived from Petri nets, and dubbed "Data Flow Petri Net".

This model offers two peculiarities. The first is that the marks are not merely objects showing the state of the parallel process represented. They also carry pieces of information such as the values of data, the destination address, the identifiers, etc. The second is that the memories are never shared. In this data flow architecture, no conflict occurs over sharing a bus or waiting for an available memory.

The paper begins with a qualitative description of the model followed by its mathematical formulation, and eventually some examples of laboratory research and implementation are described.

1. INTRODUCTION

Carrying out a parallel calculation efficiently by means of a multi-processor requires a thorough study. A model of representation of the parallel calculation is essential, in order both to express the parallelism and to determine the dynamic behaviour of the machine.

The important point to be considered when designing a multiprocessor is the relationship between the model of representation of the parallel calculation and the architecture to be implemented. This architecture must be adapted to the model of calculation and not the other way round.

Carrying out parallel programs with conventional multiprocessors meets with considerable difficulties, and does not perform as desired. These difficulties can be summarized as follows: limited competition, a complex control system, uneasy programming. They make it more difficult and complex to analyse the behaviour of the system, and to show such particular phenomena as the abortion of the calculations, the existence of permanent lockings, the re-entering of operations.

The data flow architecture is structurally different from that of conventional multiprocessors, and it does not imply additional material requirements for the execution of parallel calculation programs.

The model of representation used is derived from Petri nets (1), and dubbed "Data Flow Petri Net" (2).

## 2.  QUALITATIVE DESCRIPTION OF THE MODEL

### 2.1  Operation

An operation is carried out with an operator and a set of variables, that we represent with a Petri Net:

- $t_i$ and $t_j$ are respectively called input and output transitions of the operator $O_R$ associated to the place $p_{ij}$ .

- The places $t_i = \{p_1, p_2 \ldots p_d\}$ and $t_j = \{p_1, p_2 \ldots p_r\}$ represent respectively the data necessary to operate $O_R$ and the results that we obtain.

A place associated to a variable has only one input arc, i.e. CARD ($^\cdot$p) $\leq$ 1 and only one output arc, i.e. CARD (p$^\cdot$) $\leq$ 1 because a data is obtained by only one operator and used by a single other one.

### 2.2  Marking

A mark in a place representing a variable means that the value of the data is created in the memory. A mark in a place representing an operator means that it is activated. We suppose that a place cannot contain more than one mark, i.e. $M(p) \leq 1$.

### 2.3  Temporization

Each operation is supposed to be executed by a processor during a constant and known time $\tau$. On the Petri Net we associate to a place representing an operator a constant $\tau_i$ to indicate the time of the operation. We suppose that the time is zero for the places representing data (memory access).

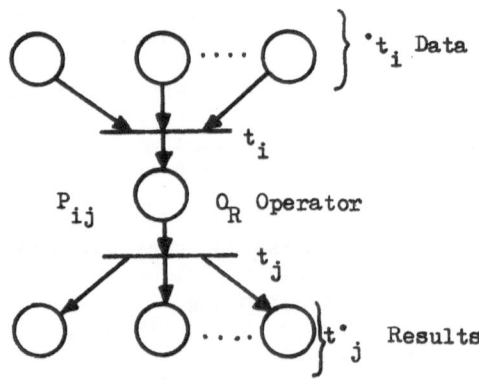

Fig. 1    Data flow Petri Net model

# 3. MATHEMATICAL DEFINITIONS

## 3.1 Two-Part Places Petri Net (TPPPN)

A TPPPN is a Petri Net $R = <P,T,\alpha,\beta>$ with

- P : a set of places divided with:

$$P_v \cup P_o = P \quad \text{and} \quad P_v \cap P_o = \emptyset .$$

$P_v$ is called set of places associated to variables.
$P_o$ is called set of places associated to operators.

- T : a set of transitions.

- $\alpha,\beta$ : incidence functions with $\forall t_i \in T$ :

$$^\bullet t_i \subset P_o \Rightarrow t_i^\bullet \subset P_v \quad \text{and} \quad ^\bullet t_i \subset P_v \Rightarrow t_i^\bullet \subset P_o$$

## 3.2 Conformable Petri Net

A Petri Net $R$ is called conformable if and only if:

- $R$ is a TPPPN.

- $\forall p \in P_v \Rightarrow CARD(^\bullet p) \leq 1$ and $CARD(p^\bullet) \leq 1$

- $\forall p \in P_o \Rightarrow CARD(^\bullet p) \geq 1$ and $CARD(p^\bullet) \geq 1$

- $\forall p_i \in P_o, p_j \in P_o$ and $i \neq j$

$$^\bullet P_i \cap ^\bullet P_j = 0 \quad \text{and} \quad P_i^\bullet \cap P_j^\bullet = 0$$

## 3.3 Temporized Petri Net (TPN)

A TPN is obtained with:

- a net $R = <P,T,\alpha,\beta>$

- a growing order $u = \{u_i\}$ $i \in N$ in such a way that to the fire of each transition suits an element of $U$.

- a set of reels $C = \{u_j - u_i\} = \{\tau i\}$

$$j > i, \quad i \in N, \quad j \in N$$

an application $\gamma$ of $P$ in $C$, in such a way that

$$\gamma(p_k) \leq (u_j - u_i)$$

$u_j$ : firing time of transition $t_j \in p^\bullet k$

$u_i$ : firing time of transition $t_i \in {}^\bullet pk$

## 3.4 Data Flow Petri Net (DFPN)

A DFPN is a 7. tuple $<R; \phi; \xi; \psi; x; 0; c>$ in which

- R is . conformable TPPPN

- $\phi/P_V : P \rightarrow X$ ; $\phi/P_0 : P \rightarrow 0$

as : $\forall~P_j \in P_o,~P_i \in P_o$ and $\phi(P_i) \equiv \phi(pj)$ ; $i \neq j$

$\qquad t_1 \in {}^\cdot P_i,~t_k \in {}^\cdot P_j,~\{\phi({}^\cdot t_1)\} \neq \{\phi({}^\cdot t_k)\}$

So, two identical operators cannot work on the same set of data.

- $\xi$ is an application $\xi : X \rightarrow M = \{ME_1 \ldots ME_u\}$

$\qquad$ as $\forall~p \in P_v,~ME \in M \Rightarrow ME = \xi(\phi(p))$.

- $M$ is called a set of memory zones.

- $\psi$ is an application $\psi: T \rightarrow C$

- $X = \{x_1, x_2, \ldots x_u\}$ is a set of variables (real, entire, logic) with values in $D_1, D_2, \ldots, D_u$ .

- $0 = \{0_1, 0_2, \ldots, 0_v\}$ is a set of operators defined as internal applications of $D_1 \times D_2 \times \ldots D_u$ .

- $C = \{c_1, c_2, \ldots, c_v\}$ is a set of conditions (predicates) on X variables.


## 3.5 Data Flow Petri Net Definitions

### 3.5.1 Temporized data flow Petri Net (TDFPN)

A Temporized Data Flow Petri Net is a net fashioned with a consistent temporized "Two-Part Places Petri Net" such as:

$\qquad \forall~p \in pv \Rightarrow \gamma(p) = \emptyset$

$\qquad \forall~p \in po \Rightarrow \gamma(p) = \tau_i \neq \emptyset$


We consider that the places associated to memories have an execution time nil.

We call time life of a mark in a place, the time during which it stays in the place, i.e. the time spent from the marking instant to the demarking instant of this place. The minimum time life of a mark in a DFPN place p is determined by the execution time associated to this place.

If the search for a result is considered to be immediate after the end of its execution, we can say that the time life of a mark is equal to the temporization time associated to this place, when representing an operator.

The time life of a mark in a place representing a data is unknown beforehand because it depends on the order of the operations.

### 3.5.2 Correct data flow Petri Net

(a) ⁻ Let us consider a Data Flow Petri Net (DFPN) R and $M_o$ its initial marking; we say that $M_o$ is correct and noted $M_o^c$ if and only if

$$\forall\ p \in P,\ M_o\ (p)\ =\ \begin{cases} 1 & \text{if } p \in P_{(v)} \mid \text{CARD } (\cdot p) = \emptyset \mid \\ \emptyset & \text{else} \end{cases}$$

Indeed, in a Data Flow Petri Net (DFPN) the initial marking corresponds to initial data.

Any marking $M'_o \subset M^c_o$ will be considered as a limited initial marking. We can define the same notion which applies to the final marking.

(b)⁻ If we consider a DFPN and Mf its final marking, we can say that Mf is correct and noted $M^c_f$ if and only if:

$$\forall\ p \in P,\ M_f\ (p)\ =\ \begin{cases} 1 & \text{if } p \in P_v \mid \text{CARD } (p\cdot) = \emptyset \mid \\ \emptyset & \text{else} \end{cases}$$

Any marking $M'f\ \ Mf$ is considered as a limited final marking.

(c)⁻ Considering R a DFPN and $M^c_o$ its initial marking, $M^c_f$ its final marking, we can say that R is correct if only

- R is safe

- $\forall\ t \in T$ is firing once only from $M^c_o$

- $\forall\ M_i \in M^c_o\ \exists\ \tau i$ in such a way that $M_i \xrightarrow{\ \tau i\ } Mf$

(d)- MARKING GRAPH

Considering R a Petri Net and an initial marking $M_o$ in such a way that consequent marking class M is ended; the marking graph of consequent marking is the turned graph (S,L) where $S = \{\vec{M_o}\}$ is the set of summits and L is the set of arcs, in such a way that:

$M_i \in \vec{M_o}$ and $M_j \in \vec{M_o}$

$\exists t_i$, $t_i$ is a R transition and $M_i \xrightarrow{\ t_i\ } M_j$

The arcs of the graph Q are labelled by the transition net.

Figures 2 and 3 show a correct Data Flow Petri Net associated to the calculation $C_{11} = a_{11}.b_{11} + a_{12}.b_{21}$ and its consequent marking graph.

4. EXAMPLES

Figures 4 and 5 show Data Flow Petri Net associated to single and multiple matrix products.

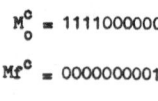

$M_0^0 = 1111000000$

$Mf^0 = 0000000001$

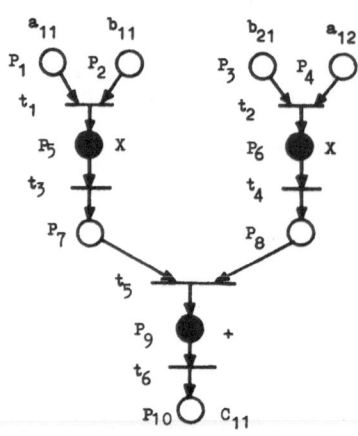

Fig. 2  Data Flow Petri Net
of the calculation:
$c_{11} = a_{11} \cdot b_{11} + a_{12} \cdot b_{21}$

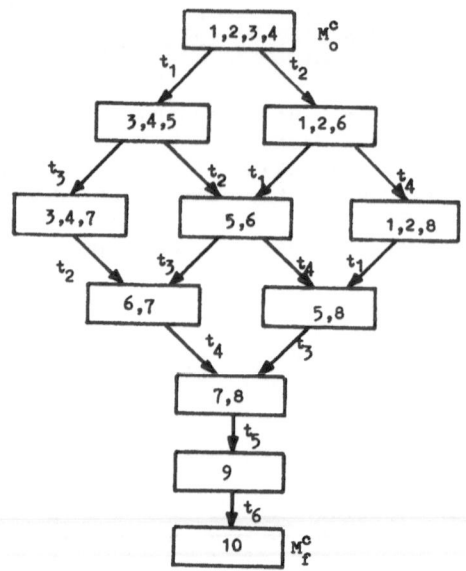

Fig. 3  Consequent marking graph
of a Data Flow Petri Net
associated to the
calculation:
$c_{11} = a_{11} \cdot b_{11} + a_{12} \cdot b_{21}$

## 5.  APPLICATIONS

Three studies have been made in the laboratory. The first two in
parallel calculation:  a single matrix multiplier (Fig. 7) and a
Lyapunov (PWM) operator (Fig. 8).

In order to reduce necessary equipment the studies have principally
treated the use of re-entering operators and the implementation of
elementary modules composed together according to the net architecture.

A memory can only be written by one processor and read by another
one.  Implementations have been obtained by using RAM memories with
separated R/W addressing.   More, calculation times are shortest because
these memories have simultaneous reading and writing operations.

Now, principal limitations are the capacities of the available
circuits (4x4; 16x4; 256x8 bits);  but modern large scale integration
techniques will allow an increase in the memory capacity of integrated
circuits, and will further simplify implementations.   However, multi-
processor machines being implemented are very simple, because no
conflict occurs over sharing a bus (no bus is shared) or waiting for an
available memory (no data is shared).

Figures 7, 8 and 9 show implementations of a single matrices
multiplier and a Lyapunov operator.

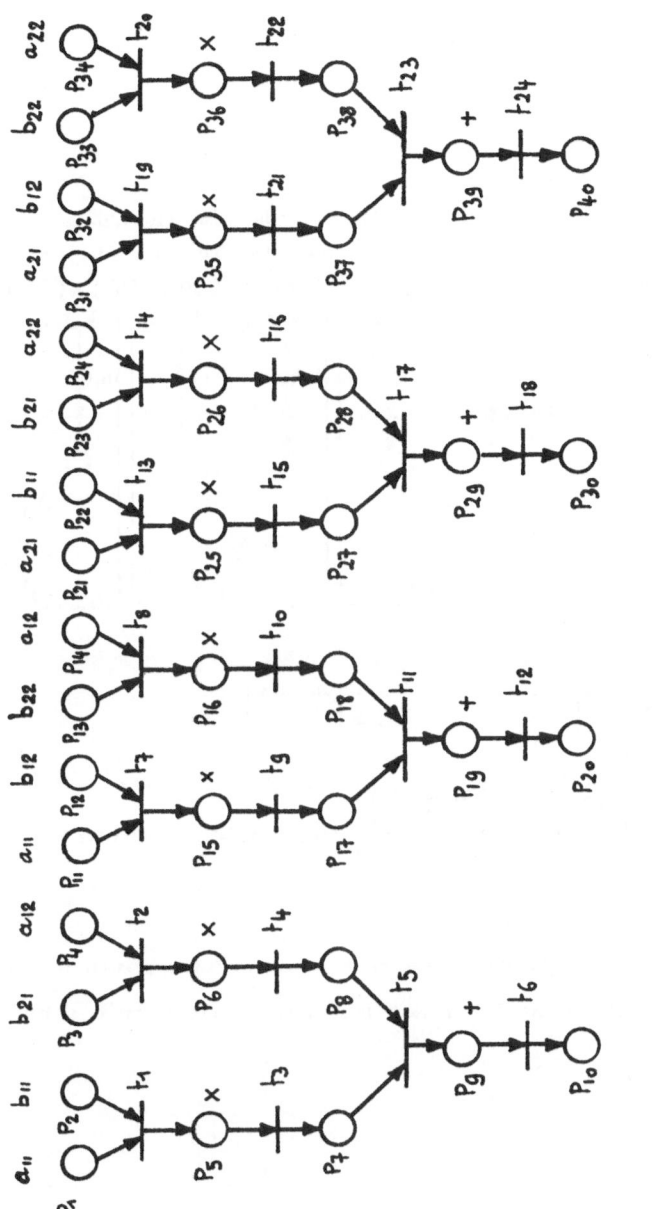

Fig. 4 Data Flow Petri Net for a single matrices product

$c_{11} = a_{11} b_{11} + a_{12} b_{21}$    $c_{12} = a_{11} b_{12} + a_{12} b_{22}$    $c_{21} = a_{21} b_{11} + a_{22} b_{21}$    $c_{22} = a_{21} b_{12} + a_{22} b_{22}$

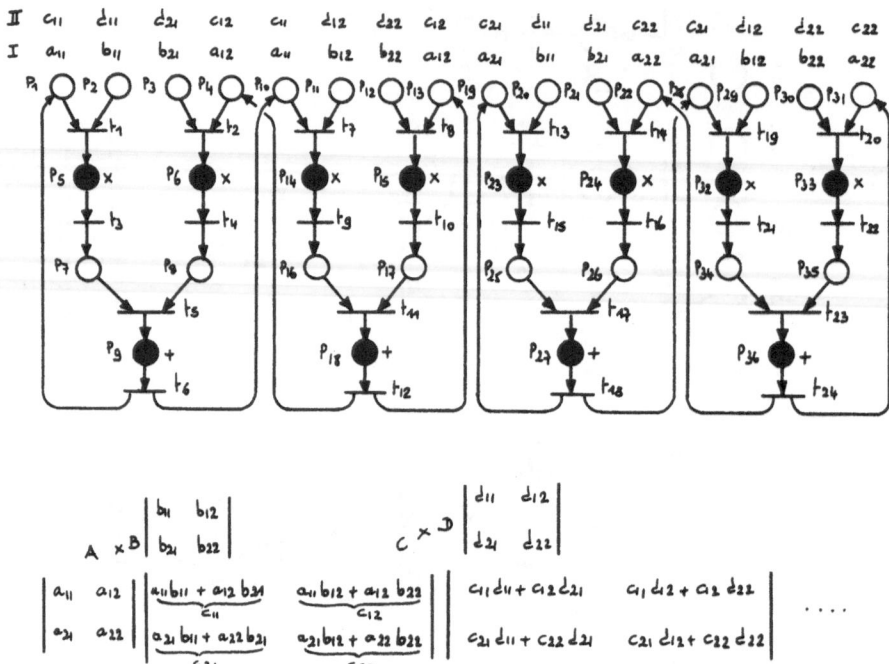

I : First data flow
II : Second data flow

x data flows are necessary to calculate x matrix products (n order matrices).

Fig. 5 Data Flow Petri Net for a multiple matrices product

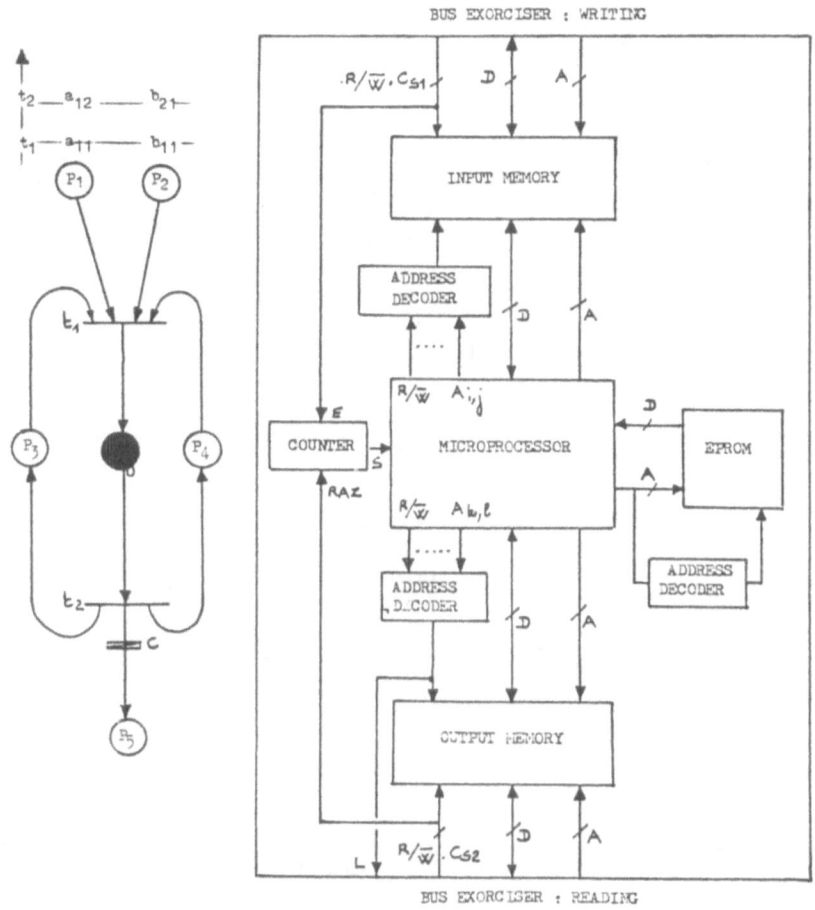

Fig. 6    Model implementation

$P_1, P_2$ :  data memories

$P_3$ :  internal register for loading the operation result

O :  treatment operator

$P_4$ :  counter which is incremented at each operation

$P_5$ :  result memory

C :  predicate (condition on $X_{p4}$) .

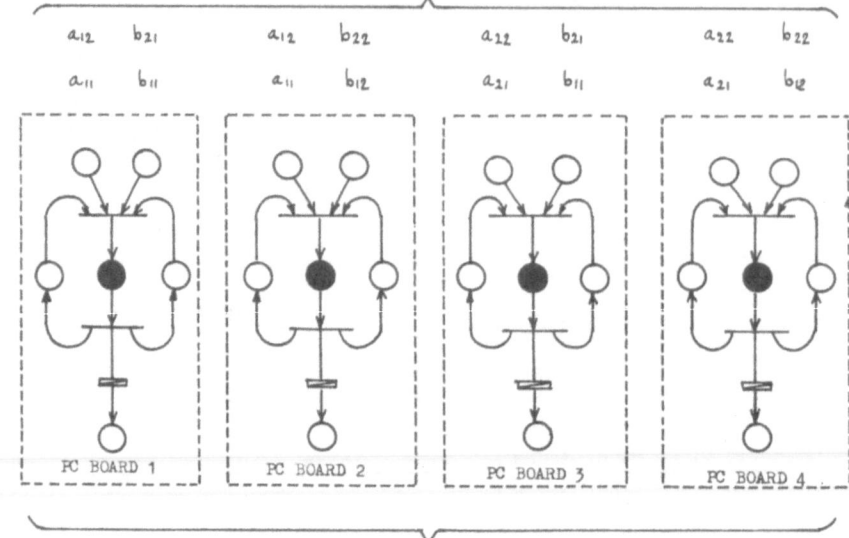

Data writing with 68000 EXORCISER BUS

| | | | | | | | | | |
|---|---|---|---|---|---|---|---|---|---|
| II | $a_{12}$ | $b_{21}$ | | $a_{12}$ | $b_{22}$ | | $a_{11}$ | $b_{21}$ | |
| I | $a_{11}$ | $b_{11}$ | | $a_{11}$ | $b_{12}$ | | $a_{21}$ | $b_{11}$ | |

| II | $a_{11}$ | $b_{22}$ |
|---|---|---|
| I | $a_{21}$ | $b_{12}$ |

PC BOARD 1   PC BOARD 2   PC BOARD 3   PC BOARD 4

EXORCISER BUS : Reading of results.

Fig. 7  Single matrices multiplier
I :   first data flow
II :   second data flow.

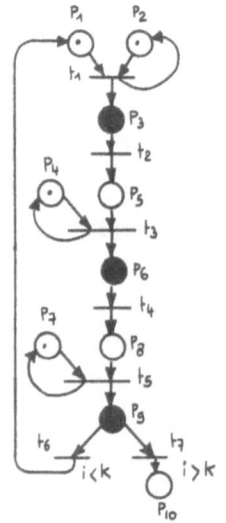

**Lyapunov operator**

$$P_{k+1} = Y \cdot P_k \cdot Y^T + Q$$

$$P_k = P_{k+1}$$

k : number of iterations required.

Fig. 8   Macro Data Flow Petri Net

248

Data writing with 68000 EXORCISER BUS

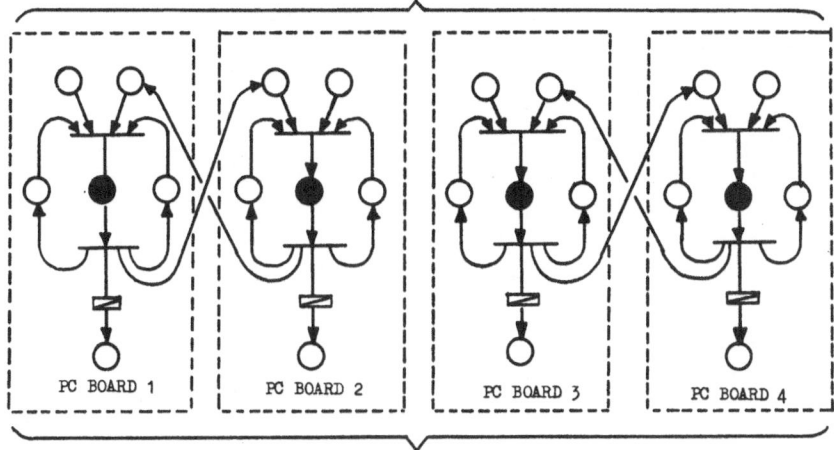

EXORCISER BUS : Reading of results.

Fig. 9    Implementation of Lyapunov operator
         (second order matrices).

BIBLIOGRAPHY

1.  G.W. Brams,  Reseaux de Petri - Theorie et pratique, Tomes 1 et 2,
        MASSON, (1983).
2.  J. Almanah,  Modelisation par reseaux de Petri a flux de donnees.
        Application a la synthese de l'operateur de Riccati rapide,
        These d'Etat MARSEILLE (1983).
3.  AFCET - Rairo,  Technique et Science informatiques.  Special
        reseaux de Petri, vol. 4, No. 1, (1985).
4.  A. Giulieri - J. Almanah,  Simulation des algorithmes de Lindquist
        et de quasilinearisation par reseaux de Petri.
        Rapport final, convention DRET 79/659 - LAIAT TOULON -
        (Juin 1982).
5.  P. Abellard,  Les systemes multiprocesseurs - Application a la
        realisation d'un biprocesseur specialise.
        LAIAT TOULON - DEA Universite de MARSEILLE. (1984).

# SYSTOLIC NETS MODELLING WITH DATA FLOW PETRI NETS

A. Giulieri, B. Barbagelata, and P. Abellard

Universite De Toulon

83130 La Garde, France

## SUMMARY

A systolic machine is a net of elementary processors in which data circulates with a constant flow. The word systol, usually used by physiologists to describe the rhythm of heart contractions, is employed to show the analogy between the operations of the heart and that of the systolic machine. Indeed, each processor pumps data from its entry to its exit, while carrying out certain operations, so that we obtain a regular data flow through the net.

The communication and calculation control in the processor is very simple. The memory place employed is little and constant, independently of the net size.

This kind of machine is interesting for two reasons: they offer an easy implementation and do not need complex commands, then they can be connected to general use machines, like Data Flow machines.

Among the possible approaches to systolic structure we have chosen their modelling with Data Flow Petri Nets. When a calculation is to be solved, first we work out calculation parallelism, then we model with Data Flow Petri Net and we synchronize the transitions.

## 1. INTRODUCTION

We use Data Flow Petri Net approach to obtain operator structure for real time calculation with systolic architecture.

Let a calculation to be carried out, we have

1 - to search for parallelism of calculation.
2 - to define elementary work of each operator.
3 - then to design the complete net to achieve the whole calculation.
4 - to define the sequence of data propagation.

## 2.  BASIC CONCEPT

Application to calculation of matrices : the product of two matrices A and B,  order NxN,  is carried out by calculating coefficients:

$$C_{ij} = \sum_{k=1}^{N} a_{ik} \cdot b_{kj}$$

First we have to define the dependence graph according to the sign evolution i,j,k, e.g.

$$A \times B \quad \begin{vmatrix} b_{11} & b_{12} \\ b_{21} & b_{22} \end{vmatrix}$$

$$\begin{vmatrix} a_{11} & a_{12} \\ a_{21} & a_{22} \end{vmatrix} \quad \begin{vmatrix} a_{11} \cdot b_{11} + a_{12} \cdot b_{21} & a_{11} \cdot b_{12} + a_{12} \cdot b_{22} \\ a_{21} \cdot b_{11} + a_{22} \cdot b_{21} & a_{21} \cdot b_{12} + a_{22} \cdot b_{22} \end{vmatrix} = c$$

So, we take 8 operators which receive $a_{ik}$, $b_{kj}$, $c_{ij}$,  carry out $c_{ij} = a_{ik} b_{kj}$,  and propagate $a_{ik}$, $b_{kj}$, $c_{ij}$ .

At the first time, we introduce $a_{11}$, $b_{11}$, $c_{11}$ in operator 1 (Fig. 1) and we connect operator 1 to operators 2,3,5.

At the second time (Fig. 2), we introduce $a_{12}$, $b_{21}$ in operator 2, $b_{12}$, $c_{12}$ in operator 3 and $a_{21}$, $c_{21}$ in operator 5.

With operator 2 we obtain $c_{11} = a_{11} \cdot b_{11} + a_{12} \cdot b_{12}$,  we connect also operator 2 to 4 and 7, operator 5 to 6 and 7.

At the third time (Fig. 3) we introduce $b_{22}$ in operator 4, $c_{22}$ in 6, and $a_{22}$ in 7.  We connect operators 4,6,7 to operator 8.  We obtain:

$$C_{12} = a_{11} \cdot b_{12} + a_{12} \cdot b_{22}$$

$$C_{21} = a_{21} \cdot b_{11} + a_{22} \cdot b_{21}$$

At the fourth time (Fig. 4), operator 8, carry out $C_{22} = a_{21} b_{12} + a_{22} b_{22}$.

We see that the calculation domain is the cube defined with

$$D = \{i, j, k, \mid 1 \leq i \leq N ; 1 \leq j \leq N ; 1 \leq k \leq N\}$$

Classic projections are:

$\pi = (0,0,1)$ or $(0,1,0)$ or $(1,0,0)$ : square net

$\pi = (0,1,1)$ or $(1,0,1)$ or $(1,1,0)$ : mixed net

$\pi = (1,1,1)$ : hexagonal net.

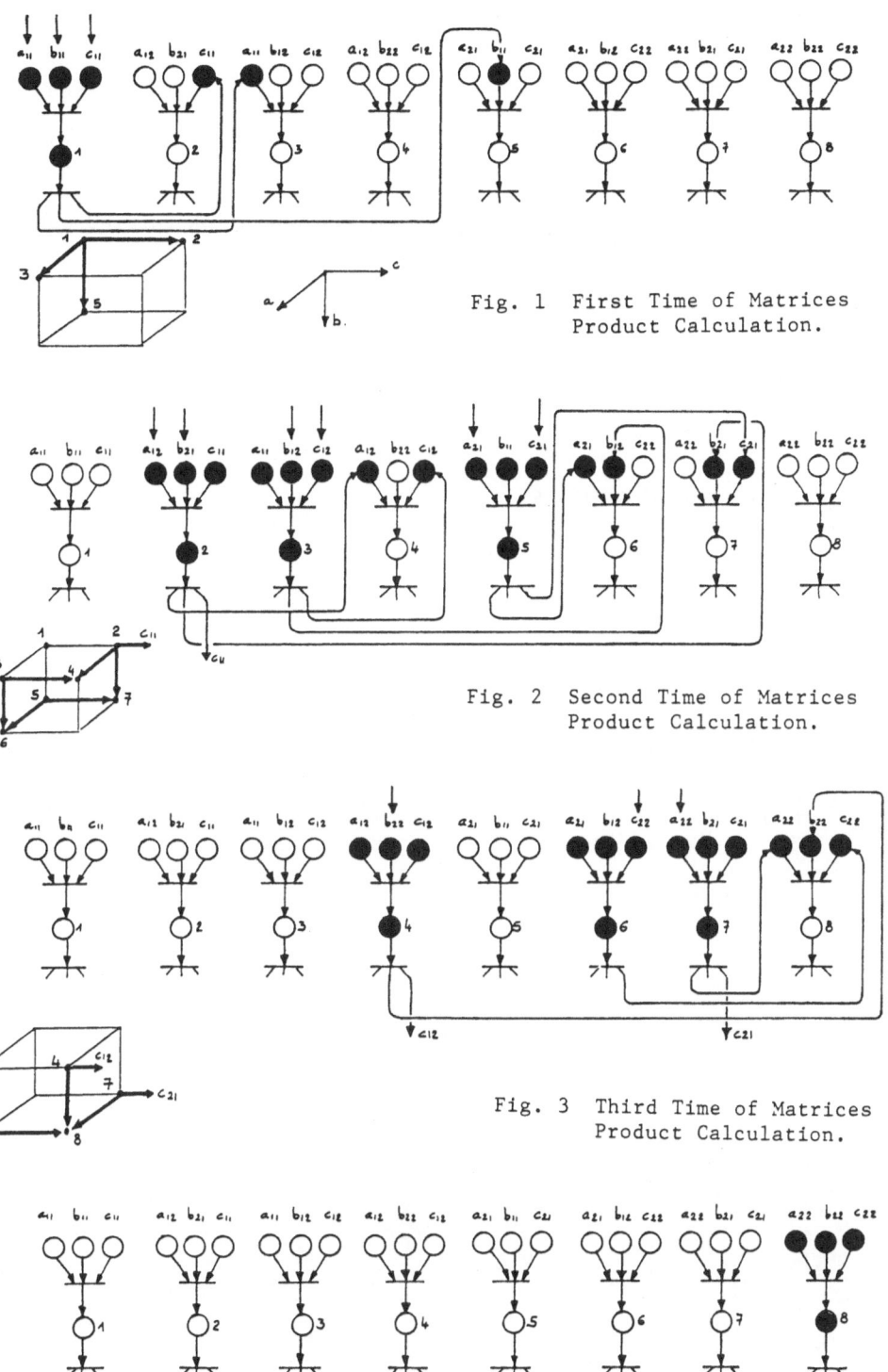

Fig. 1    First Time of Matrices
          Product Calculation.

Fig. 2    Second Time of Matrices
          Product Calculation.

Fig. 3    Third Time of Matrices
          Product Calculation.

Fig. 4    Fourth Time of Matrices
          Product Calculation.

## 3. TRANSFORMATION OF A PETRI NET INTO A SYSTOLIC NET

### 3.1 Definition

We use the following model with:

$P_1, P_2$ :  data memories

$P_3$ :  memory where the result of the operation carried out by $O_1$ is loaded

$O_1$ :  operator carrying out

$$X_{p1} X_{p2} + X_{p3}$$

and transmitting :  $X_{p1} \rightarrow P_4$

$$X_{p2} \rightarrow P_5$$

$P_4, P_5$ :  data memories .

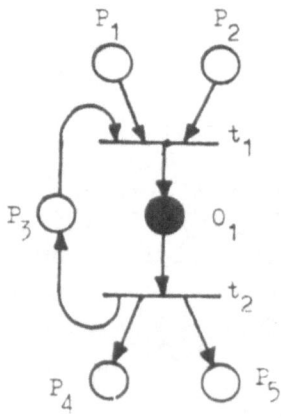

Fig. 5    Model used for the transformation
of a Petri Net into a systolic net.

## 3.2  Example

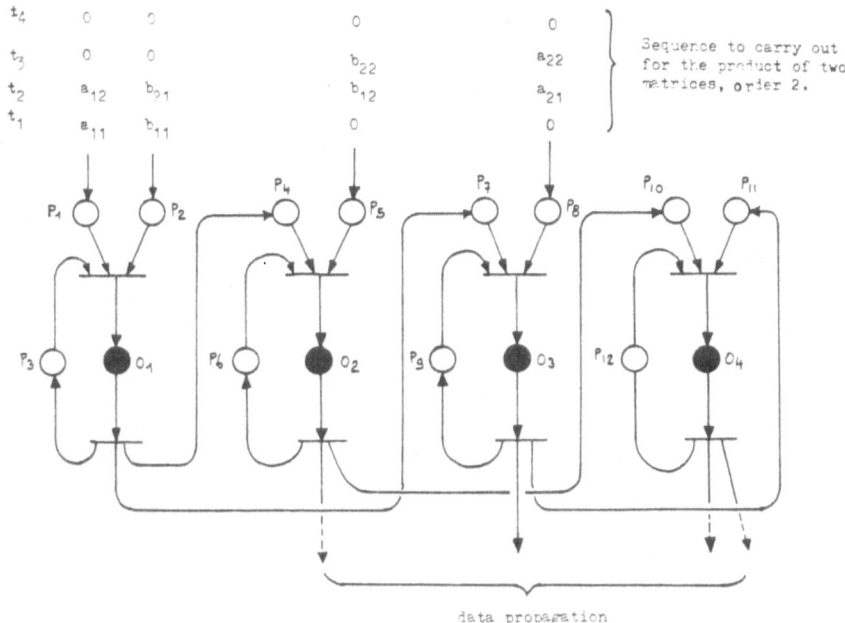

Fig. 6  Data Flow Petri Net used for the product of
two matrices order 2.

## 3.3  Petri Net

The traditional square net form is:

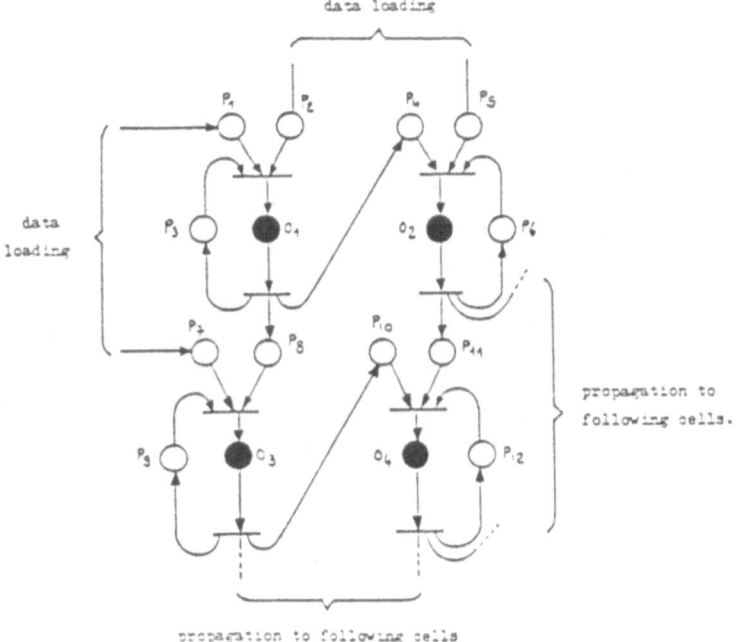

Fig. 7  Data Flow Petri Net of a systolic square net.

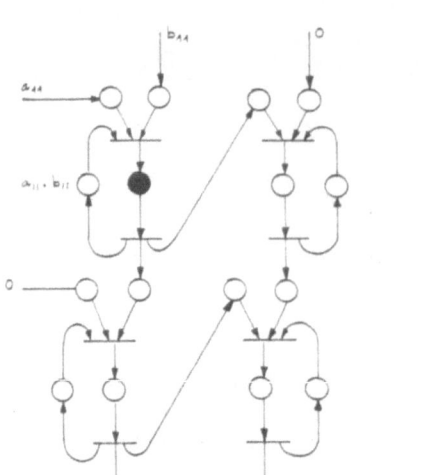

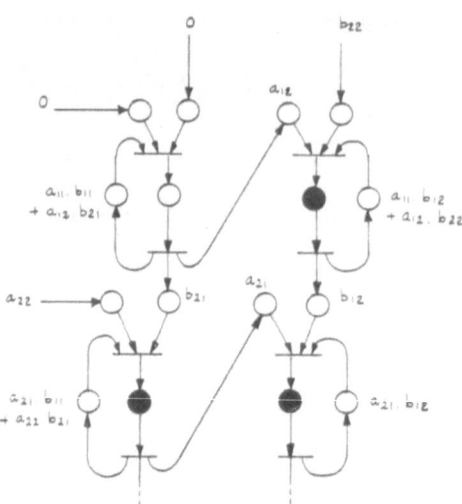

Fig. 8   Second time of the matrices
         product calculation.

Fig. 9   Third time of the matrices
         product calculation.

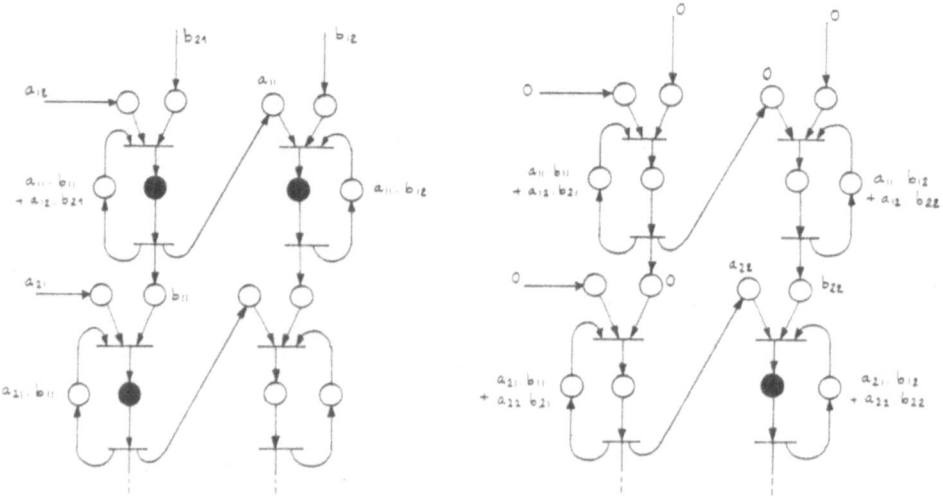

Fig. 10   Fourth time of the matrices
          product calculation.

Fig. 11   Fifth time of the matrices
          product calculation.

Data Propagation Sequence

<u>First time</u>: the whole net is cleared, i.e. $X_{p3} = X_{p6} = X_{p9} = X_{p12} = 0$.

<u>Second time</u>: (Fig. 8)

- We introduce $a_{11} \rightarrow P_1$ and $b_{11} \rightarrow P_2$
- $O_1$ carries out $a_{11}.b_{11} + 0$, loads it in $P_3$ and transmits:
  $a_{11} \rightarrow P_4$ and $b_{11} \rightarrow P_8$

<u>Third time</u>: (Fig. 9)

- We introduce $a_{12} \rightarrow P_1$; $b_{21} \rightarrow P_2$; $b_{12} \rightarrow P_5$; $a_{21} \rightarrow P_7$
- $O_1$ carries out $a_{12}.b_{21} + a_{11}.b_{11}$, loads it in $P_3$ and transmits
  $b_{21} \rightarrow P_8$, $a_{12} \rightarrow P_4$
- $O_2$ carries out $a_{11}.b_{12}$, loads it in $P_6$ and transmits $b_{12} \rightarrow P_{11}$
- $O_3$ carries out $a_{21}.b_{11}$, loads it in $P_9$ and transmits $a_{21} \rightarrow P_{10}$ .

<u>Fourth time</u>: (Fig. 10)

- We introduce $0 \rightarrow P_1$; $0 \rightarrow P_2$; $b_{22} \rightarrow P_5$ and $a_{22} \rightarrow P_7$
- $O_1$ carries out $0 + a_{11}.b_{11} + a_{12}.b_{21} \rightarrow P_3$ and transmits $0 \rightarrow P_4$ and
  $0 \rightarrow P_8$
- $O_2$ carries out $a_{12}.b_{22} + a_{11}.b_{12} \rightarrow P_6$ and transmits $b_{22} \rightarrow P_{11}$
- $O_3$ carries out $a_{22}.b_{21} + a_{21}.b_{11} \rightarrow P_9$ and transmits $a_{22} \rightarrow P_{10}$
- $O_4$ carries out $a_{21}.b_{12} \rightarrow P_{12}$ .

<u>Fifth time</u>: (Fig. 11)

- We introduce $0 \rightarrow P_1$, $0 \rightarrow P_2$, $0 \rightarrow P_5$ and $0 \rightarrow P_7$
- $O_1$ carries out $0 + a_{11}.b_{11} + a_{12}.b_{21} \rightarrow P_3$ and transmits $0 \rightarrow P_4$ and
  $0 \rightarrow P_8$
- $O_2$ carries out $0 + a_{11}.b_{12} + a_{12}.b_{22} \rightarrow P_6$ and transmits $0 \rightarrow P_{11}$
- $O_3$ carries out $0 + a_{21}.b_{11} + a_{22}.b_{21} \rightarrow P_9$ and transmits $0 \rightarrow P_{10}$
- $O_4$ carries out $a_{22}.b_{22} + a_{21}.b_{12} \rightarrow P_{12}$ .

So, we obtain the traditional following form:

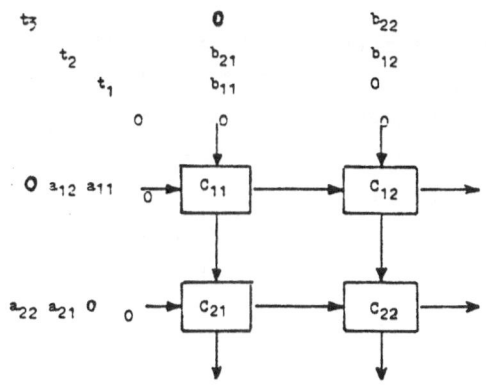

Now, let us design Data Flow Petri Net with 3 order matrices.

## 4. SQUARE NET

- At the first time $t_1$, the whole net is cleared.
- Then we introduce in the operator $C_{11}$, $a_{11}$ and $b_{11}$, and 0 in $C_{12}$, $C_{13}$, $C_{21}$, $C_{31}$
- The operator carries out $C_{11} = a_{11}.b_{11} + C_{int}(t-1) = a_{11}.b_{11} + 0$ and transmits $a_{11}$ to $C_{12}$ and $b_{11}$ to $C_{21}$
- At the second time $t_2$, $b_{21}$ is introduced in $C_{12}$ to obtain $C_{12} = a_{11}.b_{12}$ and $a_{21}$ in $C_{21}$ to obtain $C_{21} = a_{21}.b_{11}$
- These propagations continue till $t_5$. Then, the whole matrix product is finished.
- Figures 12 and 13 show a systolic square net and its data flow Petri net.

## 5. MIXED STRUCTURE

- a, b, c circulate through 3 ways.
- At the first time, $b_{11}$ and $c_{11}$ are introduced in 1 and 5 (we cleared the other cells).
- At the second time, $b_{11}$ and $c_{11}$ are propagated to 2 and 4 and we introduce $b_{12}$ and $c_{12}$ in 6 and 10.
- At the third time, $b_{11}$ and $c_{11}$ arrive in 3 where $a_{11}$ is introduced. Then calculation $C_{11} = a_{11}.b_{11}$ is carried out. At the same time $b_{21}$, $b_{13}$, $c_{21}$, $c_{13}$ are introduced in 1, 11, 5 and 15.
- Propagations are continued till the whole calculation is carried out.
- Figures 14 and 15 show a systolic mixed net and its data flow Petri net.

## 6. HEXAGONAL NET

- First we clear the whole net, and we introduce $a_{11}$, $b_{11}$ and $c_{11}$ in 5, 9, 1.
- At the second time, $a_{11}$, $b_{11}$, $c_{11}$ are propagated to 15, 17, 13.
- At the third time $a_{11}$, $b_{11}$, $c_{11}$ arrive in 19 where the product $C_{11} = a_{11}.b_{11}$ is carried out and $a_{12}$, $a_{21}$, $b_{12}$, $b_{21}$, $c_{12}$, $c_{21}$ are introduced in 4, 6, 8, 10, 12 and 2.
- At the fourth time $c_{11}$ arrives in 16, at the same time as $a_{12}$ and $b_{21}$. So the product $c_{11} = a_{11}.b_{11} + a_{12}.b_{21}$ is carried out.
- At the fifth time $c_{11}$ arrives in 7, with $a_{13}$ and $b_{31}$ . So we obtain $c_{11} = a_{11}.b_{11} + a_{12}.b_{21} + a_{13}.b_{31} \ldots$
- Figures 16 and 17 show a systolic hexagonal net and its data flow Petri net.

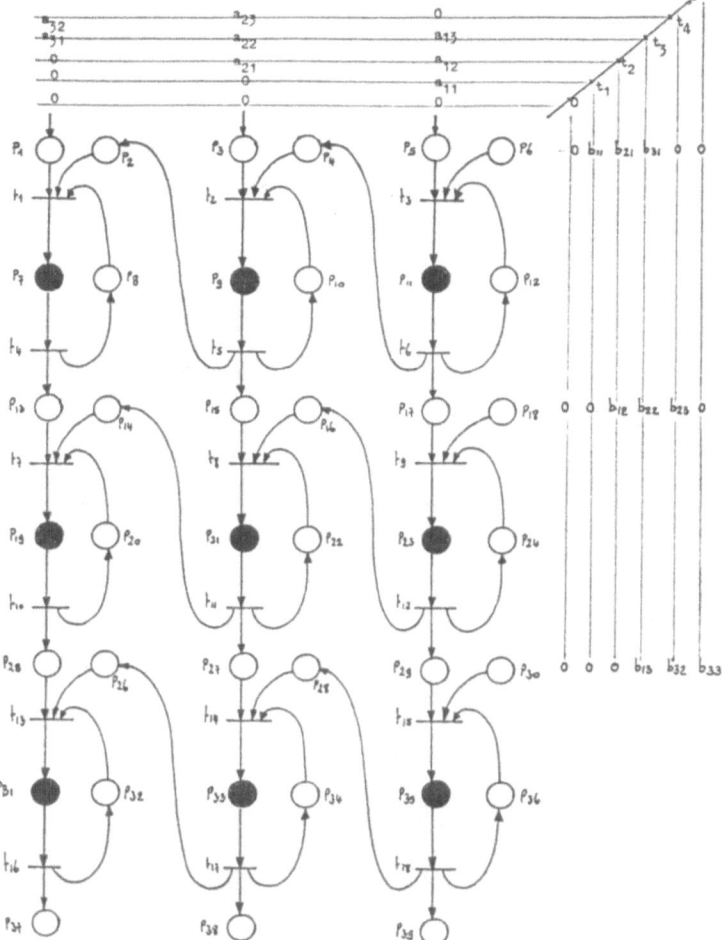

Fig. 12   Petri Net

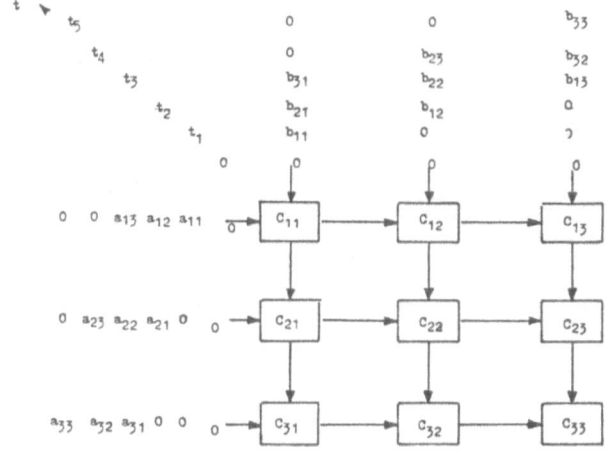

Fig. 13   Square Net

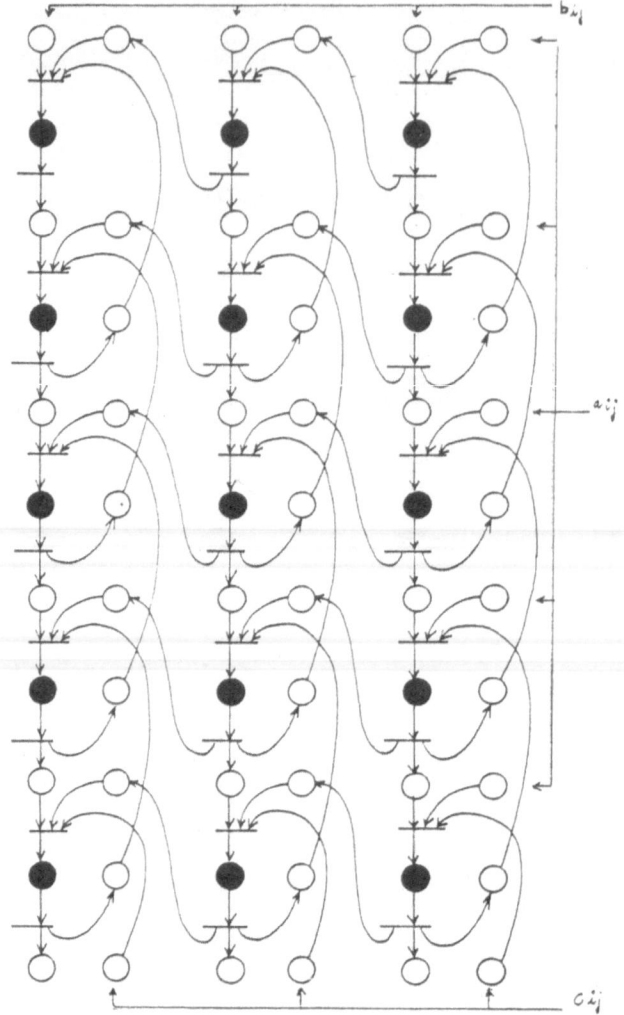

Fig. 14  Petri Net

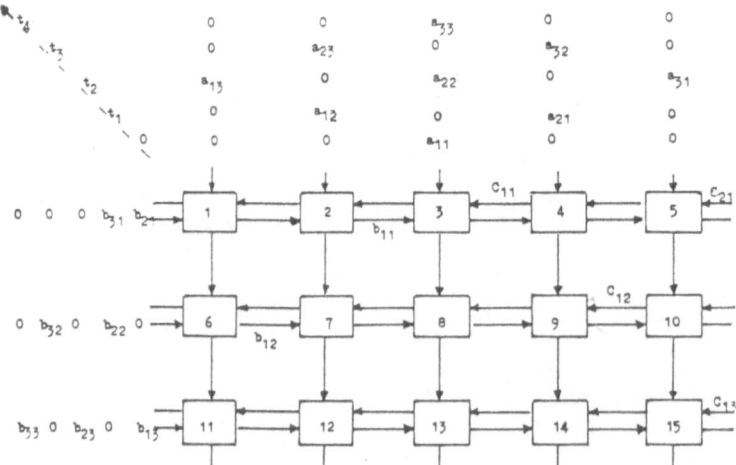

Fig. 15  Mixed Net

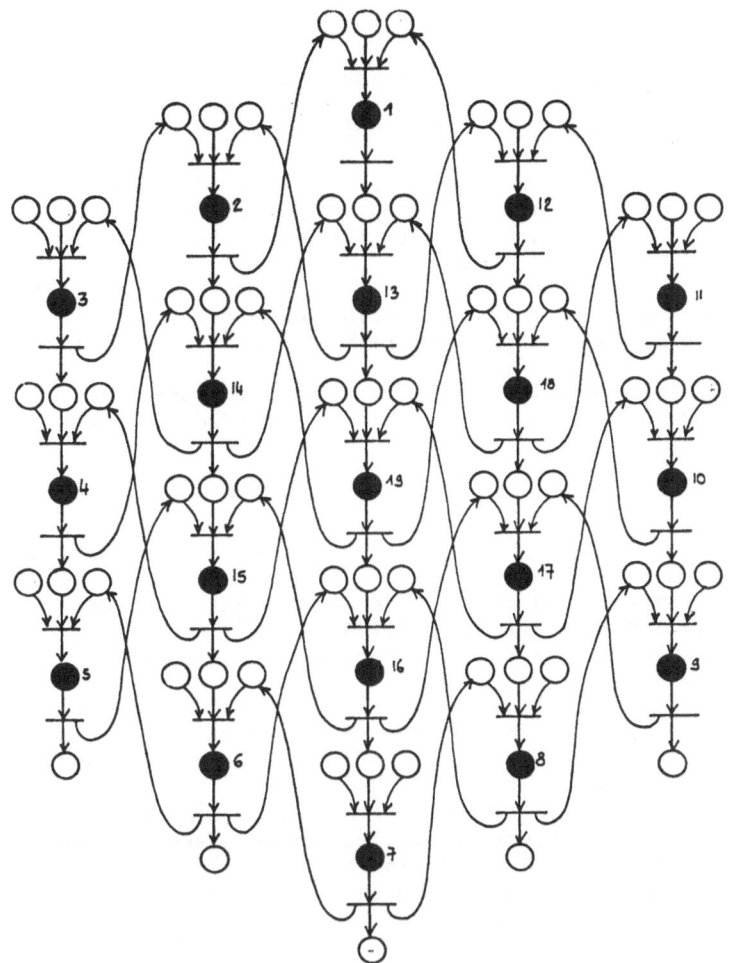

Fig. 16  Petri Net

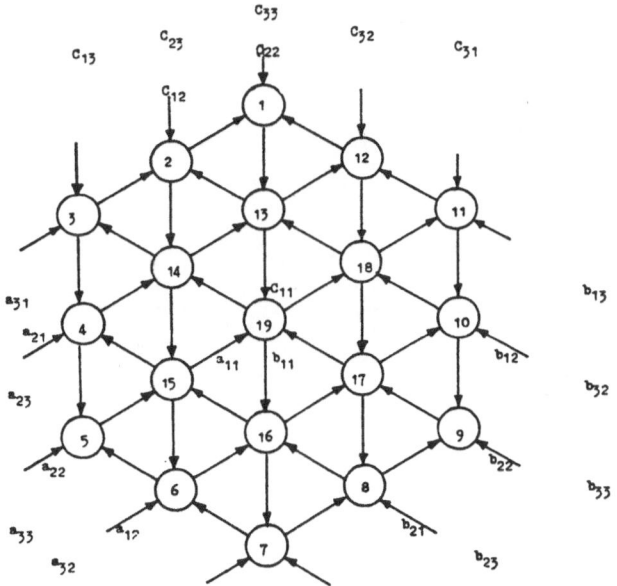

Fig. 17  Hexagonal Net

## 7. CONCLUSION

The synthesis of a Data Flow Petri Net for a model of representation and for a matrice calculation, gives us a structure with a graph of data propagation defining the calculation domain of a systolic net.

Dating from this analogy referred to P. Quinton's theory and according to the basis principle of the data propagations in a square, mixed or hexagonal systolic net, we have defined the associated Data Flow Petri Nets.

So, on those nets the characterisation of elementary cells have allowed to show a macro-structure entirely analogous to systolics.

This principle to obtain nets, without writing concurrent equations and geometrical decomposition, may easily be extended to other applications.

REFERENCES

1. J. Almanah, Modelisation par reseaux de Petri a flux de donnees.
    Application a la synthese de l'operateur de Riccati rapide.
    These d'Etat - Marseille (1983).
2. G.W. Brahms, Reseaux de Petri, Theorie et pratique.
    Tomes 1 et 2. Masson (1983).
3. H.T. Kung, "Why systolic architecture", IEEE Computer (1982).
4. A. Giulieri, H. Malki, Etude d'un operateur de Lyapunov a
    architecture systolique, LAIAT Toulon, IASTED (19-21 Juin,
    1985).
5. A. Giulieri, G. Nolibe, Etude et realisation d'operateurs special-
    ises multifonctions et application aux reseaux systoliques.
    Rapport interne LAIAT Toulon 83.268.
6. P. Quinton, The systematic design of systolic arrays.
    IRISA Report 193, (1983).

# MULTIPROCESSOR TASK SCHEDULING WITH SINGLE RESOURCE CONSTRAINTS

J. Blazewicz[*], M. Drabowski[+], K. Ecker[+], and J. Weglarz[*]

[*]Instytut Automatyki, Politechnika Poznanska, Poznan, Poland
[+]MERA-KFAP, Krakow, Poland
Technische Universitat, Clausthal-Zellerfeld, W. Germany

## ABSTRACT

An important problem arising in a multiprocessor system context is that of task allocation (scheduling) among a set of processors working in parallel. One of the assumptions commonly imposed on the processor scheduling theory is that each task is processed on at most one processor at a time. In fact, all polynomial-in-time algorithms as well as NP-completeness results for task scheduling were obtained by this assumption.

However, in recent years, together with the rapid development of microprocessor and especially multi-microprocessor systems, the above assumption has ceased to be justified in some important applications. This is especially true with some highly specialized systems where tasks may require more than one processor simultaneously. These problems create a new direction in the processor scheduling theory, in which preliminary results concerning the pre-emptive scheduling of tasks requiring one or k (k fixed, > 1) processors were obtained in [1] and [2] for the schedule length criterion.

## THE PROBLEM

In this paper we study an extension of the model defined above by assuming that besides processors, tasks require also an additional resource (e.g. channels, memory, i/o devices) for their processing. We show how this assumption affects the computational complexity of algorithms for constructing minimum length schedules. We are considering a set $T$ of $n$ tasks to be processed on a set $P$ of $m$ identical processors, $P = \{P1,...,Pm\}$. During execution, tasks may need some additional resource $R$. There are $r > 0$ copies of $R$ present in the processing system. We assume that each task $t \in T$ needs a fixed number $m'$ ($\leq m$) of processors, and a fixed number $r' \in 0..r$ of copies of the resource $R$ during its execution. Let $T(m',r')$, $m' \in 1..m$, $r' \in 0..r$, be the sub-set of all the tasks from $T$ which need $m'$ processors and $r'$ copies of $R$ during its execution.

We talk about a _feasible_ schedule for the task set $T$ if the tasks are processed in such a way that all the tasks being in execution at the same time together do not need more processors and resources than available in the system. A feasible schedule is called _preemptive_, if processing of any task can be arbitrarily prempted and continued later. A feasible schedule (preemptive or not preemptive) is called _optimal_ if it is of minimal length, i.e. the time interval between starting the first task and finishing the last one is minimal.

In [1] and [2] the problem is analysed where each task is of type $T(m',0)$, i.e. no resources are considered. It is shown that

(i)    in the case of unit execution times the problem of non-preemptive scheduling can be solved in time $O(n)$,

(ii)   in the case of arbitrary execution times the problem of preemptive scheduling is of time complexity $O(n \log n)$

     ($n$ = number of tasks to be scheduled).

We consider the problem of preemptive scheduling under the following restrictions: the task set $T$ contains tasks of type $T(m',r')$ where $m'$ is 1 or k (k fixed) and $r'$ is 0 or s (s fixed). Then $T$ is the union of the four disjoint subsets $T(1,0)$, $T(1,s)$, $T(k,0)$, $T(k,s)$. We denote these subsets by $T^O$, $T^S$, $W^O$, $W^S$, respectively. We shall present an algorithm for optimal scheduling the following sub-cases of this problem in time $O(n \log n)$:

    (a) $T^S = \emptyset$,   (b) $W^S = \emptyset$,   (c) $T^O = \emptyset$ and $W^O = \emptyset$ .

The algorithm follows an idea which was presented in [2]:

(i)     In the first step calculate a bound $C$ for the schedule length.

(ii)    Then assign all the W-tasks, i.e. all the tasks which need k (> 1) processors. The bound $C$ is chosen large enough, so that this step can always be performed.

(iii)   Next try to assign the T-tasks to those processors which are still not occupied by W-tasks. In this step it may happen that not all the T-tasks can be executed within the time limit $C$ (viz. the example below). In that case the bound $C$ has to be enlarged. If however, all the T-tasks can be scheduled feasibly, then the resulting schedule is optimal.

ALGORITHM

```
begin
INITIALIZE_C;                        {compute bound C}
repeat
   INITIALIZE_TASK_LISTS;            {task lists WL and TL are initialized}
   ASSIGN_W_TASKS;                   {WL-tasks are scheduled}
   ASSIGN_T_TASKS(OK);               {TL-tasks are scheduled}
   if not OK then RECOMPUTE_C;
   until OK;
end.
```

Procedure INITIALIZE_C;

```
begin
```
$$X^O := \sum_{T' \in T^O} t' \; ; \quad X^S := \sum_{T' \in T^S} t' \; ;$$
$$Y^O := \sum_{W' \in W^O} w' \quad\quad \sum_{W^S \in W^S} w' \; ;$$
$$Z := X^O + X^S + k * (Y^O + Y^S);$$
     {the execution times of tasks T' and W' are denoted by t' and w',

```
         resp.}
t_max   := max{t' | T' ∈ T^O ∪ T^S};
w_max   := max{w' | W' ∈ W^O ∪ W^S};
C := max{Z / m, (Y^O + Y^S) / ⌊m/k⌋, t_max , w_max , (Y^S + X^S) / r};
end;   INITIALIZE_C
```

```
procedure INITIALIZE_TASK_LISTS;

begin
WL := (W1,W2,...,Wn_1);   {list of W-tasks;
                           if W^S <> Ø then place the W-tasks (those
                           requiring resources) at first into the list
                           otherwise the order of tasks may be arbitrary}

TL := (T1,T2,...,Tn_2);   {list of T-tasks;
                           let T1,...,Tl be the resource-tasks,
                           and Tl+1,...,Tn be the non-resource-tasks.
                           n_1 + n_2 = n }

end;   {INITIALIZE_TASK_LISTS}


procedure ASSIGN_W_TASKS;

begin
        {L is the point between 0 and bound C where the last assigned
        W-task ends; viz. fig. 1.
        Within the interval [0,L), then number of processors busy with
        W-tasks is  k * (⌊Y / C⌋ + 1)), whereas in the interval [L,C),
        the number of busy processors is  k * (⌊Y / C⌋)). }
...
end;   {ASSIGN_W_TASKS}
```

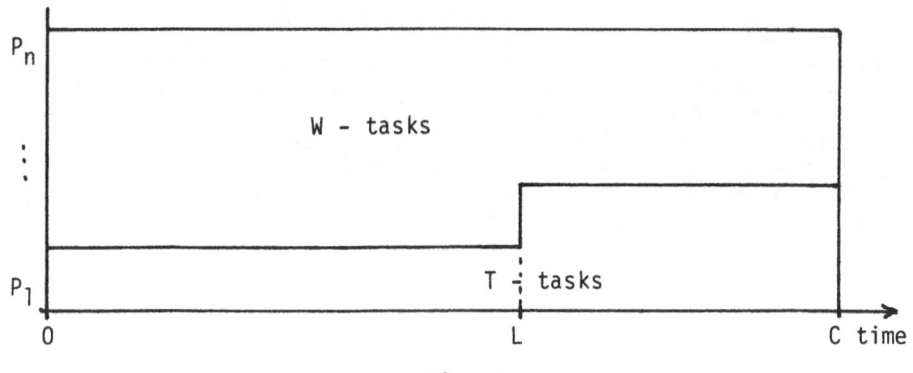

Fig. 1

```
procedure ASSIGN_T_TASKS (var OK: boolean);

begin
  {this procedure consists of two main loops;
   in the first loop assign all the resource-tasks to processors,
   considering the resource restriction.  The idea is to divide each
   task into two parts, and assign one part to the left hand side of the
   schedule (interval  [0,L)), and the other part to the right hand side;
   hereby the aim is to divide each task according to the processing
   power in the two intervals [0,L) and [L,C).  This process of task
   assignment follows exactly the algorithm given in [2] for non-
   resource tasks.
   If it is not possible to schedule all the resource-tasks then the
   bound C is too small; in that case set OK := false, otherwise set
   OK := true.}

  {If OK then execute the second loop in which the non-resource-tasks
   are assigned to the processors;  here the same algorithm applies as
   before.}
end;   {ASSIGN_T_TASKS}


procedure RECOMPUTE_C;

begin
  {This procedure is called as soon as one of the T-tasks cannot be
   scheduled within the global bound C;
   bound C is then enlarged, so that the last T-task can be scheduled,
   and this new bound is minimal.}
end;   {RECOMPUTE_C}
```

It is easy to prove that the procedure RECOMPUTE_C is called at most  m
times.  The instructions in the repeat-loop need $O(n)$ steps of time;
hence the time complexity of the main part of the algorithm results in
$O(n * m)$.


EXAMPLE

     Consider a processing system where  m=8  processors and  r=3  copies
of some resource are available;  the task set contains the four subsets
$W^0 := \{W1, W2, W3\}$,  $W^1 := \emptyset$,  $T^0 := \emptyset$,  $T^1 := \{T1, T2, T3, T4\}$;  the
W-tasks need 3 processors  (k=3),  the execution times are  w1=5, w2=3,
w3=3, t1=7, t2=6, t3=6, t4=5.  Note that the resource requirements of
the T-tasks are assumed to be 1.  In INITIALIZE_C the bound C will then
be computed to C=8.  In ASSIGN_W_TASKS the W-tasks are assigned to the
processors according to Fig. 2;  note that  L  is 3.  Considering the
still not used processors and the resource limits we calculate the ratio
of processing power for the resource tasks on each side of point L to
q1:q2,  q1=6/21, q2=15/21.  If we assign the T-tasks trying to meet this
ratio we recognize that task T4 cannot be scheduled within time limit C.
Recomputation of bound C gives then C = 35/4.  With this bound all the
tasks can be scheduled feasibly, and obviously the resulting schedule
(Fig. 3) is optimal.


266

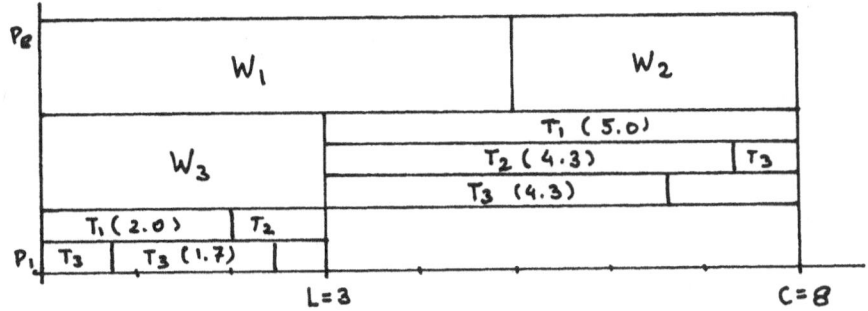

Fig. 2

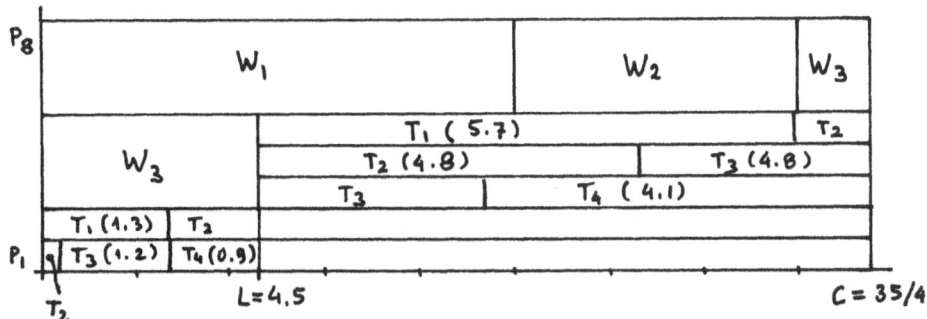

Fig. 3

REFERENCES

1. J. Blazewicz, M. Drabowski, J. Weglarz, Scheduling independent
   2-processor tasks to minimize schedule length.
   Information Processing Letters 18, pp. 267-273, (1984).

2. J. Blazewicz, M. Drabowski, J. Weglarz, Scheduling multiprocessor
   tasks to minimize schedule length.
   IEEE Trans. on Comput., to appear.

THE NUMERICAL SOLUTION OF NON-LINEAR PARABOLIC EQUATIONS

ON MIMD PARALLEL COMPUTERS

M.P. Bekakos and D.J. Evans

Dept. of Computer Studies, Loughborough University of

Technology, Loughborough, Leicestershire LE11 3TU, U.K.

ABSTRACT

This paper investigates the applicability of a new class of group explicit methods to parallel processing, the experimental vehicle being a non-linear parabolic p.d.e. of second order.

The method is briefly presented and then work is concentrated on the selection of a suitable algorithm which can effectively use the processing power of a MIMD parallel architecture and at the same time is stable and can produce highly accurate results. A numerical example is given and some impressive conclusions are derived from the experimental results and the detailed *deterministic* performance analysis in favour of the new strategy when compared with existing methods.

1.  INTRODUCTION

In recent times interest has centred on *implicit methods* for the numerical solution of the initial-boundary value problem by finite-difference /element methods. This is mainly due to stability reasons which lead to larger time-steps of integration and often increased accuracy.

The increasing availability of parallel computers, however, has meant that the *explicit methods* of solution not only offer simplicity, but also the capability that the solution can be obtained at many points at the same time. This important factor may again enable explicit methods to be competitive.

The new strategy combines stable asymmetric approximations to the partial differential equations which when coupled in groups of 2 adjacent points on the grid result in implicit equations which can be easily converted to explicit form. By judicious use of alternating this strategy on the grid points of the domain results in an algorithm which possesses unconditional stability. The merit of this approach results in more accurate solutions because of truncation error cancellations.

This paper will investigate the effectiveness of the new method to the one-dimensional non-linear parabolic partial differential equation

$$\frac{\partial u}{\partial t} = G(x, t, u, \frac{\partial u}{\partial x}, \frac{\partial^2 u}{\partial x^2}), \quad \text{for } 0 \le x \le 1, \ t \ge 0, \quad (1.1)$$

on MIMD parallel testbeds. In particular, we seek to solve the equation

$$\frac{\partial u}{\partial t} = \varepsilon \frac{\partial^2 u}{\partial x^2} - u \frac{\partial u}{\partial x}, \quad \text{for } \varepsilon > 0, \quad (1.2)$$

which has been discussed by Burgers [BURG48] as a mathematical model of turbulence and by Cole et al [COLE51] for the approximate theory for weak non-stationary shock waves in a real fluid. This equation can also be considered as a simplified form of the Navier-Stokes [AMES65] equation.

For decades this type of equation has attracted the attention of many researchers and as a result many finite-difference and finite-element methods have been proposed to solve this system. One common difficulty with the existing solutions is that as the value of $\varepsilon$ decreases a finer grid of points has to be chosen in order to obtain a reasonable accuracy. This in turn determines the problem as prohibitively time consuming and often unrealistic to solve on standard sequential computers. Since the recent utilization of parallel computers it is now possible to design effective solution methods for problems which previously were not possible to solve.

In the following sections we shall briefly investigate the existing algorithms for the solution of our problem and search for certain characteristics which are required in order to match the structure of the algorithm with the architecture of the parallel systems. Since the parallel systems selected for this investigation are of Multiple-Instruction stream, Multiple-Data stream type (MIMD), we shall require the following characteristics for the chosen solution method:

(a) Since most of the solution methods are based on finite-difference and finite-element approximations we shall look for high levels of independence between the nodes at each time-level. This independency is maximized in explicit type of solutions where all the points on a time-level can be evaluated simultaneously.

(b) The relation of the points on a time-level with other points on the previous time-level in order to seek a simple formula for all the points. Yet again, the explicit methods provide this criterion, whereas the semi-explicit methods deny the user of this very useful property.

(c) The minimization of the communication costs associated with parallel processing, which can be achieved by investigating the above criteria.

2. FINITE-DIFFERENCE SOLUTION SCHEMES

In this section we discuss the existing algorithms for the solution of equation (1.2) with the initial conditions

$$u(x, 0) = f(x), \quad 0 \le x \le 1, \quad (2.1)$$

and boundary conditions

$$\left. \begin{aligned} u(0, t) &= g_1(t) \\ u(1, t) &= g_2(t) \end{aligned} \right\}, t > 0. \quad (2.2)$$

270

The *Standard Explicit* scheme is formed by taking the central-difference approximations for $\partial u/\partial x$ and $\partial^2 u/\partial x^2$ and a forward-difference approximation for $\partial u/\partial t$. Hence, the finite-difference approximation is of the form

$$U_{i,j+1} = \varepsilon r(U_{i+1,j} - 2U_{i,j} + U_{i-1,j}) + [1 - \frac{r\Delta x}{2}(U_{i+1,j} - U_{i-1,j})]U_{i,j} \,, \qquad (2.3)$$

where $\quad U_{i,j} = U(i\Delta x, j\Delta t)$,

$$x = i\Delta x, \quad i = 0,1,2,\ldots,m$$

$$t = j\Delta t, \quad j = 0,1,2,\ldots.$$

$$\Delta x = 1/m$$

and $\quad r = \Delta t/\Delta x^2$ .

Note that $U_{i,j+1}$ is a nonlinear function in known terms $U_{i-1,j}$, $U_{i,j}$ and $U_{i+1,j}$ and therefore it can be computed directly without involving an iteration process.

Although this method is computationally simple, it suffers from a very restrictive stability condition, i.e. the value of $r$ should be $\leq 1/2$ in order to retain reasonable accuracy; in practice, however, a very small value of $\varepsilon$ will force a small value of $\Delta x$, which in turn, with $r \leq 1/2$, makes the value of $\Delta t$ too small for practical purposes. A clear advantage of this method is the independency of the points within a single time-level, which makes it naturally suitable for SIMD type architectures.

A stable *fully Implicit* scheme can be obtained by taking the backward-difference approximation to $\partial u/\partial t$ at point $(i,j+1)$ to obtain

$$U_{i,j+1} = U_{i,j} + \varepsilon r(U_{i+1,j+1} - 2U_{i,j+1} + U_{i-1,j+1})$$

$$- \frac{r\Delta x}{2}(U_{i+1,j+1} - U_{i-1,j+1})U_{i,j+1} \,. \qquad (2.4)$$

In this approximation scheme the unknowns, $U_{i-1,j+1}$, $U_{i,j+1}$, $U_{i+1,j+1}$, are involved in a nonlinear relation and therefore it must be solved iteratively.

Alternatively, approximating the partial derivatives $\partial u/\partial x$ and $\partial^2 u/\partial x^2$ by the *mean* of their central-difference approximations on the $j$th and $(j+1)$th time-level will result in a *Crank-Nicolson* type of implicit scheme; namely,

$$U_{i,j+1} = U_{i,j} + \frac{\varepsilon r}{2}[(U_{i+1,j} - 2U_{i,j} + U_{i-1,j}) + (U_{i+1,j+1} - 2U_{i,j+1} + U_{i-1,j+1})]$$

$$- \frac{r\Delta x}{4}[(U_{i+1,j} - U_{i-1,j}) + (U_{i+1,j+1} - U_{i-1,j+1})] \frac{(U_{i,j} + U_{i,j+1})}{2} \,. \qquad (2.5)$$

As can be seen, the unknowns, $U_{i-1,j+1}$, $U_{i,j+1}$ and $U_{i+1,j+1}$, are again related through a nonlinear expression which has to be solved iteratively. Although there are alternatives in which the nonlinearity of the above relation can be avoided, the implicit methods are still more expensive computationally in comparison with the explicit method. Their popularity is due mainly to their property of possessing unconditional stability, which leads to larger time-steps of integration and often increased accuracy.

Another disadvantage associated with the implicit schemes, from parallel computational aspects, is the fact that the unknowns are usually related through an expression and little or no independencies exist to be exploitable by parallel architectures. In these schemes one has to look at different labelling (colouring) techniques to see if the points in a time-step can be grouped into independent sets so that those sets can be computed simultaneously. Consequently, with the increasing availability of parallel computers and their greater throughput, the explicit methods of solution not only offer simplicity, but also the capability that the solution can be obtained at every point concurrently. This important factor reinforces the need for improved explicit procedures for utilization on parallel computers.

## 3. A NEW CLASS OF GROUP EXPLICIT (GE) METHODS

Recently, an interesting new variation of the use of Saul'yev's asymmetric equations was investigated by Evans and Abdullah [EVAN83]. The *ladder-step* formulae of these equations, for the particular case of Burgers' equation (see Bekakos [BEKA86]), are given by

$$-raU_{i+1,j+1} + (1+ra)U_{i,j+1} = (1-rb)U_{i,j} + rbU_{i-1,j} \qquad (3.1)$$

and

$$-rbU_{i-1,j+1} + (1+rb)U_{i,j+1} = (1-ra)U_{i,j} + raU_{i+1,j} \qquad (3.2)$$

where

$$a = \epsilon - \frac{\Delta x}{4}(U_{i,j+1} + U_{i,j})$$

$$b = \epsilon + \frac{\Delta X}{4}(U_{i,j+1} + U_{i,j}) \ . \qquad (3.3)$$

Hence, algorithms similar to those suggested by Larkin [LARK64] for the heat-conduction equation are possible. In particular, both the equations (3.1) and (3.2) are unconditional stable for $r>0$ and semi-explicit in the sense that if the equation (3.1) is solved in a right-to-left (RL) direction and the equation (3.2) in a left-to-right (LR) direction, then the need to solve a linear system for the solution on each line is averted.

The central theme of this new variation was not to restrict the use of the above equations solely along the x lines in the RL- and LR-directions, but to apply them to groups of two points successively along every line.

So far, we have shown that the explicit methods although very suitable for parallel processing always deny us reasonable accuracy and some stability. On the other hand, the implicit schemes offer stability, but the exploitation of these methods for parallel processing may be

272

difficult and possibly inefficient. The semi-explicit formulae above offer a trade-off between stability and the possibility of them being suitable for parallel implementation. Furthermore, it is possible to express these semi-explicit schemes in terms of pure explicit formulae to enable their efficient implementation. Of this class is the Group Explicit method which, to some extent, provides a more effective formula for implementation on parallel computers.

In order to formulate the GE equations we assume, without loss of generality, that the line segment $0 < x < 1$ is divided into an even number m of equal sub-intervals, which implies that at every time-level the number of internal points is odd, i.e. (m-1). The coupled use of Saul'yev's asymmetric equations at the points (i,j+1) and (i+1,j+1) results, in fact, in a (2x2)-set of implicit finite-difference equations which can be easily converted to explicit form, as is developed in the following.

Let us now consider any group of two points, i.e. $(i,j+1/2)$ and $(i+1,j+1/2)$, and use equations (3.1), (3.2) at these points, correspondingly, and grouped together to give the (2x2) system of equations,

$$\begin{bmatrix} 1+a_1^{(n)}r & -a_1^{(n)}r \\ -b_2^{(n)}r & 1+b_2^{(n)}r \end{bmatrix} \begin{bmatrix} U_{i,j+1}^{(n+1)} \\ U_{i+1,j+1}^{(n+1)} \end{bmatrix} = \begin{bmatrix} 1-b_1^{(n)}r & 0 \\ 0 & 1-a_2^{(n)}r \end{bmatrix} \begin{bmatrix} U_{i,j} \\ U_{i+1,j} \end{bmatrix} + \begin{bmatrix} b_1^{(n)}rU_{i-1,j} \\ a_2^{(n)}rU_{i+2,j} \end{bmatrix}$$

with                                                                                           (3.4)

$$a_1^{(n)} = \varepsilon - \frac{\Delta x}{4}(U_{i,j+1}^{(n)}+U_{i,j}) \; ; \quad a_2^{(n)} = \varepsilon - \frac{\Delta x}{4}(U_{i+1,j+1}^{(n)}+U_{i+1,j})$$

(3.5)

$$b_1^{(n)} = \varepsilon + \frac{\Delta x}{4}(U_{i,j+1}^{(n)}+U_{i,j}) \; ; \quad b_2^{(n)} = \varepsilon + \frac{\Delta x}{4}(U_{i+1,j+1}^{(n)}+U_{i+1,j}) \quad .$$

The explicit form of the above can be defined as

$$\begin{bmatrix} U_{i,j+1}^{(n+1)} \\ U_{i+1,j+1}^{(n+1)} \end{bmatrix} = \frac{1}{\Delta} \left\{ \begin{bmatrix} (1+b_2^{(n)}r-b_1^{(n)}r-b_1^{(n)}b_2^{(n)}r^2 & (1-a_2^{(n)}r)a_1^{(n)}r \\ b_2^{(n)}r(1-b_1^{(n)}r) & (1+a_1^{(n)}r-a_2^{(n)}r-a_1^{(n)}a_2^{(n)}r^2 \end{bmatrix} \right.$$

(3.6)

$$\times \begin{bmatrix} U_{i,j} \\ U_{i+1,j} \end{bmatrix} + \left. \begin{bmatrix} (1+b_2^{(n)}r)b_1^{(n)}rU_{i-1,j}+a_1^{(n)}a_2^{(n)}r^2U_{i+2,j} \\ b_1^{(n)}b_2^{(n)}r^2U_{i-1,j}+(1+a_1^{(n)}r)a_2^{(n)}rU_{i+2,j} \end{bmatrix} \right\}$$

where the inverse of the coefficient matrix for $U_{i,j+1}, U_{i+1,j+1}$ has been estimated through the use of its adjoint matrix and the determinant

$$\Delta = 1+a_1^{(n)}r + b_2^{(n)}r \quad .$$                    (3.7)

Note that the superscript n refers to the $n^{th}$ iteration number. However, despite the small time-step values $\Delta t$ the results at each time-level were so accurate that there was no need for any iteration of the non-linear problem. Furthermore, the explicit equations (3.6) are

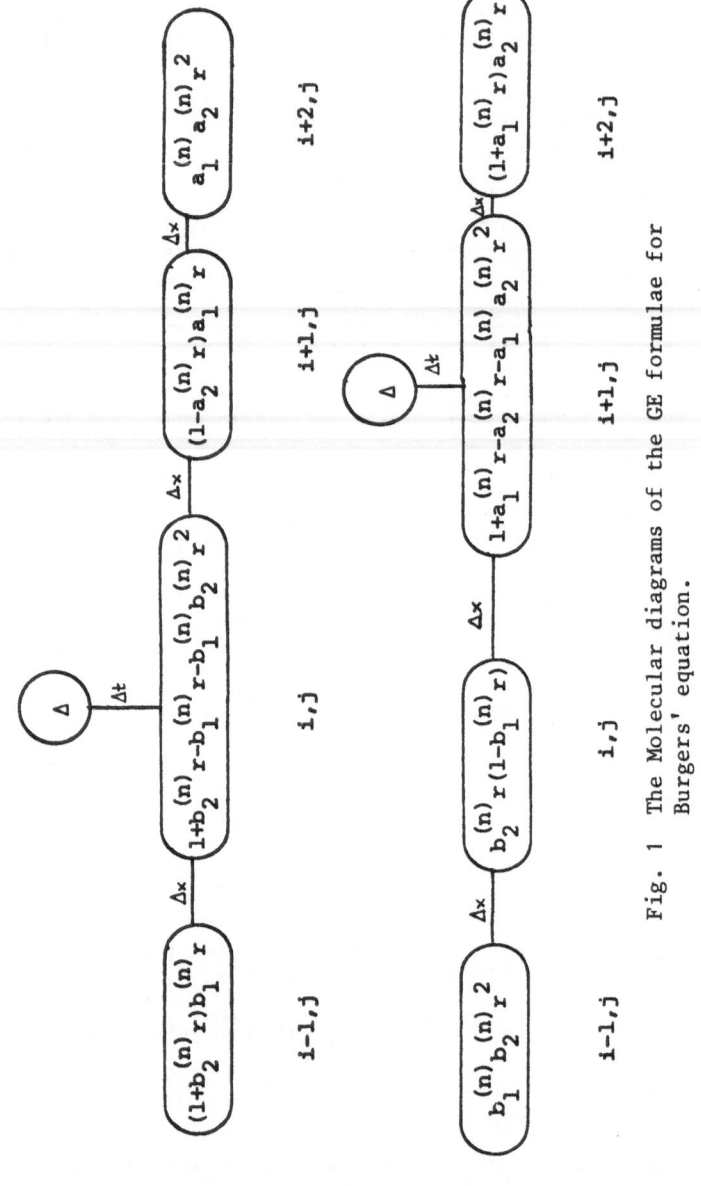

Fig. 1  The Molecular diagrams of the GE formulae for Burgers' equation.

computationally easier to handle and their computational molecular diagram is given in Figure 1.

In the following section, making use of these GE equations, we shall consider a particular scheme for this class of methods under the assumption of an odd number of sub-intervals, and this for a more computationally balanced implementation. The stability of this scheme will be seen from the numerical results exhibited, while theoretically, using the matrix method, it has been proved in [EVAN 83].

Finally, this particular GE scheme will be implemented and analysed in comparison with the Standard Explicit method (i.e. formula (2.3)), which will always form the basis for comparisons in the attempt to obtain the "true" performance results of this class of methods (see Evans and Bekakos [EVAN 85]).

## 4. THE (SINGLE) ALTERNATING GROUP EXPLICIT – (S)AGE METHOD

This scheme, for the particular case of Burgers' equation, is given by

$$(I+rG_{1,j})\underline{U}_{j+1} = (I-rG_{2,j})\underline{U}_j + \underline{b}_1$$

$$(I+rG_{2,j})\underline{U}_{j+2} = (I-rG_{1,j})\underline{U}_{j+1} + \underline{b}_2$$

$$\qquad (4.1)$$

where

$$\underline{b}_1^T = [b_2^{(n)}rU_{0,j+1}, 0, \ldots, 0, a_1^{(n)}rU_{m,j+1}],$$

$$\underline{b}_2^T = [b_2^{(n)}rU_{0,j}, 0, \ldots, 0, a_1^{(n)}rU_{m,j}]$$

$$\qquad (4.2)$$

and

$$\qquad (4.3)$$

with

$$G^{(k)} = \begin{bmatrix} a_1^{(n)} & -a_1^{(n)} \\ -b_2^{(n)} & b_2^{(n)} \end{bmatrix} \quad \text{for } k = 1,2,\ldots,\tfrac{1}{2}(m-3)$$

$$\text{and } j = 0,1,2,\ldots. \qquad (4.4)$$

$$\hat{G}^{(i)} = \begin{bmatrix} a_2^{(n)} & -a_2^{(n)} \\ -b_1^{(n)} & b_1^{(n)} \end{bmatrix} \quad \text{for } i = 1,2,\ldots,\tfrac{1}{2}(m-1),$$

This scheme diagrammatically is described by the "brick" diagram in Fig. 2.

Let us now proceed with the actual implementation of formulae (2.3) and (4.1) concentrating on their inherent parallelism in terms of the most efficient utilization of system's hardware (the *NEPTUNE* system, a *MIMD* 4-processor testbed) and software potential.

The explicit nature of the implementations offers the possibility for alternating ways of achieving the inherent parallelism. In fact, the considered each time total number of boundary and internal grid points can be divided into various task sizes to be assigned to the cooperating each time processors. The broadcasting between the working processors takes place through 'shared' real arrays storing the p.d.e.'s exact and approximate values at all the boundary and internal grid points.

Specifically, for the GE scheme, the values at the 'left' and 'right' ungrouped near boundary points, at every time-level that they occur, are computed using the Saul'yev's LR, RL asymmetric formulae, respectively. The values at all or the remaining groups of two internal points are computed using the Group Explicit finite-difference formulae. Furthermore, it should be pointed out that the exact values at the points on both boundaries, for all time-levels, as well as the exact initial values at the internal points, in either implementation, are computed separately using the chosen exact solution formula.

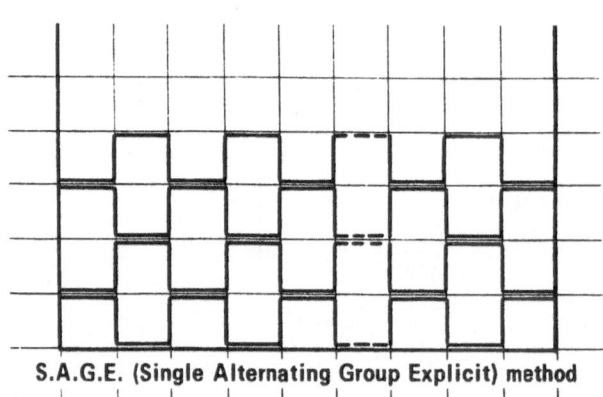

**S.A.G.E. (Single Alternating Group Explicit) method**

Fig. 2 The representative diagram of this scheme.

From the aspect of the parallel strategy in programming, both methods have been implemented in an unavoidable, but highly efficient synchronized manner which is due to the nature of the methods.

After an extensive variety of experimentation we converged to the conclusion that the implementations' *granularity factor*, or otherwise the number of processes (paths) generated at every time-level, has to be made to be equal to the number of co-operating processors.

Finally, in particular for the NEPTUNE prototype system, let us discuss some absolutely necessary specialised characteristics of it that will be mainly and repeatedly utilized in our analysis, in combination with the resource provisions of the system. More specifically, two sub-routines are available for obtaining the timing information. The routines should be embedded within a $DOALL/$PAREND sequence to force each processor to execute them. The timing is started or restarted with

CALL TIMEST

and current timing(s) is obtained using

CALL TIMOUT(ITIME),

where ITIME must be declared as a shared array of size 100 and printed out from a subsequent sequential path, its results being arranged in 8 columns.

In conclusion, to generate and terminate the parallel paths each time the $DOPAR/$PAREND construct is utilized, which is the most efficient and economical way to introduce parallelism in a program for the NEPTUNE computer.

## 5. NUMERICAL EXPERIMENTATION AND COMPARATIVE RESULTS

For our numerical experiments with Burgers' equation the following exact solution has been chosen (see Madsen and Sincovec [MADS76]),

$$u(x,t) = \frac{0.1e^{-A}+0.5e^{-B}+e^{-C}}{e^{-A}+e^{-B}+e^{-C}} , \quad 0 \leq x \leq 1, \ t \geq 0 \qquad (5.1)$$

where

$$
\begin{aligned}
A &= \frac{0.05}{\varepsilon}(x-0.5+4.95t) \\
B &= \frac{0.25}{\varepsilon}(x-0.5+0.75t) \\
C &= \frac{0.5}{\varepsilon}(x-0.375)
\end{aligned}
\left.\right\} . \qquad (5.2)
$$

This problem in practice would have a very small value of $\varepsilon$ and this in turns means that the interval [0,1] in the x direction has to be sub-divided into a very fine grid in order to obtain reasonable accuracy. In our experimental work, for a reasonably small value of $\varepsilon$, both methods have been analytically tested for a wide range of internal points and time-levels reaching the extremest figures allowed by the NEPTUNE system.

In all cases, despite the very small values of the time-step $\Delta t$, the results at each time-level are so accurate that there is no need for any iteration of the nonlinear problem.

In respect to the Number of Internal grid Points ($N_{I.P.}$) and

Time-steps experimented with, the deterministic factor for the sizes of the rectangular grids considered in the *open rectangle* $[0,1] \times [0,+\infty)$ is the grid ratio r[+] . More specifically, the grid sizes when r = 1 have been considered as the common basis to estimate the corresponding grid sizes for each individual method when r takes different values, since this is the maximum value for r satisfying the stability condition (i.e. r ≤ 1) for the 'basic' GE schemes[++]. The maximum grid size allowed by the NEPTUNE system is (1920x480), which analogously imposes the corresponding maximum grid sizes for the GE method. Certainly, the considered numbers of internal points and time-steps could equally well be in reverse order; however, this would probably lead to less accurate solutions.

The following tables, 1,2,3,4, illustrate some of the experimental results and performance measurements obtained on this MIMD system. The parameters appearing in these tables, in this priority sequence, are: Number of Internal Points, Number of utilized PROCessorS, grid ratio, Maximum Percentage Error, Maximum Absolute Error, Time-complexity (experimental), real Cost, internal acceleration (Speed-up), Efficiency, Effectiveness. In particular for the GE method, except this usual internal acceleration, we estimate two additional parameters, the Relative or normalized Speed-up ($R_{S_p}$) and the Reference internal Speed-up ($\overline{R}_{S_p}$); they represent, respectively, the ratio between the experimental time-complexity of the Standard explicit method and the experimental time-complexities of the GE method, and the Speed-ups achieved due to the increase of the grid ratio (see Bekakos and Evans [BEKA86a]).

The conclusions that arise when comparing these results are apparent and in favour of the GE scheme. The tremendous potential of this method unfolds when passing the restrictive barrier of 1 for the grid ratio and experiment with greater values. In general, from every aspect, either numerical or experimental, this method performs unbelievably. Although it is a matter of less interest to discuss the values of $S_p$ and $E_p$ parameters, however, a very close fluctuation in their values for the *(S)AGE* scheme has been observed. The revealing factors of the tremendous potential of the method are the Relative and Reference Speed-up parameters. More analytically, we note that the Relative Speed-ups are amazingly keeping up an analogous increase in values according to the increase of the grid ratio r, always being > $O(r.p)$. In actual fact, for r = 4, we reached a maximum 16.766 times acceleration compared to the Standard Explicit algorithm by only using 4 processors; it is really a very interesting matter what the result would be if we could experiment with this method for greater grid sizes (and consequently r's), unfortunately not feasible on the NEPTUNE prototype system. In other words, we have proved that this method can really accelerate in a manner independent of the system's restrictive, in terms of the number of available processors, configuration.

The explanation for this amazing result lies on the $\overline{R}_{S_p}$ parameter which exhibits a Reference internal acceleration of $O(r.p)$, reaching a maximum value of 15.357, again, for r = 4 and in a maximum utilization of the system, due to the smaller time-complexities (and consequently real Costs) caused by the proportional reduction in the total number of time-steps.

---

[+] In order to maintain the same overall time-advance length.

[++] See Bekakos and Evans [BEKA85].

Table 1 Experimental Results and Performance Measurements of the Parallel Algorithm for the Standard Explicit Method on the 'NEPTUNE' Prototype System.

| $N_{L.P.}$ | $N_{PROCS}$ | $r$ | TIME-STEPS: 120 | | | | | | | TIME-STEPS: 240 | | | | | | |
|---|---|---|---|---|---|---|---|---|---|---|---|---|---|---|---|---|
| | | | $M_{P.E.}$ | $M_{A.E.}$ | $Tc^{(e)}$ (secs) | $C_p$ | $S_p$ | $E_p$ | $F_p \cdot T_s^{(e)}$ | $M_{P.E.}$ | $M_{A.E.}$ | $Tc^{(e)}$ (secs) | $C_p$ | $S_p$ | $E_p$ | $F_p \cdot T_s^{(e)}$ |
| 240 | $\emptyset$ | 0.5 | .195053183E-01 | .149965286E-03 | 195.383 | 195.383 | 1 | 1 | 1 | .355070271E-01 | .260531902E-03 | 385.738 | 385.738 | 1 | 1 | 1 |
| | $\emptyset,1$ | | | | 100.790 | 201.580 | 1.939 | 0.969 | 1.879 | | | 198.583 | 397.166 | 1.942 | 0.971 | 1.887 |
| | $\emptyset,1,2$ | | | | 69.720 | 209.160 | 2.802 | 0.934 | 2.618 | | | 137.003 | 411.009 | 2.816 | 0.939 | 2.642 |
| | $\emptyset,1,2,3$ | | | | 52.993 | 211.972 | 3.687 | 0.922 | 3.398 | | | 104.363 | 417.452 | 3.696 | 0.924 | 3.415 |
| 480 | $\emptyset$ | 0.5 | .106825903E-01 | .106275082E-03 | 382.918 | 382.918 | 1 | 1 | 1 | .200394355E-01 | .199377537E-03 | 755.790 | 755.790 | 1 | 1 | 1 |
| | $\emptyset,1$ | | | | 197.950 | 395.900 | 1.934 | 0.967 | 1.871 | | | 392.053 | 784.106 | 1.928 | 0.964 | 1.858 |
| | $\emptyset,1,2$ | | | | 135.710 | 407.130 | 2.822 | 0.941 | 2.654 | | | 267.298 | 801.894 | 2.828 | 0.943 | 2.665 |
| | $\emptyset,1,2,3$ | | | | 102.730 | 410.920 | 3.727 | 0.932 | 3.473 | | | 202.205 | 808.820 | 3.738 | 0.934 | 3.493 |
| 960 | $\emptyset$ | 0.5 | .111256465E-01 | .106096268E-03 | 759.735 | 759.735 | 1 | 1 | 1 | .208908506E-01 | .204622746E-03 | 1498.070 | 1498.070 | 1 | 1 | 1 |
| | $\emptyset,1$ | | | | 393.510 | 787.020 | 1.931 | 0.965 | 1.864 | | | 776.815 | 1553.630 | 1.928 | 0.964 | 1.860 |
| | $\emptyset,1,2$ | | | | 268.718 | 806.154 | 2.827 | 0.942 | 2.664 | | | 530.485 | 1591.455 | 2.824 | 0.941 | 2.658 |
| | $\emptyset,1,2,3$ | | | | 203.228 | 812.912 | 3.738 | 0.935 | 3.494 | | | 401.073 | 1604.992 | 3.735 | 0.934 | 3.488 |
| 1920 | $\emptyset$ | 0.5 | .115095451E-01 | .103712082E-03 | 1513.813 | 1513.813 | 1 | 1 | 1 | .227458738E-01 | .204980373E-03 | 2987.858 | 2987.858 | 1 | 1 | 1 |
| | $\emptyset,1$ | | | | 786.073 | 1572.146 | 1.926 | 0.963 | 1.854 | | | 1548.110 | 3096.220 | 1.930 | 0.965 | 1.862 |
| | $\emptyset,1,2$ | | | | 535.260 | 1605.780 | 2.828 | 0.943 | 2.666 | | | 1054.413 | 3163.239 | 2.834 | 0.945 | 2.677 |
| | $\emptyset,1,2,3$ | | | | 404.855 | 1619.420 | 3.739 | 0.935 | 3.495 | | | 798.165 | 3192.660 | 3.743 | 0.936 | 3.503 |

(continued)

Table 1 (contd.)

| $N_{I.P.}$ | $N_{PROCS}$ | $r$ | $M_{P.E.}$ | $M_{A.E.}$ | $Tc^{(e)}$ (secs) | $C_p$ | $S_p$ | $E_p$ | $F_p \cdot T_s^{(e)}$ |
|---|---|---|---|---|---|---|---|---|---|
| 240 | ∅<br>∅,1<br>∅,1,2<br>∅,1,2,3 | 0.5 | .6706333316E-01 | .5028843388E-03 | 766.268<br>394.895<br>271.700<br>207.075 | 766.268<br>789.790<br>815.100<br>828.300 | 1<br>1.940<br>2.820<br>3.700 | 1<br>0.970<br>0.940<br>0.925 | 1<br>1.883<br>2.651<br>3.423 |
| 480 | ∅<br>∅,1<br>∅,1,2<br>∅,1,2,3 | 0.5 | .3649645530E-01 | .3631711101E-03 | 1500.133<br>775.268<br>532.595<br>402.493 | 1500.133<br>1550.536<br>1597.785<br>1609.972 | 1<br>1.935<br>2.817<br>3.727 | 1<br>0.967<br>0.939<br>0.932 | 1<br>1.872<br>2.645<br>3.473 |
| 960 | ∅<br>∅,1<br>∅,1,2<br>∅,1,2,3 | 0.5 | .4079534449E-01 | .3996491143E-03 | 2974.615<br>1541.228<br>1053.690<br>797.498 | 2974.615<br>3082.456<br>3161.070<br>3189.992 | 1<br>1.930<br>2.823<br>3.730 | 1<br>0.965<br>0.941<br>0.932 | 1<br>1.863<br>2.657<br>3.478 |
| 1920 | ∅<br>∅,1<br>∅,1,2<br>∅,1,2,3 | 0.5 | .4476282000E-01 | .4034638400E-03 | 5934.120<br>3072.303<br>2096.200<br>1588.353 | 5934.120<br>6144.606<br>6288.600<br>6353.412 | 1<br>1.931<br>2.831<br>3.736 | 1<br>0.966<br>0.944<br>0.934 | 1<br>1.865<br>2.671<br>3.489 |
| TIME-STEPS: | | | | | | 480 | | | |

Table 2 Experimental Results and Performance Measurements of the Parallel Algorithm for the (S).$A.G.E.$ Method (for $r$ = 1) on the *NEPTUNE* Prototype System.

| $N_{I.P.}$ | $N_{PROCS}$ | $r$ | M.P.E. | M.A.E. | $T_c^{(e)}$ (secs) | $C_p$ | $S_p$ | $R_{S_p}$ | $E_p$ | $F.T_p^{(e)}s$ | M.P.E. | M.A.E. | $T_c^{(e)}$ (secs) | $C_p$ | $S_p$ | $R_{S_p}$ | $E_p$ | $F.T_p^{(e)}s$ |
|---|---|---|---|---|---|---|---|---|---|---|---|---|---|---|---|---|---|---|
| 240 | ∅ | | .119335875E-01 | .866055489E-04 | 175.248 | 175.248 | 1 | 1.115 | 1 | 1 | .239556655E-01 | .175774097E-03 | 345.300 | 345.300 | 1 | 1.117 | 1 | 1 |
| | ∅,1 | | | | 89.860 | 179.720 | 1.950 | 2.174 | 0.975 | 1.902 | | | 177.320 | 354.640 | 1.947 | 2.175 | 0.974 | 1.896 |
| | ∅,1,2 | | | | 60.778 | 182.384 | 2.883 | 3.215 | 0.961 | 2.771 | | | 119.660 | 358.980 | 2.886 | 3.224 | 0.962 | 2.776 |
| | ∅,1,2,3 | 1.0 | | | 45.948 | 183.792 | 3.814 | 4.252 | 0.954 | 3.637 | | | 90.420 | 361.680 | 3.819 | 4.266 | 0.955 | 3.646 |
| 480 | ∅ | | .599190593E-02 | .599026680E-04 | 348.000 | 348.000 | 1 | 1.110 | 1 | 1 | .111145377E-01 | .111103058E-03 | 684.570 | 684.570 | 1 | 1.104 | 1 | 1 |
| | ∅,1 | | | | 178.350 | 356.700 | 1.951 | 2.147 | 0.976 | 1.904 | | | 351.420 | 702.840 | 1.948 | 2.151 | 0.974 | 1.897 |
| | ∅,1,2 | | | | 119.700 | 359.100 | 2.907 | 3.199 | 0.969 | 2.817 | | | 235.800 | 707.400 | 2.903 | 3.205 | 0.968 | 2.809 |
| | ∅,1,2,3 | 1.0 | | | 90.280 | 361.120 | 3.855 | 4.241 | 0.964 | 3.715 | | | 178.420 | 713.680 | 3.837 | 4.236 | 0.959 | 3.680 |
| 960 | ∅ | | .643055141E-02 | .641942024E-04 | 694.620 | 694.620 | 1 | 1.094 | 1 | 1 | .126221851E-01 | .126004219E-03 | 1366.180 | 1366.180 | 1 | 1.097 | 1 | 1 |
| | ∅,1 | | | | 355.440 | 710.880 | 1.954 | 2.137 | 0.977 | 1.910 | | | 695.190 | 1390.380 | 1.965 | 2.155 | 0.983 | 1.931 |
| | ∅,1,2 | | | | 238.020 | 714.060 | 2.918 | 3.192 | 0.973 | 2.839 | | | 467.260 | 1401.780 | 2.924 | 3.206 | 0.975 | 2.850 |
| | ∅,1,2,3 | 1.0 | | | 179.180 | 716.720 | 3.877 | 4.240 | 0.969 | 3.757 | | | 352.600 | 1410.400 | 3.875 | 4.249 | 0.969 | 3.753 |
| 1920 | ∅ | | .633467734E-02 | .633001328E-04 | 1391.230 | 1391.230 | 1 | 1.088 | 1 | 1 | .121325105E-01 | .121235847E-03 | 2738.860 | 2738.860 | 1 | 1.091 | 1 | 1 |
| | ∅,1 | | | | 705.080 | 1410.160 | 1.973 | 2.147 | 0.987 | 1.947 | | | 1385.270 | 2770.540 | 1.977 | 2.157 | 0.989 | 1.955 |
| | ∅,1,2 | | | | 473.460 | 1420.380 | 2.938 | 3.197 | 0.979 | 2.878 | | | 930.190 | 2790.510 | 2.944 | 3.212 | 0.981 | 2.890 |
| | ∅,1,2,3 | 1.0 | | | 356.090 | 1424.360 | 3.907 | 4.251 | 0.977 | 3.816 | | | 701.420 | 2805.680 | 3.905 | 4.260 | 0.976 | 3.812 |
| TIME-STEPS: | | | | | | 60 | | | | | | | | 120 | | | | | |

(continued)

281

Table 2  (contd.)

| $N_{I.P.}$ | $N_{PROCS}$ | $r$ | $M_{P.E.}$ | $M_{A.E.}$ | $Tc^{(e)}$ (secs) | $C_p$ | $S_p$ | $RS_p$ | $E_p$ | $F_p.T_s^{(e)}$ |
|---|---|---|---|---|---|---|---|---|---|---|
| 240 | ∅ | 1.0 | .452918150E-01 | .339627266E-03 | 685.175 | 685.175 | 1 | 1.118 | 1 | 1 |
|  | ∅,1 |  |  |  | 351.200 | 702.400 | 1.951 | 2.182 | 0.975 | 1.903 |
|  | ∅,1,2 |  |  |  | 237.675 | 713.025 | 2.883 | 3.224 | 0.961 | 2.770 |
|  | ∅,1,2,3 |  |  |  | 180.140 | 720.560 | 3.804 | 4.254 | 0.951 | 3.617 |
| 480 | ∅ | 1.0 | .219819658E-01 | .219762325E-03 | 1360.020 | 1360.020 | 1 | 1.103 | 1 | 1 |
|  | ∅,1 |  |  |  | 698.020 | 1396.040 | 1.948 | 2.149 | 0.974 | 1.898 |
|  | ∅,1,2 |  |  |  | 469.340 | 1408.020 | 2.898 | 3.196 | 0.966 | 2.799 |
|  | ∅,1,2,3 |  |  |  | 354.580 | 1418.320 | 3.836 | 4.231 | 0.959 | 3.678 |
| 960 | ∅ | 1.0 | .193599276E-01 | .193238258E-03 | 2712.530 | 2712.530 | 1 | 1.097 | 1 | 1 |
|  | ∅,1 |  |  |  | 1380.620 | 2761.240 | 1.965 | 2.155 | 0.982 | 1.930 |
|  | ∅,1,2 |  |  |  | 928.930 | 2786.790 | 2.920 | 3.202 | 0.973 | 2.842 |
|  | ∅,1,2,3 |  |  |  | 699.610 | 2798.440 | 3.877 | 4.252 | 0.969 | 3.758 |
| 1920 | ∅ | 1.0 | .232282057E-01 | .232040882E-03 | 5435.620 | 5435.620 | 1 | 1.092 | 1 | 1 |
|  | ∅,1 |  |  |  | 2765.210 | 5530.420 | 1.966 | 2.146 | 0.983 | 1.932 |
|  | ∅,1,2 |  |  |  | 1853.370 | 5560.110 | 2.933 | 3.202 | 0.978 | 2.867 |
|  | ∅,1,2,3 |  |  |  | 1393.490 | 5573.960 | 3.901 | 4.258 | 0.975 | 3.804 |

TIME-STEPS      240

Table 3  Experimental Results and Performance Measurements of the Parallel Algorithm
for the $(S).A.G.E.$ Method (for $r = 2$) on the *NEPTUNE* Prototype System.

| $N_{I.P.}$ | $N_{PROCS}$ | $r$ | $M_{P.E.}$ | $M.A.E.$ | $Tc^{(e)}$ (secs) | $C_p$ | $S_p$ | $R_{S_p}$ | $\bar{R}_{S_p}$ | $E_p$ | $F_p \cdot T_s^{(e)}$ |
|---|---|---|---|---|---|---|---|---|---|---|---|
| 240 | Ø | | .229158811E-01 | .168144703E-03 | 173.960 | 173.960 | 1 | 2.217 | 1.985 | 1 | 1 |
| | Ø,1 | 2.0 | | | 89.040 | 178.080 | 1.954 | 4.332 | 3.878 | 0.977 | 1.909 |
| | Ø,1,2 | | | | 60.200 | 180.600 | 2.890 | 6.408 | 5.736 | 0.963 | 2.783 |
| | Ø,1,2,3 | | | | 45.570 | 182.280 | 3.817 | 8.465 | 7.577 | 0.954 | 3.643 |
| 480 | Ø | | .602386892E-02 | .600218773E-04 | 346.020 | 346.020 | 1 | 2.184 | 1.978 | 1 | 1 |
| | Ø,1 | 2.0 | | | 177.500 | 355.000 | 1.949 | 4.258 | 3.857 | 0.975 | 1.900 |
| | Ø,1,2 | | | | 119.170 | 357.510 | 2.904 | 6.342 | 5.744 | 0.968 | 2.810 |
| | Ø,1,2,3 | | | | 89.510 | 358.040 | 3.866 | 8.444 | 7.648 | 0.966 | 3.736 |
| 960 | Ø | | .573427230E-02 | .573396683E-04 | 687.910 | 687.910 | 1 | 2.178 | 1.986 | 1 | 1 |
| | Ø,1 | 2.0 | | | 350.900 | 701.800 | 1.960 | 4.269 | 3.893 | 0.980 | 1.922 |
| | Ø,1,2 | | | | 235.550 | 706.650 | 2.920 | 6.360 | 5.800 | 0.973 | 2.843 |
| | Ø,1,2,3 | | | | 177.970 | 711.880 | 3.865 | 8.418 | 7.676 | 0.966 | 3.735 |
| 1920 | Ø | | .531333312E-02 | .530481339E-04 | 1375.810 | 1375.810 | 1 | 2.172 | 1.991 | 1 | 1 |
| | Ø,1 | 2.0 | | | 698.990 | 1397.980 | 1.968 | 4.275 | 3.918 | 0.984 | 1.937 |
| | Ø,1,2 | | | | 471.620 | 1414.860 | 2.917 | 6.335 | 5.807 | 0.972 | 2.837 |
| | Ø,1,2,3 | | | | 354.920 | 1419.680 | 3.876 | 8.418 | 7.717 | 0.969 | 3.757 |

TIME-STEPS: 60

(continued)

Table 3  (cont.)

| $N_{I.P.}$ | $N_{PROCS}$ | $r$ | $M_{P.E.}$ | $M_{A.E.}$ | $Tc^{(e)}$ (secs) | $C_p$ | $S_p$ | $R_{S_p}$ | $\bar{R}_{S_p}$ | $E_p$ | $F_p \cdot T_s^{(e)}$ |
|---|---|---|---|---|---|---|---|---|---|---|---|
| 240 | Ø | | .438451469E-01 | .328779221E-03 | 343.953 | 343.953 | 1 | 2.228 | 1.992 | 1 | 1 |
| | Ø,1 | 2.0 | | | 176.318 | 352.636 | 1.951 | 4.346 | 3.886 | 0.975 | 1.903 |
| | Ø,1,2 | | | | 118.973 | 356.919 | 2.891 | 6.441 | 5.759 | 0.964 | 2.786 |
| | Ø,1,2,3 | | | | 90.068 | 360.272 | 3.819 | 8.508 | 7.607 | 0.955 | 3.646 |
| 480 | Ø | | .117268600E-01 | .117063522E-03 | 679.180 | 679.180 | 1 | 2.209 | 2.002 | 1 | 1 |
| | Ø,1 | 2.0 | | | 347.780 | 695.560 | 1.953 | 4.313 | 3.911 | 0.976 | 1.907 |
| | Ø,1,2 | | | | 233.540 | 700.620 | 2.908 | 6.423 | 5.823 | 0.969 | 2.819 |
| | Ø,1,2,3 | | | | 176.780 | 707.120 | 3.842 | 8.486 | 7.693 | 0.960 | 3.690 |
| 960 | Ø | | .961446390E-02 | .961422920E-04 | 1353.130 | 1353.130 | 1 | 2.198 | 2.005 | 1 | 1 |
| | Ø,1 | 2.0 | | | 689.260 | 1378.520 | 1.963 | 4.316 | 3.935 | 0.982 | 1.927 |
| | Ø,1,2 | | | | 464.440 | 1393.320 | 2.913 | 6.405 | 5.840 | 0.971 | 2.829 |
| | Ø,1,2,3 | | | | 351.010 | 1404.040 | 3.855 | 8.474 | 7.728 | 0.964 | 3.715 |
| 1920 | Ø | | .899305195E-02 | .898241997E-04 | 2710.050 | 2710.050 | 1 | 2.190 | 2.006 | 1 | 1 |
| | Ø,1 | 2.0 | | | 1375.670 | 2751.340 | 1.970 | 4.314 | 3.951 | 0.985 | 1.940 |
| | Ø,1,2 | | | | 927.640 | 2782.920 | 2.921 | 6.397 | 5.860 | 0.974 | 2.845 |
| | Ø,1,2,3 | | | | 698.720 | 2794.880 | 3.879 | 8.493 | 7.779 | 0.970 | 3.761 |

TIME-STEPS: 120

Table 4   Experimental Results and Performance Measurements of the Parallel Algorithm for the $(S).A.G.E.$ Method (for $r = 4$) on the NEPTUNE Prototype System.

| $N_{I.P.}$ | $N_{PROCS}$ | $r$ | $M_{P.E.}$ | $M_{A.E.}$ | $Tc^{(e)}$ (secs) | $C_p$ | $S_p$ | $R_{S_p}$ | $\bar{R}_{S_p}$ | $E_p$ | $F_p.T_s^{(e)}$ |
|---|---|---|---|---|---|---|---|---|---|---|---|
| 240 | Ø | 4.0 | .447672009E-01 | .335693359E-03 | 174.062 | 174.062 | 1 | 4.402 | 3.936 | 1 | 1 |
| | Ø,1 | | | | 89.356 | 178.712 | 1.948 | 8.575 | 7.668 | 0.974 | 1.897 |
| | Ø,1,2 | | | | 60.420 | 181.260 | 2.881 | 12.682 | 11.340 | 0.960 | 2.766 |
| | Ø,1,2,3 | | | | 45.726 | 182.904 | 3.807 | 16.758 | 14.984 | 0.952 | 3.623 |
| 480 | Ø | 4.0 | .631995499E-02 | .631213188E-04 | 344.070 | 344.070 | 1 | 4.360 | 3.953 | 1 | 1 |
| | Ø,1 | | | | 177.190 | 354.380 | 1.942 | 8.466 | 7.675 | 0.971 | 1.885 |
| | Ø,1,2 | | | | 119.370 | 358.110 | 2.882 | 12.567 | 11.393 | 0.961 | 2.769 |
| | Ø,1,2,3 | | | | 90.240 | 360.960 | 3.813 | 16.624 | 15.071 | 0.953 | 3.634 |
| 960 | Ø | 4.0 | .657623261E-02 | .654459000E-04 | 686.320 | 686.320 | 1 | 4.334 | 3.952 | 1 | 1 |
| | Ø,1 | | | | 350.720 | 701.440 | 1.957 | 8.481 | 7.734 | 0.978 | 1.915 |
| | Ø,1,2 | | | | 235.620 | 706.860 | 2.913 | 12.625 | 11.512 | 0.971 | 2.828 |
| | Ø,1,2,3 | | | | 177.770 | 711.080 | 3.861 | 16.733 | 15.259 | 0.965 | 3.726 |
| 1920 | Ø | 4.0 | .512412935E-02 | .507831573E-04 | 1374.880 | 1374.880 | 1 | 4.316 | 3.954 | 1 | 1 |
| | Ø,1 | | | | 699.810 | 1399.620 | 1.965 | 8.480 | 7.767 | 0.982 | 1.930 |
| | Ø,1,2 | | | | 469.720 | 1409.160 | 2.927 | 12.633 | 11.572 | 0.976 | 2.856 |
| | Ø,1,2,3 | | | | 353.940 | 1415.760 | 3.885 | 16.766 | 15.357 | 0.971 | 3.772 |

TIME-STEPS: 60

Finally, from the pure numerical point of view, any possible doubt about the accuracy of the solution obtained through this method is proved unnecessary, since we note that the accuracy of the method generally increases along with the grid ratio r, to reach a value for the maximum absolute error (when r = 4) which is the smallest achieved in all the numerical experiments.

## 6. PERFORMANCE ANALYSIS ON THE 'NEPTUNE' PROTOTYPE SYSTEM

Since quite a few such experimental systems, otherwise called "Multi-processor testbeds" have been built to investigate algorithm performance, a comprehensive approach to performance prediction requires accurate performance frameworks in order to reduce the overall experimentation time. In accordance, most of the Multiprocessor performance frameworks have been theoretically based on statistical methods, predicting statistical mean values for performance over some time interval.

However, in real-time 'shared' memory based systems the situation is quite often different, since individual systems may involve such specialized hardware and software features that make them more enhanced than others and thus of not accurately predictable performance via the frame work of such theoretical approaches. These cases are easily taken care of in our "school" of *deterministic* performance analysis (see Bekakos and Evans [BEKA86a]), which views performance as the interaction of resources demanded by programs and provided by the Multiprocessor system, in both system dependent and independent manner; it also matches algorithm to machine trying to avoid the danger of rejecting an algorithm because it performs badly on one particular parallel system.

While the parallel algorithms design has to take into account the potential parallelism, demands for shared data and demands for synchronization, it is the last factor that is the determining feature of the design of programs for such asynchronous parallel machines as the NEPTUNE prototype system. Thus, for all parallel systems the first two features must be taken into account, but it is only for asynchronous systems that synchronization itself is a cost. Both algorithms are designed to minimize the amount of synchronization required without, however, limiting the parallelism up to four processors.

One function of synchronization can be to ensure that one set of processes terminates before the next set of processes is started, a fact which can result in 'idle' processors. In some instances, however, it can be possible to construct variants of the algorithms which allow, under weak conditions, the faster processors to move on to what were in the usual version the next set of processes.

In respect to the performance analyses of the parallel algorithms investigated herein, for a grid of size (240x480), the program dependent tables, 5, 7, 9, summarize, in this sequence, the following parameters: Grid Size, program's PHase number, the Algebraic-, (theoretical) Time-complexities per point (or group of 2 points), the Implementation cycles, the Algebraic-, (theoretical) Time-complexities per parallel path of the algorithms, the Number of parallel paths allocated per processor, the Loops of parallelism, the processing-to-access to shared data ratio and the percentage overhead, the processing-to-access to the parallel path scheduler ratio and the respective percentage overhead, the limits to performance in terms of a theoretical upper bound on the number of cooperating processors in connection with their average 'excess' access time and average access time to the Shared data resource and the cycle time of the parallel path Scheduling resource, the percentage of the average Idle

Table 5  A Program Dependent Performance Analysis of the Parallel Algorithm for the Standard Explicit Method.

| $G_S$ | PROCESSORS (p) | | | $Ac_{(i,j)}$ | $Tc_{(i,j)}^{(t)}$ (secs) | $I_{cl}$ | $Ac_p$ | $Tc_p^{(t)}$ (secs) | $N_p(p)$ | $L_O(//)$ | SHARED DATA | | PARALLEL PATH | |
| | $P_H$ | $N_t$ | $S_p$ | | | | | | | | $R_{a(s)}$ | $O_{st(s)}^{(t)}$ | $R_{a(//)}$ | $O_{st(//)}^{(t)}$ |
|---|---|---|---|---|---|---|---|---|---|---|---|---|---|---|
| 240x480 | 3 | $p<2N_{B.P.}$ | $O(p)$ $[p\|N_{B.P.}]$ | 60 flops | 0.023 | 960 | $\frac{57.6\times10^3}{p}$ flops | $\frac{22.176}{p}$ | 1 | 1 | 1:60 flops | 0.003% | 1:$\frac{57.6\times10^3}{p}$ flops | 0.005p% |
| | 4 | $p<N_{I.P.}$ | $O(p)$ $[p\|N_{I.P.}]$ | 56 flops | 0.022 | 240 | $\frac{13.44\times10^3}{p}$ flops | $\frac{5.174}{p}$ | 1 | 1 | 1:56 flops | 0.004% | 1:$\frac{13.44\times10^3}{p}$ flops | 0.023p% |
| | 5 | $p<N_{1.P.}$ | $O(p)$ $[p\|N_{1.P.}]$ | 14 flops | 0.005 | 240 | $\frac{3.36\times10^3}{p}$ flops | $\frac{1.294}{p}$ | 1 | 480 | 1:2 flops | 0.098% | 1:$\frac{3.36\times10^3}{p}$ flops | 0.093p% |

| LIMITS TO PERFORMANCE | | | | |
| $S_{d(r)}$ | $S_{d(r)}'$ | $S_{h(r)}$ | $Id_t^{(t)}$ | $W_{st}$ (secs) |
|---|---|---|---|---|
| $m_p=30,596$ | $m_p=24,509$ | $m_p=\frac{18,480}{p}$ | $\sim1.5\%$ | 0.006 |
| $m_p=28,556$ | $m_p=22,875$ | $m_p=\frac{4,312}{p}$ | | 0.005 |
| $m_p=1,019$ | $m_p=816$ | $m_p=\frac{1,078}{p}$ | | 2.913 |

Table 6  A System Dependent Performance Analysis of the Parallel Algorithm for the Standard Explicit Method.

| $G_S$ | $T_s^{(e)}$ (secs) [XPFCL] | $S_p$ ∅,1 | $S_p$ ∅,1,2 | $S_p$ ∅,1,2,3 | $\sum_{i=1}^{p} c_{y_i}$ | $t_{cy}$ (μsecs) | $Id_t^{(e)}$ | $t_b$ (μsecs) | PARALLEL PATH $O_{st(||)}^{(e)}$ | PARALLEL PATH $O_{cn(||)}^{(e)}$ | $T_s^{(e)}$ (secs) [XPFCLN] | PARALLEL CONTROL $O_{t1(||)}^{(e)}$ | $T_s^{(e)}$ (secs) [XPFCLS] | SHARED DATA $O_{t1(s)}^{(e)}$ | $W_{dc}$ (secs) |
|---|---|---|---|---|---|---|---|---|---|---|---|---|---|---|---|
| 240x480 | 766.268 | 1.940 | 2.820 | 3.700 | 5106 | ~10,800 | ~6.7% | ~686 | 0.28% | 0.42% | 765.356 | 0.12% | 764.558% | 0.1% | 55.145 |

288

Table 7 A Program Dependent Performance Analysis of the Parallel Algorithm for the $(S).A.G.E.$ Method (for $r = 1$).

| $G_S$ $P_H$ | $N_t$ $S_p$ | $Ac(i,j)$ $O(p)$ | $Tc^{(t)}_{(i,j)}$ (secs) | $I_{cl}$ | $Ac_p$ | $Tc^{(t)}_p$ (secs) | $N_p(p)$ $L_O(//)$ | SHARED DATA $R_a(s)$ | SHARED DATA $O^{(t)}_{st(s)}$ | PARALLEL PATH $R_a(//)$ | PARALLEL PATH $O^{(t)}_{st(//)}$ |
|---|---|---|---|---|---|---|---|---|---|---|---|
| 3 | $p<2N_{B.P.}$ $[p\|N_{B.P.}]$ | 60 flops  O(p) | 0.023 | $\dfrac{480}{p}$ | $\dfrac{28.8\times10^3}{p}$ flops | $\dfrac{11.088}{p}$ | 1   1 | 1:60 flops | 0.003% | $1:\dfrac{28.8\times10^3}{p}$ flops | 0.011p% |
| 4 | $p\le N_{I.P.}$ $[p\|N_{I.P.}]$ | 56 flops  O(p) | 0.022 | $\dfrac{240}{p}$ | $\dfrac{13.44\times10^3}{p}$ flops | $\dfrac{5.174}{p}$ | 1   1 | 1:56 flops | 0.004% | $1:\dfrac{13.44\times10^3}{p}$ flops | 0.023p% |
| 5 | $p<N_{I.P.}$ $[p\|N_{I.P.}]$ | 49 flops  O(p) | 0.019 | $\dfrac{120}{p}$ | $\dfrac{5.88\times10^3}{p}$ flops | $\dfrac{2.264}{p}$ | 2   120 | 1:4 flops | 0.049% | $1:\dfrac{5.88\times10^3}{p}$ flops | 0.053p% |

$G_S$ = 240x240

| LIMITS TO PERFORMANCE $S_d(r)$ | $S'_d(r)$ | $S_h(r)$ | $Id^{(t)}_t$ | $W_{st}$ (secs) |
|---|---|---|---|---|
| $m_p$=30,596 | $m_p$=24,509 | $m_p=\dfrac{9,240}{p}$ | | 0.005 |
| $m_p$=28,556 | $m_p$=22,875 | $m_p=\dfrac{4,312}{p}$ | $\approx1.5\%$ | 0.005 |
| $m_p$=2,039 | $m_p$=1,633 | $m_p=\dfrac{1,886}{p}$ | | 1.413 |

Table 8 A System Dependent Performance Analysis of the Parallel Algorithm for the $(S).A.G.E.$ Method (for $r = 1$).

| $G_s$ | $T_s^{(e)}$ (secs) [XPFCL] | $S_p$ 0,1 | $S_p$ 0,1,2 | $S_p$ 0,1,2,3 | $\sum_{i=1}^{p} c_{y_i}$ | $t_{cy}$ (μsecs) | $Id_t^{(e)}$ | $t_b$ (μsecs) | PARALLEL PATH $O_{st(ll)}^{(e)}$ | PARALLEL PATH $O_{cn(ll)}^{(e)}$ | $T_s^{(e)}$ (secs) [XPFCLN] | PARALLEL CONTROL $O_{tl(ll)}^{(e)}$ | $T_s^{(e)}$ (secs) [XPFCLS] | SHARED DATA $O_{tl(s)}^{(e)}$ | $W_{dc}$ (secs) |
|---|---|---|---|---|---|---|---|---|---|---|---|---|---|---|---|
| 240x240 | 685.175 | 1.951 | 2.883 | 3.804 | 2692 | ∼10,800 | ∼4% | ∼686 | 0.16% | 0.26% | 684.450 | 0.11% | 683.967 | 0.07% | 29.074 |

Table 9   A Program Dependent Performance Analysis of the Parallel Algorithm for the $(S).A.G.E.$ Method (for $r = 4$).

| $G_S$ | $P_H$ | PROCESSORS (p) $N_t$ | PROCESSORS (p) $S_p$ | $A_{c(i,j)}$ | $T_{c(i,j)}^{(t)}$ (secs) | $I_{cl}$ | $A_{c_p}$ | $T_{c_p}^{(t)}$ (secs) | $N_{p(p)}$ | $L_{O(//)}$ | SHARED DATA $R_{a(s)}$ | SHARED DATA $O_{st(s)}^{(t)}$ | PARALLEL PATH $R_{a(//)}$ | PARALLEL PATH $O_{st(//)}^{(t)}$ |
|---|---|---|---|---|---|---|---|---|---|---|---|---|---|---|
| | 3 | $p < 2N_{B.P.}$ $[p\vert N_{B.P.}]$ | $0(p)$ | 60 flops | 0.023 | $\dfrac{120}{p}$ | $\dfrac{7.2\times10^{3}}{p}$ flops | $\dfrac{2.772}{p}$ | 1 | 1 | 1:60 flops | 0.003% | $1:\dfrac{7.2\times10^{3}}{p}$ flops | 0.043p% |
| 240x60 | 4 | $p \leq N_{I.P.}$ $[p\vert N_{I.P.}]$ | $0(p)$ | 56 flops | 0.022 | $\dfrac{240}{p}$ | $\dfrac{13.44\times10^{3}}{p}$ flops | $\dfrac{5.174}{p}$ | 1 | 1 | 1:56 flops | 0.004% | $1:\dfrac{13.44\times10^{3}}{p}$ flops | 0.023p% |
| | 5 | $p < N_{I.P.}$ $[p\vert N_{I.P.}]$ | $0(p)$ | 49 flops | 0.019 | $\dfrac{120}{p}$ | $\dfrac{5.88\times10^{3}}{p}$ flops | $\dfrac{2.264}{p}$ | 2 | 30 | 1: 4 flops | 0.049% | $1:\dfrac{5.88\times10^{3}}{p}$ flops | 0.053p% |

| LIMITS TO PERFORMANCE $S_{d(r)}$ | $S'_{d(r)}$ | $S_{h(r)}$ | $Id_t^{(t)}$ | $W_{st}$ (secs) |
|---|---|---|---|---|
| $m_p = 30,596$ | $m_p = 24,509$ | $m_p = \dfrac{2,310}{p}$ | | 0.005 |
| $m_p = 28,556$ | $m_p = 22,875$ | $m_p = \dfrac{4,312}{p}$ | $\sim 1.5\%$ | 0.005 |
| $m_p = 2,039$ | $m_p = 1,633$ | $m_p = \dfrac{1,886}{p}$ | | 0.353 |

Table 10 A System Dependent Performance Analysis of the Parallel Algorithm for the $(S).A.G.E.$ Method (for $r = 4$).

| $G_s$ | $T_s^{(e)}$ (secs) [XPFCL] | $S_p$ | | | $\sum_{i=1}^{p} c_{y_i}$ | $t_{cy}$ (μsecs) | $Id_t^{(e)}$ | $t_b$ (μsecs) | PARALLEL PATH | | $T_s^{(e)}$ (secs) [XPFCLN] | PARALLEL CONTROL | $T_s^{(e)}$ (secs) [XPFCLS] | SHARED DATA | $w_{dc}$ (secs) |
|---|---|---|---|---|---|---|---|---|---|---|---|---|---|---|---|
| | | $\emptyset,1$ | $\emptyset,1,2$ | $\emptyset,1,2,3$ | | | | | $o_{st(//)}^{(e)}$ | $o_{cn(//)}^{(e)}$ | | $o_{tZ(//)}^{(e)}$ | | $o_{tZ(s)}^{(e)}$ | |
| 240x60 | 174.062 | 1.948 | 2.881 | 3.807 | 716 | ~10,800 | ~4.2% | ~686 | 0.17% | 0.27% | 173.880 | 0.1% | 173.690 | 0.11% | 7.733 |

time (theoretically) and the Wasted time statically. While these results are manually obtained from the program itself predicting somehow the experimental performance, Tables 6, 8, 10 are system dependent and present the "real" performance measures obtained when *running* the algorithms on the system. The estimation of most of the above figures is not simply straightforward and the reader should refer to [BEKA86a]. The additional parameters utilized in the latter tables, in the sequence of their appearance are: the uniprocessor (experimental) Time-complexity when generating the load modules using the *XPFCL* command, the achieved Speed-ups, the total number of waiting cycles because no parallel path is available, the processors cycle time, the average time wasted dynamically (experimental), the 'blocked' time, the processors static and contention (experimental) percentage Overheads when accessing the parallel path scheduling resource, the uniprocessor (experimental) Time-complexity when generating the load modules using the *XPFCLN* command, the percentage of the total Overhead due to the control of the parallel mechanisms, the uniprocessor (experimental) Time-complexity when generating the load modules using the *XPFCLS* command, the percentage of the total Overhead due to the shared data 'loading' in the shared memory module, and the Wasted time dynamically. Note that, in particular for the information obtained from the shared array ITIME, when testing both algorithms, a plethora of runs was carried out, always considering the average figures for every information taken from this array.

7. CONCLUSIVE REMARKS

The purpose of this investigation was initially to study the suitability of alternative solution methods for parabolic partial differential equations on Multiprocessor systems. This paper concentrates mainly on explicit type methods of solution whereby the simplicity of the procedures allows the user to exploit maximum parallelism with minimum overhead costs.

In the case of the implicit schemes, such as the Crank-Nicolson algorithm, the solution at each time-level is usually obtained by solving a set of linear or nonlinear systems of equations. The systems of equations encountered in such schemes have, in most cases, a banded structure as the coefficient matrix. The parallel implementation of banded linear systems of equations has been studied by Chen and coworkers [CHEN78].

As mentioned earlier, although the explicit methods are very simple and can readily be implemented on parallel computers, they require a very restrictive range for the value of r and especially so in many space dimensions.

Our study of the Group Explicit method revealed that although this scheme is more complicated to exploit parallelism, it is unconditionally stable for all values of the grid ratio parameter. Indeed, this may be a very useful property, since by selecting different values of r one can increase the value of Δt, with Δx being kept constant. In particular, by choosing r to be 4, only one quarter of the number of time-steps is required to reach the same value of t, than when r is equal to 1, and consequently the timings will be reduced by a factor of 4. There is no change in the final accuracy obtained by selecting larger values of r in order to increase the time-step Δt.

Finally, the concept of the GE method can be similarly extended to the case of a two- and three-space dimensional problem (see Abdullah [ABDU83]). For the latter case we have groups of 8 grid points taken to

293

form a 'cube', instead of the 4 points in a plane as for the two-dimensional problem. It is expected that the parallel behaviour of this class of methods will improve as the number of dimensions increases.


## 8. REFERENCES

[ABDU83]   A.R.B. Abdullah, The Study of Some Numerical Methods for
           solving Parabolic Partial Differential Equations,
           Ph.D. Thesis, L.U.T., (1983).

[AMES65]   W.F. Ames, Non-Linear Partial Differential Equations in
           Engineering, Academic Press Inc., London (1965).

[BEKA85]   M.P. Bekakos and D.J. Evans, A Basic Group Explicit
           Solution Scheme for Non-Linear Parabolic P.D.E.s on MIMD
           Parallel Systems, to appear.

[BEKA86]   M.P. Bekakos, A Study of Algorithms for Parallel Computers
           and VLSI Systolic Processor Arrays, Ph.D. Thesis, L.U.T.,
           (1986).

[BEKA86a]  M.P. Bekakos and D.J. Evans, A Model for the Deterministic
           Performance Analysis of MIMD Parallel Computer Systems,
           to appear.

[BURG48]   J.M. Burgers, Adv. Appl. Mech., vol. 1, (1948), p. 171.

[CHEN78]   S.C. Chen, D.J. Kuck and A.H. Sameh, ACM Transactions on
           Mathematical Software, vol. 4, No. 3, (1978), pp. 270-277.

[COLE51]   J.D. Cole et al, On Quasi-linear Parabolic Equations occurring
           in Aerodynamics, A. Appl. Maths., vol. 9, (1951)
           pp. 225-236.

[EVAN83]   D.J. Evans and A.R.B. Abdullah, The Group Explicit Method for
           the Solution of Burgers' Equation, Internal Report,
           Dept. of Computer Studies, No. 179, L.U.T., (March 1983).

[EVAN85]   D.J. Evans and M.P. Bekakos, An Alternative Solution Method
           for Non-Linear Parabolic Equations on MIMD Parallel
           Computers, submitted to Intern. J. of Computer
           Mathematics, (1985).

[LARK64]   B.K. Larkin, Some Stable Explicit Difference Approximations
           to the Diffusion Equation, Math. Comp., vol. 18,
           (1964), pp. 196-202.

[MADS76]   N.K. Madsen and R.F. Sincovec, General Software for Partial
           Differential Equations in Numerical Methods, for
           Differential Systems, Ed. Lapidus, L. and Schiesser, W.E.,
           Academic Press, (1976), pp. 229-242.

# AUTHOR INDEX

# SUBJECT INDEX

Abaffy-Broyden-Spedicato method, *see* ABS
ABS methods, 85-93
  class of algorithm, 90-93
ADI-system *10* Plus, 194
AGE method, *see* Method, alternating
    Group
Algorithm
  asynchronous, iterative, 234-235
  asynchronous parallel state,
    164-165, 170
  control, 164
  dynamic programming, 161-170
  iterative, 207-210, 234-235
  parallel, 123-142, 164-170
    design of, 123-142
    numerical, 123-142
  synchronous, 166, 170
Analogic AP-*400*, 192-194
Applied Dynamics International
    computer system, 194
Architecture, parallel, 93
Array processor, 185

Band filter, 217-238
Bellman's optimality principle,
    161
Benchmark model, 23-24
Bilinear continuous system
  multi-time scale, 143-159
  simulation, parallel, 143-159
Black-red decomposition method,
    236-237
Block
  interation scheme, 65-78
    convergence, 71-74
    Jacobi type, 65-78
    Newton-Raphson methods, 65-78
  lower triangular, 119
  parallel, solution methodology,
    68-74
    model equation, 68-70
Blur, 217-219, 225
Burger's equation, 274-277

Calculation modeling, parallel,
    239-249

Chaining, 187
Cluster, 31-43
Computer
  network, 6
  parallel, 161, 269-294
  system, distributed, hierarchical,
    203-215
    on-line implementation, 211-212
  time, 162
Computer Simulation Centre at the
    University of Salford,
    England, 172
Continuous bilinear system, *see*
    Bilinear continuous system
Coordination, 3, 4
  decomposition technique, 1-16
    discretization, 18
  strategy, 7
Coordinator problem, 12-13, 21,
    49-50
Coupling equation, 65-78
Crank-Nicholson scheme, 67, 271
CSPI MAP-*300* 190-192
Curse of dimensionality, 161

Data
  flow, 239-262
  multiple, 2
  stream, 184, 187
Decomposition
  algorithm, 104-106
  black-red method, 236-237
  coordination
    direct method, 8-10, 14, 17-18
    dual method, 10-14
    model, 55-62
    problem definition, 3-10
    solution methodology, 3-10
    system partitioning, 47-50
    techniques, 1-22, 24, 29
  discretization, 18
  graph-theory, 119
  hierarchical, 119-121
  red-black method, 236-237
  singular value method, 95-112
  technique, 53-55

Image restoration (continued)
processing, 97
Instruction
multiple with multiple data
(MIMD), 2
single with multiple data
(SIMD), 2
Integrated system optimization
(ISO), 203-215
and parameter estimation, 203-215

Jacobi method, 20, 65-78, 236, 237

Kalman
band filter, 221
bank design for image restoration,
adaptive, 217-238
bank, parallel, 233
Kronecker delta function, 218

Lagrangian multiplier, 10, 11, 20
Large-scale system, 31, 32, 45
partitioning into subsystems, 31
*see* Scale
Large set, of ordinary differential
equations, 45-63
*see* Differential equation, ordi-
nary, *see* Set
Linearisation, successive, 6
Linear system, *see* System, linear
Load-flow calculation, 31
Local problem, *see* Problem, local
Loft test, 28
Log-sum algorithm, 2
Low-pass filtering, 95-112
eigenfunction, 108-109
Lower block triangular system, 119
Lyapunov
equation, 219
operator, 244, 249

Macro Arithmetic Processor (MAP),
190-192 *see* CSPI MAP-*300*
Marking graph, 243
Matrix
calculation, 252-253, 256
M, 72
Method
AGE, 275
direct, 14, 18-20, 125-129
explicit, 124, 129-136, 270-272
group-, 138-142, 272-276
implicit, 124, 136, 270-272
to explicit conversion, 136-
138
Microcomputer, 203
MIMD parallel computer, 65-78,
269-294
multi-processor system, 161
solution method, 270
Motion blur, *see* Blur

Multimicroprocessor for parallel
processing, 229-238
Multiplier matrix
multiple, 246
single, 244, 245, 248
Multiprocessor, 229, 239
algorithm, 229
control, 233
and resource constraint, single,
263-267
simulation of operation, 165-166
software, 233
supervision, 233
task scheduling, 263-267
Multiprocessor Reconfigurable
Simulator (MPRS), 197
memory, common, 198
Multi-time scale bilinear system,
143-159

Navier-Stokes equation, 270
NEPTUNE system, 276-293
algorithm, parallel, for standard
explicit method, 287-292
MIMD *4*-processor test bed, 276,
278
Net modelling, 251-262
hexagonal, 258, 261
mixed, 260
square, 258, 259
structure, mixed, 258
Newton-Raphson method, 9, 20, 48,
49, 65-78

Occurrence matrix
initial, 33
square binary, 32-35
ODE, *see* Differential equation,
ordinary
Optimization of integrate system,
203-215
algorithm, 203
problem, 7, 161
discrete-time, 162
local, 213

Parallel
algorithm, numerical, design of,
123-142
calculation, modeling, 239-249
computer
facility, 161
MIMD, 65-78, 269-294
design of numerical algorithm,
123-142
estimation algorithm, 113-121
implementation efficiency, 167
processing, 17-29, 183-202,
229-238
classification, 184
multimicroprocessor for, 229-238